Springer Tracts in Modern Physics

Volume 254

For further volumes:
http://www.springer.com/series/426

Springer Tracts in Modern Physics

Springer Tracts in Modern Physics provides comprehensive and critical reviews of topics of current interest in physics. The following fields are emphasized: Elementary Particle Physics, Condensed Matter Physics, Light Matter Interaction, Atomic and Molecular Physics, Complex Systems, Fundamental Astrophysics.

Suitable reviews of other fields can also be accepted. The Editors encourage prospective authors to correspond with them in advance of submitting a manuscript. For reviews of topics belonging to the above mentioned fields, they should address the responsible Editor as listed below.

Special offer: For all clients with a print standing order we offer free access to the electronic volumes of the Series published in the current year.

Natalya A. Zimbovskaya

Transport Properties of Molecular Junctions

Springer

Natalya A. Zimbovskaya
Department of Physics and Electronics
University of Puerto Rico at Humacao
Humacao, PR, USA

ISSN 0081-3869 ISSN 1615-0430 (electronic)
ISBN 978-1-4614-8010-5 ISBN 978-1-4614-8011-2 (eBook)
DOI 10.1007/978-1-4614-8011-2
Springer New York Heidelberg Dordrecht London

Library of Congress Control Number: 2013943268

Printed on acid-free paper

Springer is part of Springer Science+Business Media (www.springer.com)

Preface

At present, we observe a long-lasting process of miniaturization of electronic devices. The ultimate limit for the miniaturization of electronic components is set by the atomic scale. However, in the case of conventional "top-down" fabrication methods and usual semiconductor materials, the smallest attainable size of the electronic components is well above this ultimate limit. Many ways for further miniaturization have been and still are suggested, including those using graphene and carbon nanotubes. The idea of using molecules as elements in electronic devices was first suggested in the seminal work by Aviram and Ratner [1], where they did envision a diode made out by assembling several diverse components into a single molecule creating a donor–bridge–acceptor combination. This work had been published in 1974 well before any clear indications within the electronic industry for the need of molecular or any other novel material technologies. However, further development of electronics brought a new interest and greatly increased research activities in this field. These activities in the field of molecular electronics reflect the convergence of two trends in the fabrication of nanodevices, namely, the "top-down" device miniaturization through lithographic methods and "bottom-up" device manufacturing through atom-engineering and self-assembly approach. The principal goal of molecular electronics is to construct electronic circuits in a "bottom-up" fashion, so that specifically designed molecules could take parts of active components as well as interconnects. Presently, molecular electronics is known to be one of the most promising developments of nanoelectronics, and the last two decades have seen an extraordinary progress in this field [2–6]. As discussed in the existing literature, the most important advantages of molecules as elements of nanodevices could be summarized as follows [5]:

- Smallness of molecules provides a faster charge transport through molecular channels and opens up opportunities for greater packing densities and higher functionalities on a chip.
- Lower heat dissipation in molecular mechanical devices which originates from typically low currents and small separation in excitations (as in conformationally mediated transitions). Switching energies of molecular mechanical devices are

expected to take on values several orders of magnitude less than those typical for the start-of-the art transistors.

- Organic chemistry and self-assembly which should be used to fabricate molecular electronic devices are significantly less expensive than the lithographic patterning typically employed in silicon-based microelectronics.
- Using molecules as building blocks of nanodevices gives means to circumvent difficulties occurring due to the necessity of patterning channels for the charge transport out of larger templates. Dealing with molecules, one may bypass these difficulties by exploiting their ability to self-assemble on various substrates.
- Potentially, molecules could be assembled in multilayers using Langmuir–Blodgett techniques. This would allow circuits including the molecules as active elements and/or interconnects to grow in the third dimension, thus creating opportunities to build up devices with complex and reconfigurable architectures.
- Silicon-based electronics has limitations caused by properties of the silicon which is characterized by a certain set of parameters that are hard to significantly change. Naturally, heating, doping, straining, and alloying of silicon offer some variability to characteristics of this material, but this variability is quite small when compared with much greater variability which organic molecules can easily accomplish just by modification of their side groups.

Successful transport experiments on molecules [7–17] confirm their significance as active elements of nanodevices. These include applications as rectifiers (molecular diodes), field effect transistors (molecular triodes) switches, memory elements, and sensors. However, it is necessary to remark that, to a significant extent, the advantages of molecule-based electronics still remain potentialities rather than established facts. In many cases the molecules do not really behave as anticipated, and performance of existing molecular-based devices needs significant improvement to make them suitable for industrial applications. This means that further and deeper understanding of processes determining transport characteristics of molecules is necessary to achieve. This is especially important in view of the fact that in the last two decades this particular research field has been suffering through a state of flux which was ranging from exaggerated optimism to clearly expressed pessimism. Currently, the situation is improving. Recent experiments are reporting higher yields and become reproducible between diverse research groups. Also, better agreement between theory and experiment is being achieved [6, 18]. All these emphasize the above-mentioned importance of thorough analysis of the physics underlying electron transport through molecular junctions. Detailed understanding of the electron transport at the molecular scale is a key step to device designing and controlling. Theory of electron transport through MMMs is being developed in the last two decades, and main transport mechanisms are currently elucidated in general terms. However, progress of experimental capabilities in the field of molecular electronics brings new theoretical challenges causing further development of the theory.

The key element and basing block of molecular electronic devices is a junction including two metal electrodes (leads) linked by a molecule. Usually, the electrodes are microscopically large but macroscopically small contacts which may be connected to a battery to provide the bias voltage across the junction. Accordingly, most of the theoretical and experimental studies so far have been concentrated on various aspects of electron transport through such systems known as metal–molecule–metal (MMM) junctions. A molecule included into the junction may be treated as a quantum dot coupled to the charge reservoirs. The discrete character of energy spectrum on the dot (molecule) is combined with nearly continuous energy spectra on the reservoirs (leads) occurring due to their comparatively large size, and this combination determines transport properties of the junction. To a considerable extent, transport characteristics of a certain MMM depend on the composition and structure of the molecular linker. It opens opportunities to take advantage of the variability of chemical compounds to design MMMs with the desired properties for use as elements in molecular electronic devices, so the molecular electronic research often requires participation of chemists as well as physicists. Therefore, molecular electronics is a multidisciplinary research field. Also, the coupling between the molecule and the leads significantly affects the transport properties of MMMs. As a part of a MMM, the molecule is turned into a open system electronically hybridized with the contacts, often making it difficult to ascertain where the molecule ends and the leads begin. Significant hybridization may occur even in the case of well-defined molecules put in contact with metal electrodes. As a result, molecular linkers in MMM junctions often behave similar to other open systems such as carbon nanotubes.

In practical molecular junctions the electron transport is always accompanied by nuclear motions in the environment. Accordingly, the MMM conduction is affected by the coupling between electronic and vibrational degrees of freedom. Nuclear motions underlie the interplay between the coherent electron tunneling through the junction and inelastic thermally assisted hopping transport [2]. Also, electron–phonon interactions may result in polaronic conduction [19, 20], and they are directly related to the junction heating [21] and to some specific effects such as alterations in both shape of the molecule and its position with respect to the leads [22, 23]. The effects of electron–phonon interactions may be manifested in the inelastic tunneling spectra (IETS), which present the second derivative of the current in the MMM d^2I/dV^2 versus the applied bias voltage V [24–26]. The inelastic tunneling spectroscopy may be a valuable method for identification of molecular species within the conduction region, especially when employed in combination with scanning microscopy. However, one may remark that the theory suggested so far needs further development to ease its application to practical MMMs.

To a significant extent, the specifics of the interplay between different transport mechanisms depend on the molecule size. Usually, electron transport through small individual molecules is nearly ballistic, and it may be reasonably considered as coherent resonance tunneling. As the size of the relevant molecule increases, the contribution from stochastic nuclear motions to the molecular conduction strengthens, destroying the coherence and removing the interference effects. Exploring electron transport through macromolecules such as proteins and DNA, one must emphasize incoherent scattering and consider thermally assisted intermolecular hopping as the predominating transport mechanism. Accordingly, there exist a hierarchy of models suitable for analyzing the transport properties of MMM junctions. The hierarchy is built up based on the fact that the increasing degrees of dephasing in the transport require decreasing amounts of quantum mechanics in the corresponding theoretical model. So, as the dephasing strengthens one must move from multielectron rate equations and coherent quantum kinetics to semiclassical Boltzmann approach and then to classical diffusion equations.

As well, transport characteristics of molecular junctions could show effects originating from electron–electron interactions, which give rise to such interesting quantum transport phenomena as Coulomb blockade and Kondo effect. Both suppression of the electron transport through molecular junctions at low values of the bias voltage occurring due to the charging energy in the molecule (Coulomb blockade) and the increase in the electrical conduction of the junction near zero bias voltage (Kondo effect) were observed in experiments on molecular and carbon nanotube junctions [3, 8, 27–33]. Theory of electron transport through MMMs and quantum dots taking into account electron–electron interactions on the dot is being developed (see, e.g., [34–37] and references therein). However, this theory is not completed so far, and it still meets with unresponded challenges.

Presently, significant attention of the research community is being given to studies of transport properties of magnetic molecules and molecular clusters. Electron–electron correlations leading to Coulomb blockade and Kondo physics and ferromagnetic many-body correlations play the predominating part in transport properties of these molecules and molecular complexes. Break-junction experiments and modeling of the electron tunneling through Mn_{12} complexes were reported [38]. The effect of spin blockade when a molecule traps an electron and blocks the current because transitions out of the trapping state are forbidden due to the spin conservation rules is also being studied [39, 40]. Another interesting subject is the electron transport through molecular networks made out of metal nanoparticles linked by molecules [40–42]. These systems reveal significant potentialities for nanoelectronic applications. To properly study electron transport in the networks one may treat them as sets of MMM junctions, each including two metal clusters connected by a molecular linker. An important issue in these studies is the effect of the electron structure of the clusters (nanoelectrodes) which in this case cannot be ignored. Another important aspect is to develop an approach enabling to compute the network conductance taking into account its geometry, which is a nontrivial and currently unresolved task.

The significance of electron spin in semiconductor physics was recognized in 1990s, thus making an opening for a new research field commonly called spin-electronics or spintronics. The electron spin in semiconductors can preserve coherence over long times and distances which offer opportunities to use it as the logic bit in memory and logic devices. The usefulness of electron spin is emphasized by the fact that its orientation may be manipulated by means of electric field. This is based on spin–orbit interaction, which is an intrinsic property of electronic structures. The spin Hall effect utilizing spin–orbit interaction instead of the magnetic field gradient gives a good example of all-electrical spin manipulation in semiconducting materials. However, the spin–orbit interactions play a double part. Giving the means to manipulate the electron spin direction by the external electric field, these interactions also serve as the main source of spin dephasing, thus destroying spin coherence and shortening spin relaxation times.

The recent progress in molecular electronics gave rise to the prospects of molecular spintronics. These prospects are appealing because the spin–orbit interactions in carbon-based organic molecules are known to be weak, which results in extremely long spin relaxation times. At the same time, the electron spin direction may be manipulated by the external magnetic field and by intrinsic magnetic moments of the ferromagnetic leads or those characterizing the molecular linker (in the case of magnetic molecules used as such). Presently, various aspects of spin transport through MMM junctions are being intensively studied both theoretically and experimentally.

To quantitatively analyze transport properties of metal–molecular junctions, one needs to have detailed knowledge concerning the electron band structure of these systems. Electronic structure plays a crucial part in determining important transport properties of molecules in both ballistic and diffusive transport regimes. When the transport is strongly inelastic (diffusive regime), the molecular conductance is related to the mobilities in the transport channels. In turn, these mobilities are determined by the scattering times and the effective masses of charge carriers in the channel. All these parameters take on values controlled by the specifics of the electron band structure through the electron density of states associated with the channel. In the case of ballistic transport, the current through the molecule is determined by the quantum mechanical electron transmission which explicitly depends on the energies related to molecular orbitals.

The electron structure of the considered system could be analyzed using different ways depending on its specifics. For instance, in studies of junctions where the electrodes are linked with the graphene nanoribbon, one may employ a phenomenological approach to describe the ribbon electronic structure, for within such approach one could successfully capture its most important features stipulated by the edge chemistry and the roughness of the ribbon. However, while considering proper MMM junctions, one needs to employ advanced computational methods to analyze the electronic structure of molecular linkers which includes effects due to exchange-correlation interactions as well as those originating from the interaction with contact atoms.

Being a very interesting research subject by itself, the studies of electron transport through molecules are equally important due to essential similarities between the latter and long-range electron transfer chemical reactions, which are already studied for several decades. These reactions could be developed in donor–bridge–acceptor molecular systems. The donor is some molecule (reductant) or a part of a macromolecule, which donates an electron to the acceptor (oxidant) via the molecular bridge. The long-range electron transfer plays an essential part in biological processes such as signal transduction across membranes, photosynthesis, enzyme catalysis, and some others [43]. Historically, studies of the long-range electron transfer strongly contributed to the "birth" of molecular electronics.

Similarities between the electron transport through MMMs and long-range electron transfer reactions were repeatedly emphasized in theoretical works (see, e.g., [44–52]). Thus the studies of electron transport through molecular junctions may bring new insight into the nature and characteristics of the long-range electron transfer reactions.

Molecular electronics is a diverse and rapidly growing field. Currently there exists a multitude of works reporting the results of both theoretical and experimental studies of transport properties of molecules and MMMs. The purpose of this book is to give an overview of the main physical mechanisms controlling the transport and the main characteristics of the latter. As far as possible we avoid the detailed descriptions of computational formalisms commonly used to theoretically analyze electron transport through molecules as well as experimental techniques. These methods and formalisms are described elsewhere. In the first chapter of this book we introduce basic concepts, which enter into a description of the electron transport through molecular junctions and briefly describe relevant experimental methods. In the next chapter we describe theoretical methods commonly used to analyze the electron transport through molecules. The next three chapters contain a description of various effects manifested in the electron transport through MMMs as well as the basics of density-functional theory and its applications to electronic structure calculations in molecules. Some nanoelectronic applications of molecular junctions and similar systems are discussed in the last chapter. In the Appendix, several MATLAB codes used to solve some problems discussed in the book are presented.

The author is sincerely grateful to all colleagues with whom she collaborated during the years given to studies of various aspects of electron transport through molecules, namely: G. Gumbs, N.J. Pinto, A.T. Johnson, M.R. Pederson, A. Blum, and B. Ratna. Also, the author is pleased to thank S. Datta and M.A. Rather for helpful and elucidating discussions, and G.M. Zimbovskiy for his help with the manuscript preparation.

Humacao, PR Natalya A. Zimbovskaya

Contents

1 **General Description** 1
1.1 Conduction and Electron Current Through a Single Molecule 1
1.2 Gating and Electric Potential on the Bridge 8
1.3 Electron–Electron Interactions: Coulomb Blockade 12
1.4 Dephasing 16
1.5 Electron–Phonon Interactions 18
1.6 Spin Transport 23
1.7 Building of Metal–Molecule–Metal Junctions 28
1.8 Characterization of Molecular Junctions 33

2 **Transport Theory** 39
2.1 Retarded and Advanced Green's Functions for Electrons Traveling Through a MMM Junction 39
2.2 The Green's Function for the Atomic Wire; an Analytical Example 47
2.3 Electron Transmission and Landauer Expression for the Current Through a MMM Junction 52
2.4 Electron Transport Through a Junction as a Multichannel Scattering Problem 56
2.5 Nonequilibrium Green's Functions Formalism 64
2.6 Rate Equations 73

3 **Ballistic Transport** 79
3.1 Charge Transfer and Electrostatic Potential Distribution in Unbiased Junctions 79
3.2 Electric Charge and Potential Distribution Over a Biased MMM Junction: Current Rectification 84
3.3 Fano Resonances 90
3.4 Simmons Model for Electron Transport Through Molecular Insulators 93
3.5 Conformational Gating of Molecular Bridges 99

3.6 Effect of the Electronic Structure of the Leads on the Electron Transport Through a Junction: Negative Differential Resistance 102
3.7 Coulomb Blockade and Charge Oscillations: NEGF Approach 105
3.8 Coulomb Blockade: Multielectron Master Equations Approach 110
3.9 Transport Through Magnetic Molecules 115
3.10 Kondo Effect 119
3.11 Kondo Effect in Nonequilibrium Quantum Dots 126

4 Inelastic Transport 133
4.1 Vibration-Induced Features in the Electron Conductance and Current Through MMM Junctions 133
4.2 Vibration-Induced Features in the Inelastic Electron Tunneling Spectra (IETS) 139
4.3 The Effect of Molecular Vibrations on the Coulomb Blockade and Kondo Anomaly 148
4.4 Dissipative Transport: Direct Coupling of the Molecular Bridge to the Phonon Bath 153
4.5 Dissipative Transport: Indirect Coupling of the Molecular Bridge to Thermal Phonons 158
4.6 Current Fluctuations 163
4.7 Heat Transfer, Thermal Conductance, and Thermopower 168
4.8 Polaron Effects: Hysteresis, Switching and Negative Differential Resistance 175
4.9 Molecular Junction Conductance and Long Range Electron Transfer Reactions 178

5 Electronic Structure Calculations in Molecules 183
5.1 Kohn–Sham Equations 183
5.2 Approximations for the Exchange-Correlation Energy Term 192
5.3 Long-Range Interactions in the DFT 196
5.4 Excitation Energies 200
5.5 Time-Dependent DFT 204
5.6 Dielectric Response of Fullerenes 207
5.7 Density-Functional-Based Investigations of Molecular Magnets 212
5.8 Photoexcitation of the Light-Harvesting Carotenoid-Porphyrin-C_{60} Molecular Triad 217
5.9 Specifics of the Electron Structure Calculations for MMM Junctions 221

6 Nanoelectronic Applications of Molecular Junctions 231
6.1 Field-Effect Transistors 231
6.2 Single-Molecule Diodes 239
6.3 Quantum Dots and Carbon Nanotubes as Switches and Memory Elements 242
6.4 Molecular Switches and Memories 248

6.5 Chemoselective Sensors 255
6.6 Thermoelectric Devices 262
6.7 Molecules on Silicon Surfaces 268

7 Conclusion 273

A MATLAB Codes Used to Generate Text Figures 277

References 307

Index 335

Chapter 1
General Description

1.1 Conduction and Electron Current Through a Single Molecule

We start our analysis of MMMs transport properties by adopting an extremely simplified model for the junction. We consider the molecule (presented as a set of energy levels) placed in between two leads (left L and right R). The leads are treated as free electron reservoirs with nearly continuous energy spectra. Currently, we omit from consideration electron–electron correlations (Coulomb interactions) and electron–phonon interactions. The effects of these interactions on the electron transport are discussed in the next chapters. Thus we consider all atomic nuclei in the molecule to be fixed to their equilibrium positions, which corresponds to the Born–Oppenheimer approximation. This is a reasonable first approximation appropriate at low temperatures, when the thermal energy $k_{\rm B}T$ is much smaller than the molecular levels broadening originating from the coupling of the molecule to the leads, which is characterized with the energy $\Gamma = \Gamma_L + \Gamma_R$. Within this approximation, the molecule energy levels are time independent.

Energy levels of a molecule coupled to the leads of a MMM differ from those of a single molecule of the same kind. The difference arises due to the changes in the electrostatic potential inside the molecule. These changes occur as a result of the charge transfer between the molecule and the leads, and their intensity is determined with the coupling strength. When a molecule is weakly coupled to the contacts, the charge transfer is insignificant, and the molecular orbitals remain nearly undisturbed. The stronger is the coupling, the more noticeable shift it gives to the molecular orbitals [53–63]. An extremely fruitful concept of an "extended molecule" was developed to take into account the effect of the molecule coupling to macroscopic metal leads. This concept was first suggested by Xue et al. [58] and is commonly adopted ever since. The concept is based on the very reasonable assumption that due to the screening in the metal contacts, the changes in the charge density and electric potential induced by the attached molecule are mostly localized

N.A. Zimbovskaya, *Transport Properties of Molecular Junctions*, Springer Tracts in Modern Physics 254, DOI 10.1007/978-1-4614-8011-2_1,

over small regions on the contacts surfaces. These regions may include only a few atoms. Therefore, one may separate out the "extended molecule" consisting of the molecule itself and those atoms from the lead surfaces which are significantly disturbed by the molecule. The effect of the molecule on the remaining parts of the leads may be taken into account by introducing an extra electrostatic potential, which provides the redistribution of the electric charge outside the extended molecule.

So, speaking of molecular orbitals one must realize that due to the coupling of the molecule to the contacts, these energy levels are shifted from their positions inherent to the single molecule. Also, the coupling to the contacts results in the broadening of molecular orbitals. To take this into account in a simplest way one may replace a discrete level with the energy E_0 with a Lorentzian density of states $D(E)$ depending on energy E :

$$D(E) = \frac{1}{2\pi} \frac{\Gamma}{(E - E_0)^2 + (\Gamma/2)^2}. \tag{1.1}$$

Here again, $\Gamma = \Gamma_L + \Gamma_R$, and $\Gamma_{L,R}$ are the parameters which have the dimensions of energy and characterize the coupling of the molecule to the contacts.

So long as there is no bias voltage applied across the junction, the latter remains in equilibrium characterized by the equilibrium Fermi energy E_{F}. The location of the Fermi energy is the most important factor which determines the conduction through the molecular bridge at low bias voltage. In bulk intrinsic semiconductors the Fermi level is situated in the middle of the bandgap separating the valence band from the conduction band. When the material is doped, the actual position of the Fermi level is shifted from the intrinsic level position, and the shift is controlled by the degree of doping. In molecules one meets a similar situation. In this case the part of the intrinsic Fermi level is taken by the so-called charge neutrality level (CNL). The CNL position is determined by the requirement that the number of states below it must be equal to the number of electrons in the molecule. According to this rule, the charge neutrality level of the junction is located in between the highest occupied molecular orbital (HOMO) and its lowest unoccupied orbital (LUMO). The position of the Fermi level with respect to CNL depends on the amount of electric charge, which comes to the molecular linker from the leads in the process of establishing the equilibrium. The intensity and character of the charge transfer between the molecule and the contacts in the MMM junction control the location of the Fermi level in the same way as doping in semiconducting materials is doing. The charge transfer process is driven by the difference between the CNL and the chemical potential of the contacts being simultaneously opposed by the Coulomb cost of charging up the molecule. The precise location of the Fermi energy is very difficult to discover because of its high sensitivity to the electronic correlations, polarization and image charges on the electrodes, as well as the precise geometry of the contacts between the molecule and the leads (Fig. 1.1). Usually, in transport calculations concerning practical systems, E_{F} is assumed to be located someplace in between HOMO and LUMO as well as CNL, and it is treated as a fitting parameter,

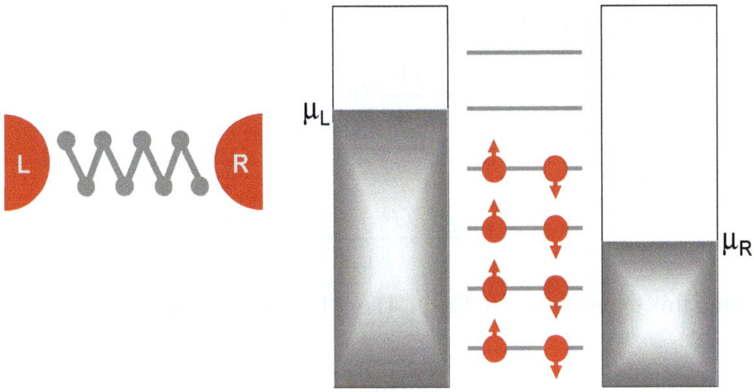

Fig. 1.1 Schematic drawing of a metal–molecule–metal junction (*left*). The molecule is represented by a set of spin-degenerated energy levels (*right*). When the bias voltage is applied across the junction, the levels situated between μ_L and μ_R serve as channels for the electron transport

whose value is chosen within certain reasonable limits to provide better agreement with experimental transport characteristics (see e.g. [64]).

When the bias voltage is applied across the junction, it keeps the left and right contacts at different electrochemical potentials μ_L and μ_R. Then the electric current appears in the junction, and molecular energy levels located in between the electrochemical potentials μ_L and μ_R take the major part in maintaining this current. Electrons from occupied molecular states tunnel to the contacts in accordance with the voltage polarity, and the electrons from one electrode travel to another one using unoccupied molecular orbitals as intermediate states for tunneling. So, molecular orbitals work as channels for the electron transmission. One may illustrate this using a very simple model for the MMM junction [4]. Within this model, the molecular bridge is simulated by a single spin-degenerated energy level with the energy E_0. An electron put on the level escapes into the leads at the tunnel rates $\Gamma_\beta/\hbar$ where the index β is labeling left and right contacts. When the level remains in the equilibrium with the contact β, the number of electrons on the level equals

$$n_\beta = 2 f_\beta(E_0) \tag{1.2}$$

where $f_\beta(E_0)$ is the Fermi distribution function:

$$f_\beta(E) = \left\{ 1 + \exp\left[\frac{E - \mu_\beta}{kT}\right]\right\}^{-1} \tag{1.3}$$

Under nonequilibrium conditions, the occupation of the level (n) could be found by equating the current I_L which flows through the left metal–molecular contact:

$$I_L = \frac{e\Gamma_L}{\hbar}(n_L - n) \tag{1.4}$$

to the current I_R flowing through the right contact:

$$I_R = \frac{e\Gamma_R}{\hbar}(n - n_R). \tag{1.5}$$

This gives

$$n = 2\frac{\Gamma_L f_L(E_0) + \Gamma_R f_R(E_0)}{\Gamma_L + \Gamma_R} \tag{1.6}$$

Substituting this expression into Eq. (1.4) or (1.5), we obtain

$$I = \frac{2e}{\hbar}\frac{\Gamma_L \Gamma_R}{\Gamma_L + \Gamma_R}\{f_L(E_0) - f_R(E_0)\}. \tag{1.7}$$

Here, e is the electron charge.

This result confirms that the current flows only in the bias voltage window where the difference of the Fermi functions takes on nonzero values. The bridge energy level E_0 may serve as the transport channel when it is situated in this window but not otherwise. Also, the expression (1.7) shows the current to be inversely proportional to $\hbar/\Gamma_{\text{eff}}$ given by the relation

$$\frac{\hbar}{\Gamma_{\text{eff}}} = \frac{\hbar}{\Gamma_L} + \frac{\hbar}{\Gamma_L}. \tag{1.8}$$

So, the quantity $\hbar/\Gamma_{\text{eff}}$ has the physical meaning of the overall escape time for the electron from the channel. Within the adopted simplified model, the transit time inside the channel is zero and the resistance of the junction comes solely from the contacts between the leads and the molecular bridge.

Deriving the expressions (1.6) and (1.7) we neglected the broadening of the bridge level due to its coupling to the leads. To take this effect into account we need to multiply the right sides of Eqs. (1.6) and (1.7) by the Lorentzian expression (1.1) and to carry out integration over energies. Also, in the expressions for the Fermi functions in the integrands, the bridge level energy E_0 must be replaced by the variable E. The results are

$$n = \frac{1}{\pi}\int \frac{\Gamma_L f_L(E) + \Gamma_R f_R(E)}{(E - E_0)^2 + (\Gamma/2)^2} dE, \tag{1.9}$$

$$I = \frac{e}{\pi\hbar}\int \frac{\Gamma_L \Gamma_R}{(E - E_0)^2 + (\Gamma/2)^2}\{f_L(E) - f_R(E)\} dE. \tag{1.10}$$

This expression agrees with the well-known Landauer formula for the electron tunnel current [4]:

$$I = \frac{e}{\pi\hbar}\int T(E)\{f_L(E) - f_R(E)\} dE \tag{1.11}$$

where the electron transmission function $T(E)$ has the form

$$T(E) = \frac{\Gamma_L \Gamma_R}{(E - E_0)^2 + (\Gamma/2)^2}. \tag{1.12}$$

To further clarify the physical meaning of the expression for the current one may rewrite it as follows:

$$I = \frac{e}{\pi\hbar} \int T(E)\{f_L(E)[1 - f_R(E)] - f_R(E)[1 - f_L(E)\}dE. \tag{1.13}$$

Here, the integrand is explicitly presented as the difference between two fluxes, one being directed from the left to the right and another one flowing in the opposite direction. We remark that the expression (1.12) for the electron transmission does not take into account the effects of the electrostatic potential which appears in the conducting channel due do combined influence of the bias voltage and electron–electron interactions. These effects and alterations which they bring to the electron transmission are discussed in the following section. Nevertheless, simplified expressions (1.9)–(1.13) allow to illustrate some important characteristics of electron transport. For instance, it follows from Eq. (1.10) that there exists the upper limit for the molecular conduction. In agreement with the Landauer's result [65] the maximum conductance of a channel for the electron transmission with a single spin-degenerated energy level equals

$$G_0 = \frac{e^2}{\pi\hbar} = (25.8\,\mathrm{k}\Omega)^{-1} \tag{1.14}$$

where again e is the electron charge and $\hbar$ is the Planck's constant. This is a truly remarkable result. First, it proves that the minimum resistance of a molecular junction $R_0 = G_0^{-1}$ cannot become zero. In other words, one never can short-circuit a device operating with quantum channels. Secondly, it shows that the minimum resistance of a MMM junction is independent of the length of the molecular linker. This gives grounds to believe that the minimum resistance R_0 is associated with the points of contact of the molecule with the electrodes.

Conductance g in practical quantum channels associated with molecular orbitals can take on values significantly smaller than G_0, depending on the delocalization of molecular orbitals participating in the electron transport, as well as on the quality of molecule–electrodes contacts. Also, it strongly depends on the scattering of electrons traveling through the molecular bridge. The total resistance $r = g^{-1}$ includes contributions from contact and molecular resistances and could be presented as follows [66]:

$$r = \frac{1}{G_0}\left(1 + \frac{1-T}{T}\right) \tag{1.15}$$

where the electron transmission coefficient T usually takes on values less than unity.

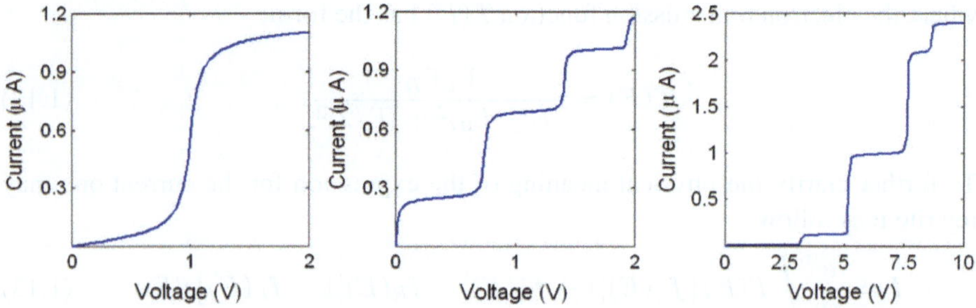

Fig. 1.2 Electron current through a single-site bridge symmetrically coupled to the leads (*left panel*), through a chain including five hydrogen-like atoms (*middle panel*), and through a hydrocarbon chain coupled to copper electrodes (*right panel*). The curves are plotted assuming $T = 30\,\text{K}$, $E_0 = -0.2\,\text{eV}$ (*left panel*); $E_0 = 0$, $b = 0.2\,\text{eV}$ (*middle panel*)

To theoretically analyze current–voltage characteristics of practical molecular junctions, one needs to employ one of the existing computational methods combined with the electron band structure calculations. However, one may visualize general features of current–voltage curves without recourse to computations. As the magnitude of the bias voltage (V) rises, the difference between μ_L and μ_R increases. This brings increase in the number of molecular orbitals located between μ_L and μ_R. These orbitals take part of channels for the electron transport; therefore, the appearance of an extra channel causes a sudden jump in the current. As a result, low-temperature I–V curves look as sequences of steps. At higher temperatures the plateaus become washed out and the I–V curves may lose their step-like character.

As illustration, we present two examples of current–voltage curves. In the first example we mimic a molecular linker in a MMM junction as a one-dimensional chain of identical hydrogen-like atoms with nearest neighbors interactions. We assume that there exists only one electron state per isolated atom with the energy E_0 and that the coupling between the neighboring sites in the chain is characterized by the parameter b. Such a model was suggested and theoretically analyzed by D'Amato and Pastawski [67] and then used in numerous works on the electron transport through molecules (see, e.g., [45, 46]). Assuming that the bias voltage is symmetrically distributed between the electrodes:

$$\mu_{L,R} = E_{\text{F}} \pm \frac{1}{2}|e|V, \tag{1.16}$$

it was shown that for a single-site chain the I–V curve displays a step-like profile with the step at $V = 2E_0$. This is presented in Fig. 1.2 (left panel). When the chain includes several atoms, it is characterized by the set of states (orbitals), and their number coincides with the number of atoms in the chain. Now, the current–voltage curve exhibits a sequence of steps as shown in the middle panel of Fig. 1.2. The longer is the chain, the more steps appear in the I–V curves. The second example concerns the electron transport through a hydrocarbon chain placed between copper electrodes. The relevant molecular energy levels were computed using the concept of an extended molecule, which included four copper atoms at each side of the chain. The current–voltage characteristic is shown in the right panel of Fig. 1.2. Again, we observe the stepwise curve where the steps are associated with the set of transport channels provided by the molecular orbitals.

Conductance of molecular junctions strongly depends on the length of the molecule linking the leads. As follows from the theory confirmed by numerous experiments, the electron transmission through a molecule is proportional to $\exp(-\beta L)$, where L is the molecule length and β is the decay constant taking on different values for different kinds of molecules. As discovered in the experiments, the decay constant takes on values within the range 1–5 nm^{-1} [44, 68]. Therefore, according to Eq. (1.15) the tunnel conductance of molecules which are few nanometers long is much smaller than G_0.

For long molecular linkers, as well as in the case of long-range electron transfer reactions, the electron transport through a molecule may be treated as a sequence of tunnelings between potential wells. Each well represents one atomic orbital of the molecular bridge [69–71]. Within this approach, every atom is represented by a set of sites corresponding to its states. Sometimes, it is more convenient to subdivide molecular bridges in certain atomic groups such as methyl groups CH_2, instead of separating out single atoms. Then any such group can be represented as a set of sites, each corresponding to a molecular orbital for the group. Using the accepted model where electron–electron and electron–phonon interactions are omitted and neglecting other factors which may change the energy of traveling electron (such as electron–photon interactions), one may reasonably treat the electron transport through the junction as its tunneling between atoms and atomic groups along the molecular bridge. Correspondingly, the electron transmission through the molecule becomes expressed in terms associated with the barriers separating the sites. For instance, using a simple model describing a molecule as a set of $N + 1$ identical sites separated by identical potentials barriers, it was shown [72] that the electron transmission through this system is given by the expression

$$T = \frac{4\delta^N}{(\lambda_+^N + \lambda_-^N)^2}. \tag{1.17}$$

Here, δ is the electron transmission through a single barrier and $\lambda_\pm = 1 \pm \sqrt{1 - \delta}$.

Considering the electron transport through a molecule as a sequence of tunnelings between the sites referred to certain atoms, one may visualize the motion of the traveling electron along the molecule. This gives grounds to further explore the matter looking, for instance, for the preferable pathways for the electron transport. The concept of preferable or primary pathways was suggested two decades ago in the course of studying of long-range electron transfer reactions in proteins [73], and it was and still is used in studies of electron transfer and electron transport through big molecules such as proteins and DNA. In whole, the described way to visualize the electron motion through the molecular bridge is a useful complement to the principal model where the molecule is represented by the set of energy levels.

1.2 Gating and Electric Potential on the Bridge

As known, active electronics is mostly based on gate-controlled devices. Therefore, the ability to gate molecules is one of the critically important issues presently facing the molecular electronics. Putting aside for a while the discussion of relevant technical and experimental problems, we concentrate on qualitative analysis of the gate-controlled electron transport through MMM junctions. So, we introduce the third terminal (gate) along with already present electrodes, which are often called the source and the drain terminals. The schematics of the gated MMM junction is presented in Fig. 1.3. Applying the electrostatic potential V_G to the gate terminal, one may shift energy levels of the molecular linker from their original positions, thus changing transport properties of the system. When the gate is sufficiently strongly coupled to the molecule, the shift may become significant enough to move some energy levels out of/into the window determined by the bias voltage applied between the source and drain. Thus the number of conducting channels could be altered which affects the transport properties of the junction. This explanation is appropriate when we assume that the bias voltage is distributed between the leads and the positions of the molecule energy levels remain more or less undisturbed by the bias. We may consider another arrangement which is closer to typical experimental conditions by assuming that one of the electrodes is kept at fixed potential. Then the applied bias is moving up/down the electrochemical potential of the remaining electrode and the energy levels of the molecular bridge. The action of the gate is to oppose this bias-induced shifts keeping the bridge levels positions less dependent

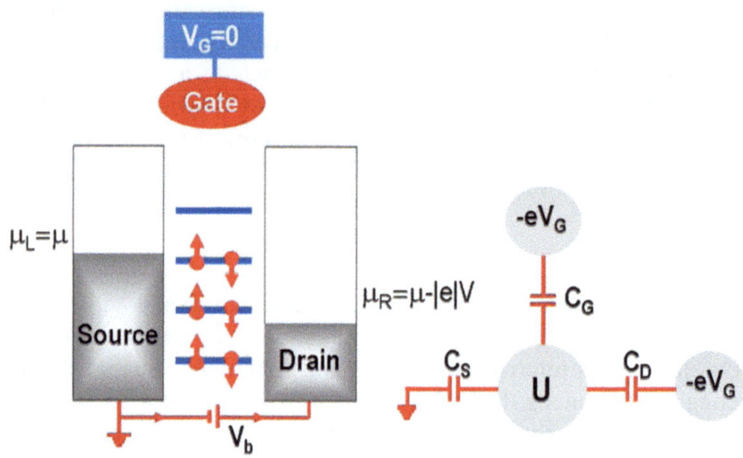

Fig. 1.3 *Left panel*: schematics of a metal–molecule–metal junction where the bias voltage V_b is applied in such a way that it provides the electron transport from the source (left lead) to the drain (right lead). The voltage V_G applied to the third terminal (gate) may shift energy levels of the molecular bridge, thus changing transport properties of the system. *Right panel*: a capacitive circuit model for the electric potential energy U on the bridge

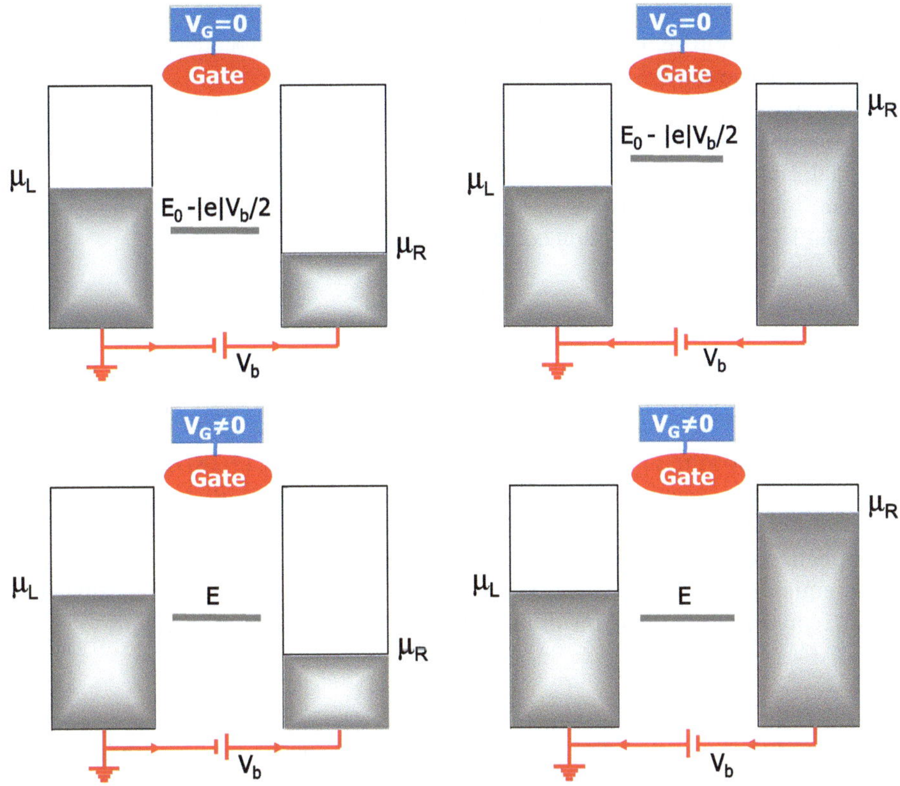

Fig. 1.4 Schematics of the energy-level diagram for the forward and reverse bias voltage plotted assuming that the left electrode is grounded in the absence (*top panels*) and in the presence (*bottom panels*) of the potential on the gate terminal

of the bias. The ultimate effect of the gating is the same as before. Strong gating may change the number of conducting channels participating in the transport for a certain bias voltage polarity.

For illustration, we again turn to the simple model of a junction where the molecule is represented by a single level, and we assume for certainty that the left electrode is grounded. At first we suppose that there is no potential on the gate. Then we may expect (as a first approximation) the molecular level to be shifted halfway between μ_L and $\mu_R - |e|V$ by the bias voltage V for both positive and negative bias voltage polarities (see Fig. 1.4). This means that the current will flow through the molecular bridge for either bias polarity with an equal magnitude. However, if we apply a sufficient potential on the gate, so that the bridge level becomes independent of the bias voltage, a very different picture emerges, as shown in the bottom panels of Fig. 1.4. Now, the bridge level is located between μ_L and $\mu_R - |e|V$ for positive bias ($V > 0$), so it serves as the conducting channel for this bias polarity. However, in the case of reversed polarity ($V < 0$), the bridge level falls out of this gap, and it ceases to participate in transport. As a result, the current–voltage curve should be very asymmetric showing the effect of current rectification.

The net electrostatic potential over the junction V_{tot} is formed by the combined contributions from the bias between the source and drain, the potential on the gate terminal, and the potential which appears as a result of the charge transfer between the source and drain electrodes and the molecule. When the molecule is neutral, the potential V_{tot} obeys the Poisson equation of the form

$$\nabla(\epsilon \nabla V_{\text{tot}}) = 0. \tag{1.18}$$

Here, ϵ is the relative dielectric permittivity of the system which could vary in space. Assuming again that the left (source) electrode is grounded, one may put boundary conditions as $V = 0$ (source electrode), $V = V_{\text{G}}$ (gate), and $V = V_{\text{D}}$ (drain electrode). Following [4, 5], we visualize the solution of Eq. (1.18) by introducing a model where the junction is simulated by the capacitive circuit shown in the right panel of Fig. 1.3. The total electric charge in the system $Q = CV_{\text{tot}}$ where C is the effective capacitance of the circuit is shown in Fig. 1.3. It could be obtained by summing up the capacitances $C_{\text{S}}, C_{\text{D}}$, and C_{G} corresponding to the capacitors which represent the source, the drain, and the gate electrodes, respectively:

$$C = C_{\text{S}} + C_{\text{D}} + C_{\text{G}}. \tag{1.19}$$

As the total charge is distributed among the elements of the junction (the bridge is supposed to be neutral), one obtains

$$CV_{\text{tot}} = C_{\text{D}} V_{\text{D}} + C_{\text{G}} V_{\text{G}}. \tag{1.20}$$

The corresponding potential energy U equals

$$U = -eV_{\text{tot}} = \frac{C_{\text{D}}}{C}(-eV_{\text{D}}) + \frac{C_{\text{G}}}{C}(-eV_{\text{G}}). \tag{1.21}$$

Now, we assume that the molecular linker in the junction is charged due to the presence of Δn electrons which arrived there from the leads in the process of the charge transfer. Then Eq. (1.20) is to be modified to take into account this charge on the bridge. As a result, we have

$$CV_{\text{tot}} = -e\Delta n + C_{\text{D}} V_{\text{D}} + C_{\text{G}} V_{\text{G}}. \tag{1.22}$$

Using this expression as well as the expression for the effective capacitance (1.19), we may present the potential energy in the form

$$\tilde{U} = U + \frac{e^2}{C}\Delta n. \tag{1.23}$$

By this equation, one introduces an important constant $U_0 = e^2/C$ which shows how much the potential energy changes when we bring one extra electron on the bridge. This quantity is called the single-electron charging energy.

So, two terms in Eq. (1.23) are corresponding to the contributions from the bias and gate voltages (U) and from the electron–electron interactions ($\frac{e^2}{c}\Delta n$), respectively. This electrostatic energy influences the electron density of states on the molecular bridge. As a zero approximation, we may replace $D(E)$ given by Eq. (1.1) by $D(E - \tilde{U})$ where

$$D(E - \tilde{U}) = \frac{1}{2\pi} \frac{\Gamma}{(E - \tilde{U} - E_0)^2 + (\Gamma/2)^2}. \tag{1.24}$$

Using the modified expression for the electron density of states, one may rewrite the expression for the level occupation and the electron current through the junction as follows:

$$n = 2 \int \frac{\Gamma_L f_L(E) + \Gamma_R f_R(E)}{\Gamma_L + \Gamma_R} D(E - \tilde{U}) dE, \tag{1.25}$$

$$I = \frac{2e}{\hbar} \int \frac{\Gamma_L \Gamma_R}{\Gamma_L + \Gamma_R} D(E - \tilde{U}) \{ f_L(E) - f_R(E) \} dE. \tag{1.26}$$

Unlike the previously presented Eqs. (1.10) and (1.11), these two expressions are coupled to each other because the electrostatic energy $\tilde{U}$ given by Eq. (1.23) includes the term which depends on the level occupancy "n." Therefore, the solution of Eqs. (1.23), (1.25), and (1.26) must be found by means of a self-consistent iterative method. As suggested by Datta [4], one can carry out their computations in two steps. First, Eqs. (1.23) and (1.25) are to be simultaneously solved bringing the converging sequence of iterations for the energy $\tilde{U}$, and after that the converged energy is to be employed to calculate the current using Eq. (1.26).

The appearance of the electrostatic energy $\tilde{U}$ in the expression for the electron current may lead to significant changes in transport properties of the considered junction. As discussed before, the combined action of bias and gate potentials (described by the term U in Eq. (1.23)) may result in current rectification. Also, it may bring about some other effects. Among them there is the so-called negative differential resistance (NDR). The NDR appears as the decrease of the current accompanying the increase of the bias voltage occurring within a certain range of the bias variation. The second term in Eq. (1.23) originating from electron–electron interactions is responsible for screening at the contacts between the molecule and the leads, and it opposes to the charge transfer processes. Under certain conditions, this opposition may cause blocking the contacts from adding electric charge to the molecular bridge, which results in a prominent insulating behavior of the junction known as Coulomb blockade.

Beyond the simple model adopted here, one needs much more sophisticated formalisms to analyze the effects of the gate voltage and Coulomb interactions. These formalisms are discussed in the following chapters. However, even the simplified model employed here clearly indicates the necessity of self-consistent iterative approach to solving the coupled equations for the current, electrostatic potential distribution and the occupation of the bridge level, inherent to advanced formalisms.

1.3 Electron–Electron Interactions: Coulomb Blockade

Single-electron charging effects in small-area tunnel junctions (including MMMs) may significantly affect transport characteristics, giving rise to a variety of observable phenomena such as Coulomb blockade and Kondo anomaly in the conductance through the junction. This happens due to the small electric capacitance of these junctions, which allows their electrostatic energy to be noticeably changed when even a single electron is tunneling through. The electron–electron interaction energy for a collection including N electrons is proportional to the number of interacting pairs:

$$U_{\rm ee}(N) = \frac{1}{2} U_0 N(N-1) \tag{1.27}$$

where U_0 is the average interaction energy per pair. If one adds an extra electron to the collection, this energy receives a correction $\Delta U(N) = U_{\rm ee}(N+1) - U_{\rm ee}(N) = U_0 N$, so U_0 may be treated as the single-electron charging energy. The effects of electron–electron interactions become important when the charging energy U_0 is greater than the thermal energy $k_{\rm B}T$ and the energy Γ which characterizes the coupling of the molecule to the leads and the consequent broadening of molecular energy levels. Electron–electron interactions contribute to the energy spectrum of a molecule. The molecule energy levels E_k differ from the "bare" energies E_k^0 computed ignoring interactions among the electrons. A simple approximation for the relation between E_k and E_k^0 was suggested by Beenakker [74]:

$$E_k = E_k^0 + U_0 N \pm \frac{1}{2} U_0 \tag{1.28}$$

where N is the number of electrons on the molecule and $\pm$ stands for adding/removing an electron. As for the constant U_0, in earlier works [74, 75] it was approximated in the same way as in Eq. (1.23), namely, $U_0 = e^2/C$ where C is the electrostatic capacitance of the quantum conductor such as a molecule or a quantum dot. This approximation is still used in some works (see, e.g., [76]). When the number of electrons significantly exceeds unity, one may omit the last term in Eq. (1.28) and express the correction to the "bare" energy E_k^0 arising due to electron–electron interactions by means of a self-consistent potential U_{SCF} (see e.g. [64]). Within the Hartree approximation,

$$U_{\mathrm{SCF}} = U_H = \frac{e}{2} \int d\mathbf{r}' \frac{n(\mathbf{r}')}{|\mathbf{r} - r'|}. \tag{1.29}$$

Here, $n(\mathbf{r})$ is the electron density, and the integration is carried out over the volume occupied by the system. As known, actual interaction energy takes on smaller values than those predicted by Eq. (1.29) because the electrons correlate their motions to avoid each other. As a result, the potential U_{SCF} acquires a negative term, which partially balances the Hartree term.

In general, the self-consistent field models reduce the effects of electron–electron interactions to the shifts of molecular orbitals determined by the potential U_{SCF}. We may illustrate this using a simple example of a MMM junction where the molecule is represented by a single energy level. This junction is described by the well-known Anderson Hamiltonian [77]:

$$H = H_L + H_R + H_M + H_T \tag{1.30}$$

where the terms H_β ($\beta = L, R$) correspond to the left and right lead, respectively; H_M describes the molecular bridge; and the term H_T takes into account the coupling of the molecule to the leads. The electron–electron interactions are included into the term H_M, which may be written in the form

$$H_M = \sum_\sigma E_0 c_\sigma^\dagger c_\sigma + U_0 c_\uparrow^\dagger c_\uparrow c_\downarrow^\dagger c_\downarrow. \tag{1.31}$$

Here, $c_\sigma^\dagger, c_\sigma$ are the creation and annihilation operators for the electrons on the molecule, E_0 is the "bare" energy for the single molecular orbital, and U_0 is the single-electron charging energy. Within the SCF approach, E_0 is replaced by

$$E_\sigma = E_0 + U_0 \langle n_{-\sigma} \rangle \tag{1.32}$$

where the occupation number $\langle n_{-\sigma} \rangle = \langle c_{-\sigma}^\dagger c_{-\sigma} \rangle$. Then the Hamiltonian H_M may be reduced to the form

$$H_M = \sum_\sigma E_\sigma c_\sigma^\dagger c_\sigma \tag{1.33}$$

Significant efforts were applied to find suitable expressions for U_{SCF} beyond Hartree approximation, and various models were developed [55, 56, 64, 78–83]. Applications of these models for electron–electron interactions to studies of molecular conduction bring controversial results. The models based on the self-consistent field approach appear fairly successful in describing some characteristics of the electron transport through molecular junctions, especially those observed at room

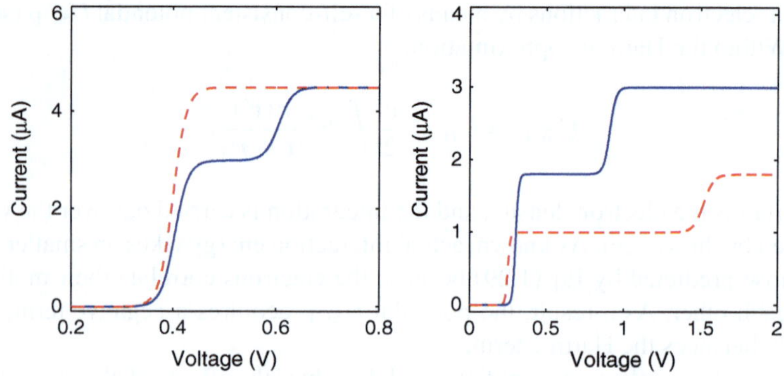

Fig. 1.5 Coulomb blockade in the MMM where the molecule is represented by a single orbital. The current–voltage curves are plotted assuming $T = 30\,\text{K}$, $U_0 = 0.5\,\text{eV}$, $E_0 = -0.2\,\text{eV}$. *Left panel*: I–V curves for a junction symmetrically coupled to the leads computed using the master equations (*solid line*) and the SCF approach (*dashed line*). *Right panel*: I–V curves for an asymmetrically coupled junction. The curves are plotted assuming $\Gamma_L = 2\Gamma_R$ (solid line) and $\Gamma_L = 4\Gamma_R$ (*dashed line*)

temperatures [84]. However, these models fail to adequately describe another important characteristics of electron transport including Coulomb blockade. As stated by Muralidharan, Ghosh, and Datta [83], the self-consistent field models in general are insufficient to properly treat nonequilibrium electron correlation effects regardless of specific improvements in the computations of the renormalized energy levels of the molecule. Better results in theoretical analysis of electron–electron correlations were achieved using "master equations" that is rate equations for the occupation probabilities $P(k, N)$ of many-body electron states $|k, N\rangle$ with the energies $E_k(N)$ [81, 83–85]. The rate equations express these probabilities in terms of transition rates between many-body states differing by a single electron.

The approach based on "master equations" brings good results in analyzing the electron transport through MMM junctions and other similar systems when the linker is weakly coupled to the leads, so that the charging energy U_0 is much greater than the coupling strength Γ. Within this regime the Coulomb blockade occurs, and the master equations approach allows to adequately describe this phenomenon, whereas self-consistent field models do not. For instance, it is known that the current–voltage characteristics for a junction including a single-site bridge exhibit two steps (see Fig. 1.5). The steps correspond to the gradual adding/removing of two electrons with the different spin orientations from the spin-degenerated bridge level. Assuming the symmetric contacts of the bridge with the leads ($\Gamma_L = \Gamma_R$), the first step in the current–voltage curve must be two times higher than the second one. This result has a clear physical sense. Obviously, one may put the first electron of an arbitrary spin orientation to the empty level on the molecular bridge. However, the spin orientation of the second electron put to the bridge level is determined by the spin orientation of the first one. So, there are two ways to add/remove the first electron to/from the empty/filled level on the bridge but only one way to add/remove the second one. We may remark that when $\Gamma_L \neq \Gamma_R$, the difference in the step heights is determined by the relation between the coupling parameters. For strongly asymmetric junctions, the steps become nearly equal in height. The master equation-based theory confirms this result. On the contrary, the self-consistent field models do not show two steps at all, or they bring the wrong values for the ratio of the steps heights, as was discussed in detail in [83].

However, the master equation-based methods also have limitations. While they bring sound results in analyzing electron–electron interactions induced effects on the transport characteristics of weakly coupled MMMs, their generalization to the case of stronger coupled junctions is not justified. For instance, these methods fail to theoretically describe Kondo peak in the molecular conduction. More sophisticated methods employing various modifications of the nonequilibrium Green's function formalism (NEGF) are developed to better analyze electron–electron interaction effects on the transport through molecular junctions and another systems including a quantum dot coupled to the charge reservoirs. These studies started by Meir et al. [86, 87] are still in progress (see, e.g., [34–37, 88–92]). These formalisms are discussed in the next chapter. Here, we briefly discuss the Hamiltonian for the junction taking into account electron–electron interactions. This Hamiltonian has the form (1.30). Assuming that the molecular linker is characterized by the set of orbitals and electrons belonging to different orbitals may interact as well as those belonging to the same energy level, we may write

$$\begin{aligned} H_M =& \sum_{k\sigma} \left(E_k^0 c_{k\sigma}^\dagger c_{k\sigma} + U_{0;kk} c_{k\sigma}^\dagger c_{k\sigma} c_{k,-\sigma}^\dagger c_{k,-\sigma} \right) \\ &+ \frac{1}{2} \sum_{k,k';\sigma,\sigma'\ (k\neq k')} U_{0;kk'} c_{k\sigma}^\dagger c_{k\sigma} c_{k',\sigma'}^\dagger c_{k',\sigma'}. \end{aligned} \tag{1.34}$$

Here, the third term describes Coulomb interactions between electrons associated with different energy levels. Usually, the leads are treated as reservoirs of noninteracting electrons, which results in the following expressions for the terms H_β:

$$H_\beta = \sum_{r\sigma} \epsilon_{r\beta\sigma} c_{r\beta\sigma}^\dagger c_{r\beta\sigma} \tag{1.35}$$

where $\epsilon_{r\beta\sigma}$ are the single-electron energies in the electrode β for the electron states r, σ and $c_{r\beta\sigma}^\dagger$, $c_{r\beta\sigma}$ denote the creation and annihilation operators for the leads. It is commonly assumed that the energy levels $\epsilon_{r\beta\sigma}$ are uniformly spaced over the ranges corresponding to the electron conduction bands of the metallic electrodes for both spin orientations, and the spacing between two adjacent levels is much smaller than the band width. The last term in the Hamiltonian (1.30) describing the coupling of the leads to the molecular linker has the form

$$H_T = \sum_{k,r,\beta,\sigma} \tau_{kr\beta\sigma}^* c_{r\beta\sigma}^\dagger c_{k\sigma} + H.C. \tag{1.36}$$

Here, the coupling parameters $\tau_{kr\beta\sigma}$ describe the couplings of the r, σ electron states belonging to the lead β to the k, σ energy level of the molecule.

1.4 Dephasing

In the previous analysis, we neglected all scattering events which may occur when electron is traveling along the molecular bridge assuming that such events solely happen at the contacts between the molecule and the electrodes. Under these assumptions, the phase of an electron remains conserved as it travels between the electrodes. Correspondingly, the electron current through the junction is described by the Landauer formula (1.11). However, electron transport though practical molecular junctions is always somewhat affected by phase-breaking scattering events. In general, electronic dephasing occurs due to entanglement between the electronic degrees of freedom and these of an external object (such as a phonon, photon, or spin) provided that the dynamics of this object is governed by the thermal environment. When an electron is scattered by such an object, the phase information transferred from the electron is rapidly dissipated due to the coupling of the scatterer to the thermal bath. This leads to the loss of the phase coherence.

The simplest way to analyze electron transport in the presence of dephasing was suggested by Buttiker [72], who has proposed to treat each scattering center as a virtual contact which draws zero net current. Introducing local distribution functions for the electrons on these virtual contacts $f_i(E)$, one may write the expression for the current I_i at an arbitrary terminal "i" summing up the currents flowing there from all remaining terminals (real as well as virtual) included into the system. To further simplify the following analysis we exclusively concentrate on the dephasing and momentum scattering and ignore electron transitions between different energy levels on the molecule. Then the electron transport remains elastic although incoherent, and contributions to the current I_i from the diverse terminals could be described by the expressions similar to the Landauer formula:

$$I_i = \frac{e}{\pi\hbar} \sum_j \int dE T_{ij}(E) \big[f_i(E) - f_j(E) \big]. \tag{1.37}$$

Implying that the net current associated with every virtual terminal must equal zero, we can derive the system of linear algebraic equations including the transmission coefficients $T_{ij}(E)$ and electron distribution functions on the terminals $f_j(E)$. Solving these equations we express the distribution functions for the scattering centers (not necessary Fermi–Dirac) in terms of Fermi functions associated with actual terminals. Assuming that the system includes only two terminals with the corresponding Fermi distribution functions $f_L(E)$ and $f_R(E)$ and a single-breaking scatterer with the electron distribution function $f_s(E)$, we get

$$f_s(E) = \frac{T_{LS}(E)f_L(E) + T_{SR}(E)f_R(E)}{T_{LS}(E) + T_{LR}(E)}. \qquad (1.38)$$

Substituting this result into Eq. (1.37) we may reduce the expression for the electron current to the form given by Eq. (1.11) where the electron transmission $T(E)$ is presented as follows:

$$T(E) = T_0(E) + \frac{T_{LS}(E)T_{SR}(E)}{T_{LS}(E) + T_{SR}(E)}. \qquad (1.39)$$

Here, $T_0 \equiv T_{LR}(E)$ is the transmission corresponding to the coherent electron tunneling through the molecule, and the second term describes the effect of dephasing. This result gives grounds for interpretation of the electron transport through the considered system as a combination of coherent tunneling through a scattering-free channel with the resistance proportional to $T_{LR}^{-1}(E)$ connected in parallel with a scattering-dominated channel represented by two resistors connected in series with the resistances proportional to $T_{LS}^{-1}(E)$ and $T_{SR}^{-1}(E)$, respectively [5].

In principle, the loss of coherence does not necessary lead to energy dissipation. One may consider the above-discussed situation when electrons traveling through the MMM junction suffer dephasing, but their transport remains solely elastic, so that the changes in the electron energies are forbidden. However, in practical systems this situation is highly unlikely to occur. Usually the dephasing is accompanied by the energy dissipation. The usual and most frequent source of electron phase breaking occurring in the process of electron transport through molecules is electron scattering by thermally activated nuclear motions. Since the combined electron–phonon system evolves coherently [93], the phase information which is transferred from the scattered electrons to the environment should be eventually restored after a certain time interval τ_r (which is an analog of the Poincare recurrence time in dynamical systems). The characteristic time τ_r rapidly increases as the number of the environmental degrees of freedom enhances. In practical MMM junctions, τ_r is much greater than other characteristic times associated with the electron transport. This makes the loss of phase coherence practically irreversible.

The importance of dephasing/dissipative effects depends on several factors. Among these factors, the temperature and the size and complexity of the molecular bridge are predominating. The temperature determines the intensity of stochastic nuclear motions, and the molecule size determines the contact time, that is, the time for electron to travel through the junction and to undergo the effects of the environment. As shown by Buttiker and Landauer [94], the contact time is proportional to the number of sites on the molecular linker which provides the intermediate states for the electron tunneling. For small molecules and at low temperatures the contact time is shorter than characteristic times of the stochastic fluctuations in the environment, and the dephasing remains rather weak. Under these conditions one may expect only small extra broadening of the molecular energy levels to occur due to the effect of the phonons representing the thermal environment and, correspondingly, treat the electron transport as nearly coherent.

For longer molecules and at higher temperatures the effect of these thermal phonons grows stronger. When the dissipative effects are strong enough, they may bring essential changes into the conduction mechanisms causing the transition from the elastic transport by the electron tunneling to the thermally assisted hopping transport mechanism. The latter is typical for disordered materials with localized electron states such as conducting polymers [95, 96] and for carbon nanotubes [97]. When the hopping transport mechanism works, the thermal phonons act as a source of the electrical conductivity. Accordingly, the hopping contribution to the electron transmission increases as temperature rises. On the contrary, the tunnel transmission through the junction reduces when the temperature goes up. Another significant difference between the tunnel and hopping transmission is manifested in their respective dependencies of the molecule length L. As mentioned above, the tunneling transmission exhibits the exponential fall as L increases. The hopping transmission decreases significantly slower, being inversely proportional to the length L [70]. The crossover from tunneling to thermally assisted transport could not be identified with the loss of coherence. Strictly speaking, thermal-assisted electron transport from a lead to a molecular bridge may be followed by its coherent propagation along the molecule. However, while these crossovers are not identical, they occur under similar conditions. So, when thermally activated transport predominates, one may expect the decoherence within the molecular bridge to be significant.

Early measurements of transport characteristics of metal–molecular junctions were carried out for systems with short molecular linkers. In these systems, the electron–molecule contact time took on values within the subfemtosecond range even at room temperatures. This implied a weak effect of thermal phonons on the traveling electrons resulting in coherent transport. The loss of coherence was provided by gating which reduced the injection gap [9]. In this way, the contact between the traveling electron and the molecule was prolonged giving sufficient time for the decoherence to develop. As an example, the temperature dependencies of electron current through a MMM junction with a tercyclohexylidene molecule serving as a linker are shown in Fig. 1.6. The presented curves give evidence of the crossover from the tunneling to the thermally activated transport. The dependence of molecular conductance from the length of the molecule may be easier observed on large molecules. An interesting demonstration of the transition from tunneling to thermally activated hopping was seen in junctions with DNA linkers [30]. It was demonstrated that an exponential length dependence typical for electron tunneling through short DNA segments was replaced by an inverse length dependence observed on longer segments (see Fig. 1.7).

1.5 Electron–Phonon Interactions

Electron transport through practical molecular junctions is always affected by nuclear motions. The interplay between electronic and nuclear dynamics is a significant factor with important, sometimes crucial, implications for the transport

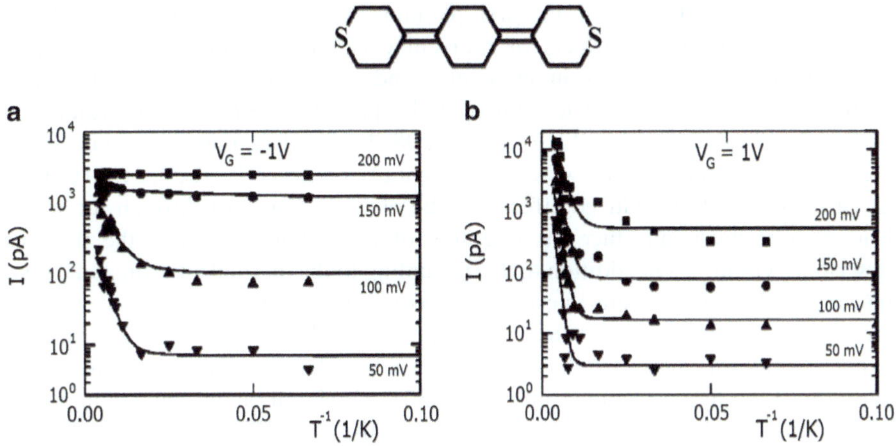

Fig. 1.6 A crossover from tunneling to thermally activated transport through sulfur end-functionalized tercyclohexylidene molecules linking gold electrodes [9]. Reprinted with the permission of the American Chemical Society (ACS)

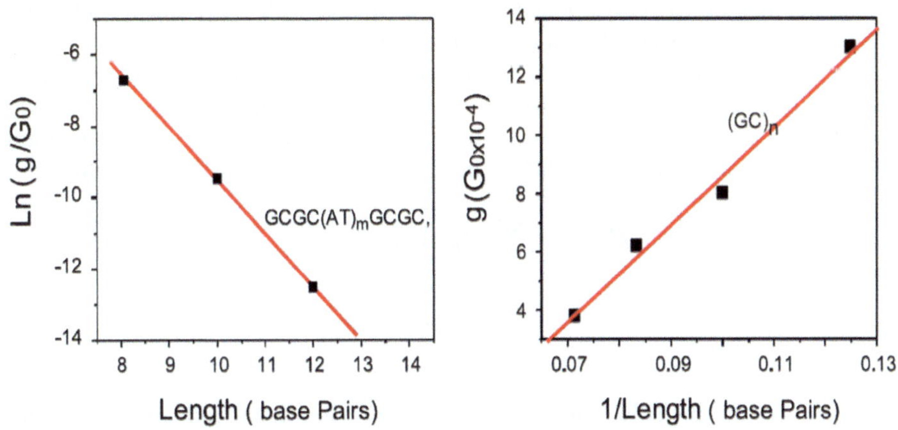

Fig. 1.7 Turnover from tunneling to hopping observed in short strand duplex DNA molecules (after [30]). Reprinted with the permission of the ACS

characteristics. As discussed in the previous section, stochastic fluctuations of the electrostatic potential originating from thermally activated motions of ions cause loss of coherence and energy dissipation in the process of electron transport. Electron coupling with a phonon thermal bath is directly related to the issue of junction heating and, consequently, junction ability to remain stable while the ambient temperature rises making nuclear motions more intensive. This, in turn, raises the important problem of heat conduction by molecular junctions and, in general, of thermoelectric properties of such systems. At present, various aspects of heat transfer through MMM junctions are being intensely studied.

Besides the influence of thermal phonons, electrons traveling through the junction experience the effects of molecular vibrations. Theoretical studies of electron–vibron interactions in molecules and other similar nanosystems (e.g., carbon nanotubes) were carried out over the past decade by a large number of authors (see, e.g., [19, 20, 22, 24–26, 34, 98–109]. Also, manifestations of the electron–vibron interactions were experimentally observed [2, 8, 23, 29, 110–117]. To analyze vibration-induced effects on the electron transport through molecular bridges, one must introduce the coupling of molecular orbitals to the phonons describing vibrations in addition to their coupling to the thermal environment. Also, one must take into account interactions between the phonons associated with the thermal nuclear motions and those associated with the molecule vibrations. As a result, extra terms appear in the expression for the Hamiltonian of the junction. In addition to the terms included into Eq. (1.30), the Hamiltonian now contains two extra terms describing the phonons, which have the form [25, 26]

$$H_{\text{ph}} = \sum_l \hbar\Omega_l a_l^\dagger a_l + \sum_m \hbar\omega_m b_m^\dagger b_m \tag{1.40}$$

where the first addend corresponds to vibronic modes with the frequencies Ω_l, and the second one describes the set of phonons representing the thermal environment whose frequencies are denoted as ω_m. Accordingly, $a_m^\dagger, a_m$ and $b_m^\dagger, b_m$ are creation and annihilation operators for the vibrational and thermal phonons, respectively. Besides, one must include into the Hamiltonian terms $H_{\text{e-ph}}$ and $H_{\text{ph-ph}}$ describing electron–phonon and phonon–phonon interactions. Following [25], we write

$$\begin{aligned} H_{\text{e-ph}} = & \sum_{k,\sigma,l} \Lambda_{kl} A_l c_{k\sigma}^\dagger c_{k\sigma} \\ & + \sum_{k,\sigma,m} \lambda_{km} B_m c_{k\sigma}^\dagger c_{k\sigma}; \end{aligned} \tag{1.41}$$

taking into account solely interaction between vibrational modes and thermal phonon bath, we have

$$H_{\text{ph-ph}} = \sum_{lm} M_{lm} A_l B_m. \tag{1.42}$$

Here, $A_l = a_l^\dagger + a_l$, $B_m = b_m^\dagger + b_m$, and the parameters Λ_{kl}, λ_{km}, and M_{lm} have the dimensions of energy and represent electron–phonon and phonon–phonon coupling strengths.

As follows from Eq. (1.40), the vibrational contribution to the bridge Hamiltonian is represented by a set of harmonic normal modes. The electron–phonon interaction terms given by Eq. (1.41) are taken to be linear in the phonon coordinate operators A_l and B_m. The term $H_{\text{ph-ph}}$ is bilinear in these coordinates. The bilinear form for this term is convenient because it yields an exact expression for the self-energy of vibrational phonons appearing as a result of their interaction with thermal phonon bath. However, the bilinear form given by Eq. (1.42) may appear to be unappropriate in situations when one considers molecular interactions with condensed environments. In these situations, it may happen that the Debye cutoff frequency characterizing the environment is smaller than frequencies of vibrational modes. Relaxation of molecular vibrations is then caused by multiphonon processes that result from nonlinear interactions. One may mimic the multiphonon process by introducing a continuous spectral function $\rho_{\text{ph}} \sim \dfrac{\omega}{\omega_c} \exp\big[-\omega/\omega_c\big]$ describing thermal phonons [93].

The total Hamiltonian of the considered system may be obtained by combining terms (1.30) and (1.40)–(1.42). In solving different problems the phonon-related terms may be distributed between the exactly solvable (zero-order) part of the total Hamiltonian and its perturbation part in different ways. This especially applies to the terms $H_{\text{e-ph}}$ and $H_{\text{ph-ph}}$. These terms could be included into the perturbation part along with the term H_T associated with the coupling of the molecular bridge to the leads [see Eq. (1.36)]. Also, under certain conditions, these terms could be included into zero-order Hamiltonian bringing renormalizations of electron energies and of frequencies of vibrational modes. For instance, within the often used polaron model the vibrational contribution to $H_{\text{e-ph}}$ is replaced by a diagonal term

$$\tilde{H}_{\text{e-ph}} = \sum_{k,\sigma} \Lambda_k A_k c_{k\sigma}^{\dagger} c_{k\sigma} \tag{1.43}$$

whereas the contribution from the thermal phonons is omitted. Within this model, the energies of the bridge levels E_k get corrections ΔE_k called reorganization energies or polaronic shifts. The expression for the polaronic shift has the form

$$\Delta E_k = \sum_k \frac{(\Lambda_r)^2}{\hbar \Omega_r}. \tag{1.44}$$

The values of polaronic shifts provide good estimations for the electron–vibron coupling strengths. Particular manifestations of electron–vibronic effects in the transport characteristics of a MMM junction are determined by the relation of the relevant energies. Putting aside electron–electron and phonon–phonon interactions, these are the coupling strengths of the molecule to the leads $\Gamma_{L,R}$, electron–phonon coupling strengths Λ_{kl} corresponding to the interactions of traveling electrons with vibrational phonons, thermal energy $k_{\text{B}}T$ which determines the effects of the thermal environment, and the value of the energy gap E_g between HOMO and LUMO and the nearest lead chemical potential. When the molecule is weakly coupled to the leads ($\Gamma_{L,R} \ll \Lambda, \hbar\Omega$) and the temperature is low ($k_{\text{B}}T \ll \Gamma_{L,R}$), the electron transport through the junction may give rise to a strong vibrational excitation, and one may expect vibration-induced features to appear in the current–voltage characteristics [107, 110]. For stronger coupled junctions, electron–vibron interactions are not revealed in the current–voltage curves themselves, but they remain manifested in the inelastic tunneling spectra of the traveling electrons [25].

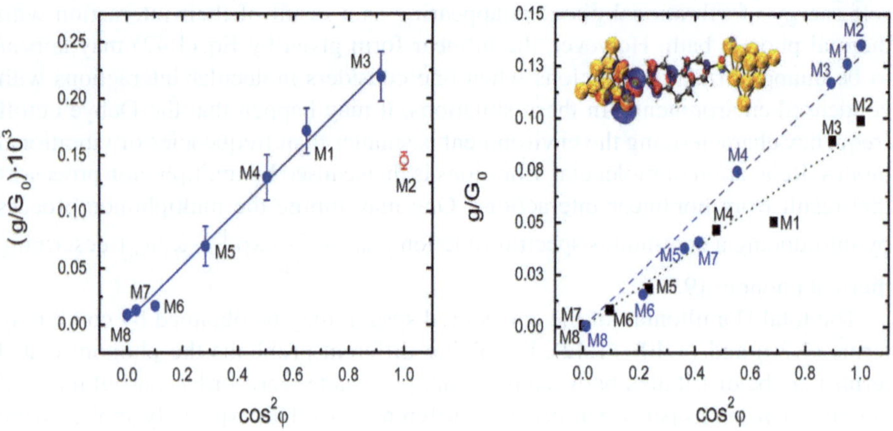

Fig. 1.8 *Left panel*: experimentally determined conductance of thiol-terminated biphenyl molecules as a function of torsion angle φ. The *solid line* is the linear fit to all studied molecules excluding the molecule M2. *Right panel*: conductance vs $\cos^2\varphi$ computed using DFT assuming atop–atop (*circles*) and bridge–bridge (*squares*) bonding configurations where the sulfur is bonded to one or two gold atoms on the leads, respectively (after [118]). Reprinted with the permission of ACS

These manifestations could differ depending on the relation between the energy gap E_g and the energies of the relevant vibrational phonons. When E_g significantly surpasses the latter as well as the corresponding coupling strengths ($E_g \gg \hbar\Omega, \Lambda$), one observes an off-resonance IETS signal (or set of such signals). When E_g becomes small (either due to a large bias voltage applied across the system or due to a suitable gating), the conditions for resonance tunneling of electrons through the junction become satisfied. This may lead to qualitative changes in the IETS signals which may display features associated with the vibrational structure of intermediate molecular ions [26].

In addition, some important transport characteristics of MMM junctions strongly depend on nuclear conformations of the molecular bridges. The effect of molecular geometry on the conductance was observed in experiments [14, 118]. Different configurations of the bridge ions could be considered as "frozen" vibrations. Correspondingly, their effect on electronic properties of the junctions could be treated within the formalism describing electron–vibron interactions. Configurational modulations of electronic properties may lead to changes in the conductance spectrum arising due to displacements of nuclear coordinates which are similar to the well-known geometry dependencies of optical spectra. These changes in the molecular conductance could be quite strong. For instance, it was reported that the conductance of biphenyl molecules was decreased by a factor of 30 following the variation of torsion angle between the phenyl rings from 0° to 90° [118]. These results are reproduced in Fig. 1.8. Within an extreme situation, a MMM junction may exhibit a nonzero conductance only provided that the bridge has a certain particular configuration being insulating otherwise.

The dependence of molecular conductance of the arrangement of ions belonging to the molecule is a breakdown of the Condon approximation. Within this approximation, the electron transport is assumed to be independent of the molecular geometry. Nevertheless, these dependencies are experimentally observed as a phenomenon known as stochastic switching. The stochastic switching could be qualitatively explained as follows. In molecular junctions, changes in the geometry in which the electron transport occurs modulate the electronic Hamiltonian. This in turn leads to variations in the conductance. When the geometric changes are given rise by random excitations, the conductance variations also appear in a random, stochastic fashion. Recent studies clearly demonstrate that modulations in the geometry of interfacial atoms in MMM junctions may result in changes in the conductance ranging from factors of several fold up to factors as large as 1,000. This may happen when the coupling strength along the electrode/molecule tunneling direction changes [59]. The stochastic switching phenomenon, as well as other manifestations of the electron's interactions with the molecules vibrational modes, is discussed in more detail in the following chapters.

1.6 Spin Transport

Recently, molecular spintronics has attracted significant attention of the research community due to its promising potential in designing of high-speed and low-energy consuming quantum electronic devices (see, e.g., [119–128]). The molecular spintronics started from the pioneering experiment of Tsukagoshi and coworkers [129], who injected spin-polarized electrons into a carbon nanotube and found that the spin diffusion length in this system was as large as 130 nm. However, potentialities of the electron spin in bringing memory and logic functionalities on the same nanodevice were already recognized in semiconductor physics at the time when the Tsukagoshi's experiment had been performed. The electron spin may be treated as the ultimate logic bit. As was demonstrated in experiments on semiconductors, it preserves coherence over long times and distances [130, 131]. This offers excellent opportunities for the electron spins to be used in semiconductor-based quantum logic devices first suggested in [132].

Electron spin directions may be manipulated using magnetic fields and the effects of spin–orbit interactions, as had been first proposed by Datta and Das [133]. The spin–orbit interaction is a relativistic effect, which couples the spin degree of freedom of an electron to its orbital motion. In general, the spin–orbit Hamiltonian H_{SO} can be written in the form

$$H_{SO} = V_{SO} \mathbf{S} \cdot \mathbf{L} \tag{1.45}$$

where $\mathbf{S}, \mathbf{L}$ are the spin and angular momentum of the electron, respectively. The factor $V_{SO}(\mathbf{r})$ is related to the electrostatic potential controlling the electron motion. Thereby, the spin–orbit interaction may be a useful handle for manipulating the electron spin direction by external electric fields and/or gate voltages. It is responsible for the spin Hall effect [134–137], which is a good example of the electrical spin manipulation. Another important interaction which affects the spin transport is the so-called hyperfine interaction between electron and nuclear spins. Assuming that a single electron with the spin $\mathbf{S}$ interacts with a nucleus, whose spin equals $\mathbf{S}_N$, we may present the corresponding Hamiltonian in the form

$$H_{\text{hyp}} = \alpha_{\text{hyp}} \mathbf{S} \cdot \mathbf{S}_N \tag{1.46}$$

where α_{hyp} is the hyperfine coupling strength. In certain semiconductors, the hyperfine interaction was shown to be useful for controlling nuclear spins by optically polarized electron spins [138]. At the same time, both spin–orbit and hyperfine interactions are very important sources of spin dephasing [139]. These interactions may significantly reduce both durations and distances of spin coherence, thus worsening the conditions for operations of spintronic devices.

Metal–molecular junctions possess all important properties which spin-electronics needs. Electronic structure and conductivity of molecules are controllable, so the linkers with desirable properties could be designed and inserted into the MMM junctions. Both paramagnetic and ferromagnetic molecules are available. Also, the molecules may be attached to the leads in diverse manners, which allows to manipulate transport properties of the junction more efficiently than those of semiconductor-based nanodevices. Valence band spin splitting in various semiconductors is significantly greater than in carbon-based materials including organic molecules commonly used as linkers in MMM junctions [119]. This indicates a relative weakness of the spin–orbit interactions in these materials. Also, most of the molecules used for spin transport are π-conjugated molecules where the electrons mostly travel through molecular states localized near carbon atoms and the predominating isotopic form of carbon (^{12}C) is not hyperfine active. So, intrinsic molecular properties of molecular linkers could provide weak spin–orbit and hyperfine interactions, if necessary. Weakness of these interactions creates opportunities to prolong the spin lifetime and to increase the spin diffusion length. This is important in designing devices using an electron spin as the logic bit.

The prototype of all spin devices is the spin valve, which behaves as a spin polarizer/analyzer. It was demonstrated that MMMs may exhibit the spin-valve-like behavior. One of the first attempts to analyze a molecular spin polarizer was carried out by Pati and coworkers [140, 141]. They considered a junction where a molecule was sandwiched in between two magnetic clusters. The clusters were coupled to nonmagnetic leads (see Fig. 1.9). This junction provides a difference between the electric currents I_σ corresponding to the different spin orientations. The spin polarization P of the junction may be defined as follows:

$$P = \frac{I_\uparrow - I_\downarrow}{I_\uparrow + I_\downarrow}. \tag{1.47}$$

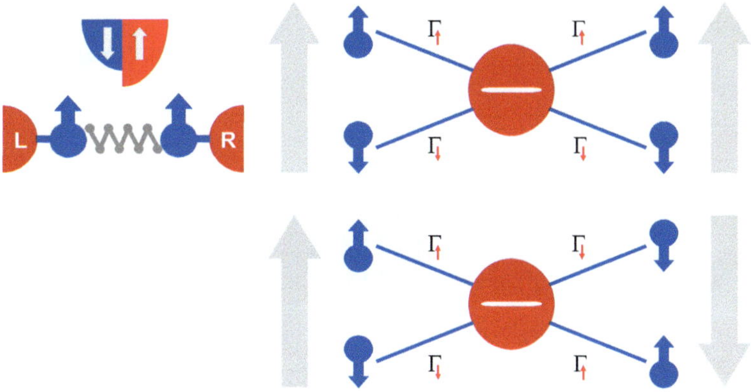

Fig. 1.9 *Left panel*: schematics of spin-splitted manifolds with different occupations typical for the conduction band of a magnetic metal (*top*) and the molecular spin polarizer studied in [140,141] (*bottom*). *Right panel*: schematic drawing which illustrates electron transport through a junction with magnetic leads for different orientations of magnetic moments on the leads

More efficient molecular spin valves are the junctions with ferromagnetic leads, and they were studied in numerous works (see, e.g., [119, 120, 125–127, 142, 143]. Spin transport through such a junction is schematically illustrated in the right panel of Fig. 1.9 where the molecule is represented by a single state. As known, the Fermi level of magnetic metals such as Fe, Ni, and Co is located in that region of density of states where d states are predominating. These states are spin split, and difference Δ between two split states takes on considerable values of the order of 1 eV. As a result, the Fermi surfaces for electrons with different spin orientations differ, and one of the two spin-split manifolds (majority band) is more occupied than the other (minority band). Also, the parameters $\tau_{kr\beta\sigma}$ describing the coupling of the electron states on the ferromagnetic leads to the molecule states differ for different spin orientations of the electrons belonging to the leads. These two factors are responsible for the difference in the currents I_σ and, consequently, for the spin polarization created by the junction. As shown in Fig. 1.9, the junction spin polarization strongly depends on the mutual orientations of the magnetic moments on the leads. We illustrate this neglecting for simplicity of all interference effects. We introduce hopping parameters $\Gamma_{\uparrow\beta}$ and $\Gamma_{\downarrow\beta}$ ($\beta = L, R$) to characterize the coupling of the leads to the molecular state. For the same lead these parameters take on different values determined by the spin orientation of an electron participating in the transport and by the magnetic moment of the lead ($\Gamma_{\uparrow\beta}$ and $\Gamma_{\downarrow\beta}$ correspond to the parallel and antiparallel spin orientation, respectively). Assuming that the junction is symmetrically coupled ($\Gamma_{\sigma L} = \Gamma_{\sigma R} \equiv \Gamma_\sigma$) and the magnetic moments of the leads are parallel, one may conclude that the transmission of majority electrons is proportional to $\Gamma_\uparrow^2$ and that of minority electrons is proportional to $\Gamma_\downarrow^2$ [119]. When the magnetic moments of the leads are antiparallel, the transmission does not depend on the spin orientation of a traveling electron and could be

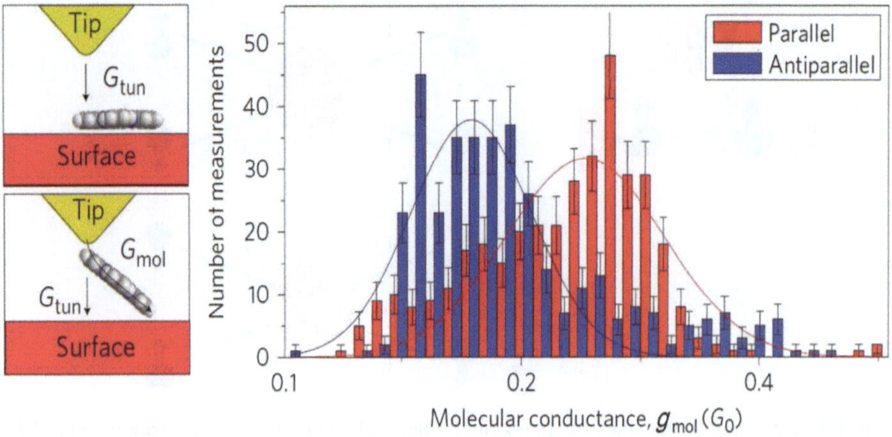

Fig. 1.10 *Left panel*: schematics showing the formation of a Co–hydrogen–phthalocyanine–Co junction. *Right panel*: histograms of molecular conductances for parallel and antiparallel orientations of magnetic moments on the electrodes observed in experiments (after [142]). Reprinted with the permission of the Nature Publishing Group (NPG)

estimated as $\Gamma_\uparrow\Gamma_\downarrow$. Obviously, this gives evidence of different spin polarizations of the junction for the two considered cases. We remark that the best efficiency of a molecular spin valve shown in the right panel of Fig. 1.9 should be achieved when only electrons with a certain spin orientation could travel through the molecular bridge. Then the current in the case of antiparallel magnetic moments of the leads vanishes, and the effect of giant magnetoresistance (GMR) may be observed. The spin polarization of such a junction with parallel orientations of magnetic moments equals 1, so it behaves as an ideal spin polarizer.

Recently, the GMR effect was reported in MMM systems consisting of two ferromagnetic electrodes linked by a hydrogen phthalocyanine molecule [142]. These molecules were first adsorbed onto cobalt islands set on the copper surface, and then the contact with STM tip positioned above the aromatic side groups of the molecules was established. This is schematically illustrated in Fig. 1.10. Measurements of the molecular conductance for parallel (g_P) and antiparallel (g_{AP}) orientations of magnetic moments on the electrodes showed a significant difference between them corresponding to GMR $= (g_P - g_{AP})/g_{AP} \approx 60\,\%$.

Transport studies in metal–molecular junctions with magnetic leads should be based on the corresponding Hamiltonian for the junction. Neglecting for simplicity the effects of phonons and putting aside for a while the spin–orbit interactions, we may present the Hamiltonian in the form (1.30) where the term H_M describing the molecule itself is determined by Eq. (1.34). Also, the expressions for H_β hold on their form (1.35) where

$$\epsilon_{r\beta\sigma} = \epsilon_{r\beta} + \sigma M_\beta. \tag{1.48}$$

Here, M_β is the magnitude of magnetization of the corresponding lead and $\epsilon_{r\beta}$ are eigenenergies of the Hamiltonian H_β for a nonmagnetic lead. Now, the transfer Hamiltonian could be presented as follows:

$$H_T = \sum_{k,r,\beta,\sigma} \tau^*_{kr\beta\sigma} \left(\cos\left[\frac{\theta_\beta}{2}\right] c^\dagger_{r,\beta,\sigma} - \sigma \sin\left[\frac{\theta_\beta}{2}\right] c^\dagger_{r,\beta,-\sigma} \right) \times \exp\left[\frac{i\sigma\varphi_\beta}{2}\right] c_{k\sigma} + H.C. \tag{1.49}$$

where the angles $\varphi_\beta, \theta_\beta$ determine the position of the magnetization M_β in the coordinate system whose z axis is directed in parallel with the electron spin. Including into consideration the spin–orbit interactions one needs to further change the Hamiltonian of the junction. As shown in [143], the spin–orbit interaction causes significant changes in the transfer Hamiltonians which accept the form

$$H_L = \sum_{k,r,\sigma} \tau^{*(L)}_{kr\sigma} \left(\cos\left[\frac{\theta_L}{2}\right] c^{\dagger(L)}_{r,\sigma} - \sigma \sin\left[\frac{\theta_L}{2}\right] c^{\dagger(L)}_{r,-\sigma} \right) \times \exp\left[\frac{i\sigma\varphi_L}{2}\right] c_{k\sigma} + H.C. \tag{1.50}$$

$$H_R = \sum_{k,r,\sigma} \tau^{*(R)}_{kr\sigma} \left(\cos\left[\frac{\theta_R}{2}\right] c^{\dagger(R)}_{r,\sigma} - \sigma \sin\left[\frac{\theta_R}{2}\right] c^{\dagger(R)}_{r,-\sigma} \right) \times \exp\left[\frac{i\sigma\varphi_R}{2}\right] \exp\left[-i\sigma\Phi_{\rm SO}\right] c_{k\sigma} + H.C. \tag{1.51}$$

The factor $\exp\left[-i\sigma\Phi_{So}\right]$ describes the spin precession originating from the spin–orbit interaction.

Also, the spin–orbit interaction leads to the emergence of a new term in the junction Hamiltonian. According to [143] this term is described by the expression

$$H_{\rm SO} = \sum_{k,k'} t_{k,k'} c^\dagger_{k,\sigma} c_{k',-\sigma} + H.C. \tag{1.52}$$

where the parameters $t_{k,k'}$ obey the relationship $t_{k,k'} = -t_{k',k}$, which means that all diagonal terms t_{kk} vanish. Off diagonal terms describe spin flips between different energy levels on the molecular bridge caused by spin–orbit interactions. However, according to Eq. (1.52), the molecular states remain at least twofold degenerated due to the symmetry properties of the parameters $t_{k,k'}$. This guarantees that in equilibrium the molecular linker is not spin polarized. So, the expressions (1.30), (1.34), (1.46), and (1.50)–(1.52) describe the MMM junction including a nonmagnetic molecule linking ferromagnetic leads.

Presently, significant interest of the research community is concentrated on transport properties and spintronic potentialities of magnetic molecules, that is, the molecules with a global net spin **S** [144]. As mentioned before, carbon-based molecules commonly used as linkers in MMM junctions do not possess a noticeable spin. The occurrence of the net molecular spin is caused by the transition metal ions inserted into an original carbon-based molecules [145]. As an example one may cite Mn_{12}, $(Mn_{12}O_{12}(CH_3COO)_{16}(H_2O)_4)$ molecules, whose transport properties were experimentally studied in [38, 39] and/or vanadium-iron-cyclopentadienyl and similar molecular complexes studied by Wu and coworkers [127]. Specific features of magnetic molecules originate from the fact that the occurrence of a nonzero molecular spin **S** lifts the degeneracy of spin-multiplet states and introduces the magnetic quantum number M_S which labels the energy levels $(-S < M_S < S$ where S is the molecular spin magnitude). In the absence of external perturbations, the magnetic molecule may remain in a certain M_S state. However, a perturbation creates mixing between M_S states. Consequently, transitions between various M_S levels become possible, and they obey selection rules determined by the particular symmetry of the perturbation. Transport calculations considering electron transport in magnetic molecules were carried out in several works (see, e.g., [146–150]. Theoretical studies of the electron transport through magnetic molecules are not completed so far. One of the reasons is that the results of such calculations crucially depend on the characteristics of the electron structure of considered molecules. The latter may be obtained in the course of *ab initio* computations, which had been and are being carried out [151–155].

There exists an interesting possibility for designing of all-molecular spin valves where both spin injector and spin detector are incorporated into the molecule itself and the sole function of the leads is to serve as electron reservoirs. Such a junction resembles that is analyzed in [140, 141]. It was suggested [156] that all-molecular spin polarizers may be engineered using dicobaltocene molecules, each including a pair of Co ions separated by a spacer. Among other intriguing phenomena related to the molecular magnetism, we should mention the effects occurring due to the interaction of a magnetic molecule with a surface of a ferromagnetic film where the molecule is positioned [157]. Also, an induced magnetic moment could appear on an originally nonmagnetic molecule/carbon nanotube placed on such a film [158, 159]. Finally, ferromagnetism was reported in molecular complexes made out of gold nanoparticles and organic molecules that is in the systems consisting of nonmagnetic elements [160–162]. Presently, these and other phenomena are being investigated.

1.7 Building of Metal–Molecule–Metal Junctions

In the following two sections we briefly discuss experimental issues. The experimental realization of metal–molecule–metal junctions has been provided by the development of both adequate chemical methods for preparation of molecular/electrode interfaces appropriate for forming metal–molecule contacts and nanoscale characterization techniques. Vital to experimental studies of metal–molecular junctions is the ability to form atomic-size metallic contacts. Commonly, such contacts are created by growing molecules on the metal substrate and using a scanning probe as a top contact [82, 163]. In some experiments a metal cluster or a mercury droplet were grown on the top of the molecule. This allows both ends of the molecule to be chemisorbed to the metal contacts avoiding large vacuum barriers typical for scanning tunneling microscopy (STM) probing techniques [164]. Also, the contacts may be formed by mechanically breaking a thin metal wire [the so-called mechanically controlled break junctions (MCBJ)] and by electromigration. When the electromigration method is employed, a metal wire is prepared on the substrate and then broken by driving a strong electric current through it. In addition, atomic-size contacts may be formed electrochemically, by immersing the electrodes together with a counter electrode into an electrolyte solution. The formation of an atomic-size contact between a molecule and electrodes is illustrated in Fig. 1.11. The metal–molecule contacts could be established when the molecule possesses anchor groups at the ends which are designed to bond to the metal contacts. With gold electrodes which are often used in building up MMM junctions, the anchor groups are usually thiol ($-s$-) and/or amino ($-NH-$) groups [165]. The thiol (sulfur) groups at the ends of the molecule work as "alligator clips" attaching the molecule to the gold atoms of the substrate. Regretfully, the thiol-gold bonds are not sufficiently strong to prevent molecules from diffusing on the substrate surface.

This molecular diffusion brings significant variations into the molecular conduction, for the hybridization between the relevant orbitals of sulfur and gold (which varies as the molecule moves) controls the charge transfer between the gold electrodes and the molecular bridge [82, 166]. In this respect, amino groups are better fitted, for they create well-defined bonds with gold atoms with fewer variants of placing [14, 15]. Practically, the MMMs could be made out by immersing an already broken (mechanically or by the electromigration) metallic atomic-size contact into the solution containing potential molecular linkers as shown in Fig. 1.11. When these molecules are in the solution, their anchor groups are protected by certain protection groups. When a molecule approaches the metallic contacts, these protection groups detach themselves from the molecule ends and the anchor groups attach the molecule to the metal contacts. Mechanically or otherwise controlled break junctions allow sampling of a few molecules due to the small cross-sections of the metal contacts. However, the molecule could position itself in between the electrodes in many different ways as illustrated in Fig. 1.12. This brings strong variations in the values of coupling parameters. As a result, transport properties of MMM junctions with identical leads and molecular linkers could vary [167]. Actually, it is difficult to reliably control the quality of the contacting process and even to ensure that the molecules actually do bridge the gaps between the leads and become strongly bounded to both sides.

Due to the relative feasibility of manufacturing of self-assembled thiol-gold structures, gold electrodes are commonly used in MMM junctions. However, some work concerning the junctions with nonmetallic electrodes was reported.

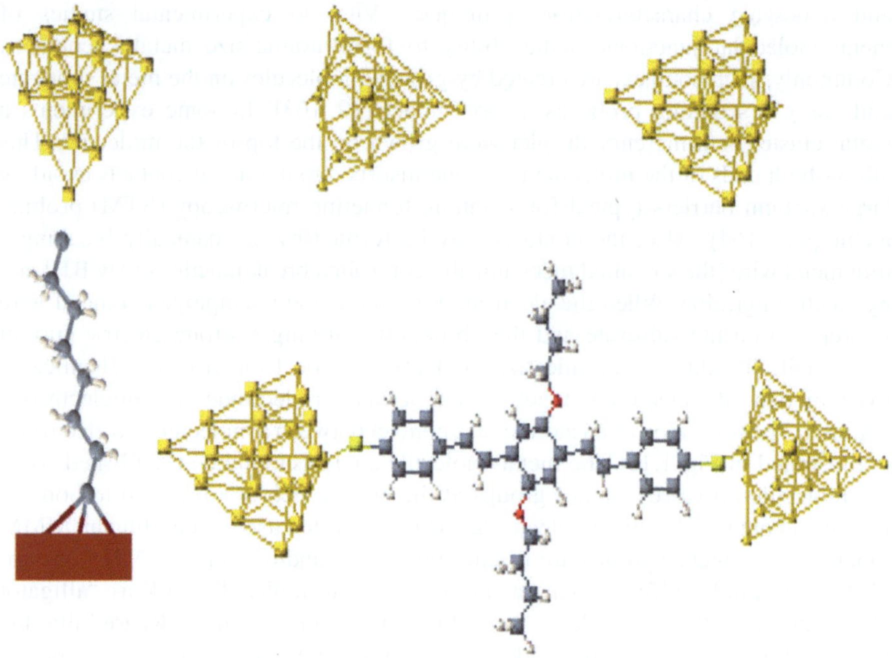

Fig. 1.11 Schematics of a metal–molecule–metal junction formation using a scanning probe as a top atomic-size contact (*left*). Mechanically controlled break-junction formation (*right*)

Experiments were carried out on the junctions with silicon and carbon electrodes [168–173]. Transport in semiconductors is dominated by their bandgap; therefore, one may expect significantly richer transport behavior of the junctions where semiconductors are involved than in the case of MMMs. Also, molecule–semiconductor bonds often are of the covalent nature. This implies potentialities for reducing geometric variability in the junction. This is important for the variability creates difficulties in the interpretation of experimental data concerning characterization of the molecular junctions with metallic leads. On the other hand, it is easier to assemble metal–molecule structures than those requiring semiconductor–molecule bonding. However, studies of molecule–semiconductor junctions continue and bring interesting results.

Numerous experiments have been carried out using self-assembled monolayers on suitable substrates. The counter electrodes included in these systems could be developed using nanopores and/or nanowells with evaporated metal top contacts. These architectures are compatible with planar processing and suitable for circuit-level integration. Since these systems include a contact which is not prefabricated, there is a possibility of pinholes and shortcuts through the molecular layers due to electromigration. If this happens, the conductance through the system may be determined by these flaws, affecting the transport properties of the junction. The likelihood for shortcuts is aggravated when the metal substrate is polycrystalline, so

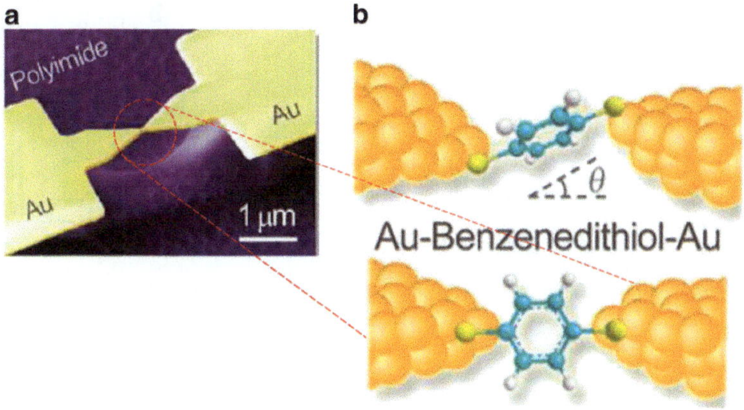

Fig. 1.12 Schematics illustrating multivariance in the molecule bonding to the leads in the process of fabrication of a single-molecule break junction (after [164]). Reprinted with the permission of the ACS

that the molecules (which follow the substrate geometry during the assembly) may be poorly packed leaving gaps open to the shorts. Other methods of building up assembled molecular monolayers include the so-called metallic flakes, metal dots, and indirectly deposited metals [174–179]. Finally, these exist crossed-wire devices (see Fig. 1.13). Some of them are suitable for characterization of self-assembled monolayers (SAM), and other include Langmuir–Blodgett films with static contact geometries which are more fitting for memory applications.

Molecular self-assemblies are quite appealing, as a way to create a multitude of identical molecules, although the time required to assemble a monolayer consisting of a thousand of molecules (i.e., a micron-sized structure) is still too long for industrial applications. Presently, significant progress is made to improve the quality of the substrate on which the molecules are grown. It becomes possible to create relatively large areas of single-crystal gold where the molecules may be orderly self-assembled [180]. During the last decade encouraging results relating to the top contact of large-area molecular junctions where "the linker" is an assembly including a large molecules were reported [181, 182]. It was suggested to use top contacts made out of conducting polymer (polyethylenedioxythiophene, also known as PEDOT) films. The flexibility which is an inherent property of polymer materials allowed these films to conform to the junction shapes over large areas, thus presenting gaps between the molecules and contact.

One of the important issues in molecular electronics is the gateability of MMM junctions. The effective gate control is essential for use of molecular junctions as building blocks in various nanoelectronic devices. In general, the problem of gate control in miniaturized devices is recognized as one of most significant challenges facing the semiconductor-based electronics. The root of the difficulties is that the efficient gate control could be maintained only when the gate electrode is placed very close to the transport channel. As the device size is getting smaller,

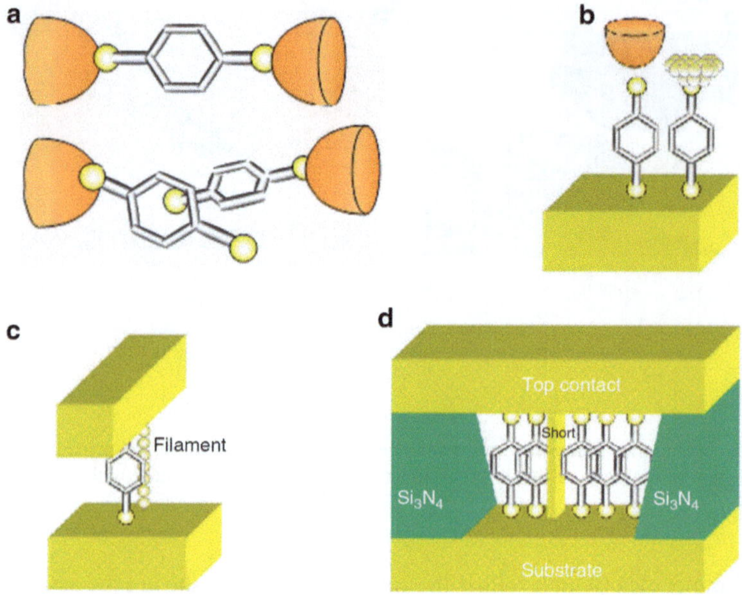

Fig. 1.13 Schematics of two-terminal MMM junctions: (**a**) break junctions, (**b**) nanoscale probes, (**c**) crossed wires as contacts, (**d**) large-area junction with evaporated top contact (courtesy of Dr. A. Ghosh)

the transport channel shortens, requiring the corresponding miniaturization of the gate. However, the insulating oxide which is the necessary part of the gate whose purpose is to present the leakage of the current from the transport channel to the gate electrode cannot scale below a certain limit without becoming transparent to tunneling electrons. This problem is not easily resolved or circumvented in molecular electronics. The extremely small size of molecules taking part of transport channels in MMM junctions requires equivalently thin oxides to be used in building gate electrodes, and this facilitates the electron tunnelings from the molecule to the gate. Also, due to the small distances separating the leads in the MMM junction, one usually needs significant (a few volts) gate voltage to efficiently control the electronic energy levels of the molecule in the junction which gives rise to additional difficulties. Presently, some experts question the very possibility of finding a satisfactory way to provide gating for a molecule in a solid-state environment [5]. However, one may circumvent the difficulties by placing the molecule into liquid ambience [183, 184]. Under these conditions the large dipole moments of the surrounding solvent could help to maintain the gating control.

Up to present, the most of experiments were carried out using a single MMM junction or a self-assembled monolayer. Now, several research groups are analyzing two- or three-dimensional networks of metal nanoparticles linked with molecular entities [40–42, 185, 186]. Two examples of molecular networks are shown in Fig. 1.14. Transport through such a network is easier to measure than through a

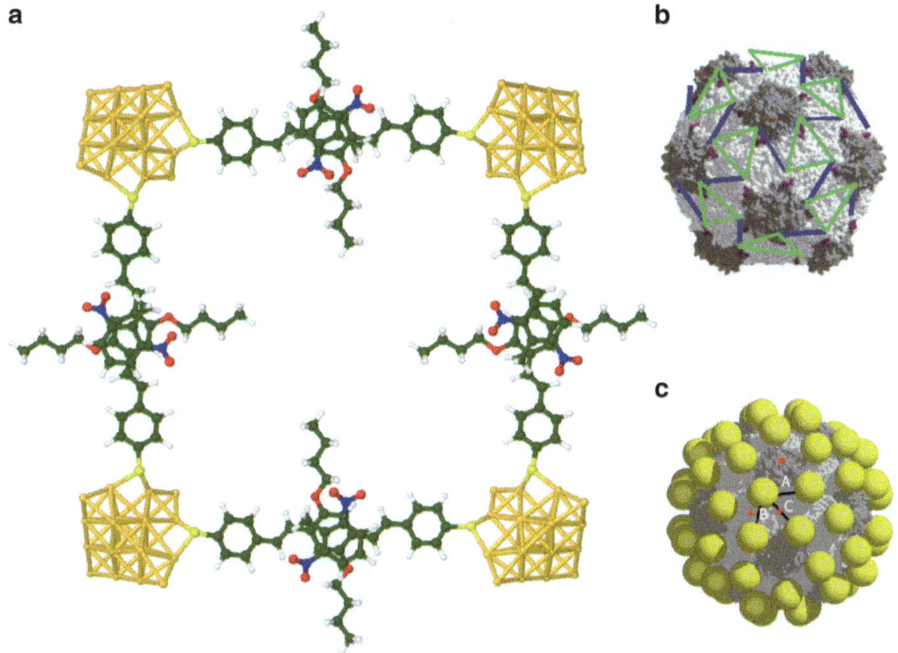

Fig. 1.14 Examples of a two-dimensional (*left*) and three-dimensional (*right*) molecular network built out of gold nanoparticles linked by oligophenylenevinylene molecules (courtesy of Dr. T. Baruah and A. Blum)

single junction. However, the process of interpretation of the results meets some difficulties. The root of these difficulties is the uncertainty in the relation of the measured network characteristics and the characteristics of the network elements. One may expect that some kind of averaging of the characteristics automatically occurs when the whole network is measured which facilitates the interpretation of the results. However, the nature and specific features of this averaging need further analysis.

1.8 Characterization of Molecular Junctions

There are various ways to measure current flowing through molecular junctions. Different methods have different advantages and deficiencies. Some approaches are scalable and integrable, while others are suitable for imaging and characterization of junctions linked by single molecules. Also, some methods allow scrupulous control, and another contain possibilities for high-throughput testing. Numerous experimental studies of the electron transport through molecules (and some of them very important) were carried out using the STM. In these experiments the molecule is placed on a metal surface, and a scanning tunneling microscope tip is approached to it, as shown in Figs. 1.11 and 1.13b. In general, there exists a significant vacuum gap between a non-bound end of the molecule and the STM tip. This makes the system an atypical MMM junction, for only one end of the molecular bridge is attached to the electrode. Nevertheless, the STM type measurements have been crucial in studying transport properties of molecules [111, 163, 164, 187–189]. The presence of a gap between the STM tip edge and the molecule leads to the specific voltage distribution across the junction. The major part of the bias voltage drop occurs in this gap, which justifies the Tersoff–Hamann picture [190, 191] implying that junction conductance is proportional to the local density of the electron states at the Fermi level at the tip position. In other words, in the experiments using the STM geometry the voltage applied across the junction only slightly affects the molecular orbitals, and they could be probed by appropriately positioning the tip edge.

The STM disadvantage is that minute variations in the tip position with respect to the molecule may bring significant changes in the measured current. Also, a typical STM molecular interface is physisorbed, reducing the current due to the intervening vacuum barriers. Break junctions are able to provide stronger coupling between the molecule and the contacts than that typical for STM experiments, thus increasing the conduction, but they always run the risk of nonideal bounding, as discussed before. An interesting opportunity is provided by using crossed wires brought together by Lorenz attractive forces as contacts for a MMM junction. Deposited cross wires with a Langmuir–Blodgett film in between can be used to probe single molecules (see Fig. 1.13c). However, such a system may suffer from electromigration and filament formation [5].

Studying an ordinary MMM junction where both ends of the molecule are attached to the leads, one may start from measuring current–voltage characteristics [82, 163]. To make these measurements more efficient, the third terminal is often introduced into the system as a supplement to the leads. As discussed before, the gate electrode is used to apply the gate voltage which directly affects molecular orbitals shifting them with respect to the chemical potentials of the leads. As the electron transport through the junction occurs via HOMO or LUMO levels, the shifting of the molecular orbitals by the gate voltage may significantly change the transport properties of the junction (see, e.g., [8, 9, 19, 27, 183, 184, 192]).

In an ideal experiment, one would be studying a single molecule with the well-defined geometry situated between the leads. However, atomistic properties of the contacts between the molecule ends and the leads often remain uncertain in practical experiments. At the same time, theoretical models used to analyze transport properties of a particular junction must adopt some specific atomistic picture of these contacts. Theoretical reconstructions of the atomic-size contacts between the leads and the molecular linker may appear unrealistic, leading to disagreements between the theory and the experimental observations even in the cases when the atomistic properties of the contacts are correctly established. Moreover, often it is difficult to determine either a single molecule really had made a contact with the leads, thus forming the MMM junction. Under these conditions, it is not surprising that transport properties for apparently identical junctions vary. Therefore, it has become

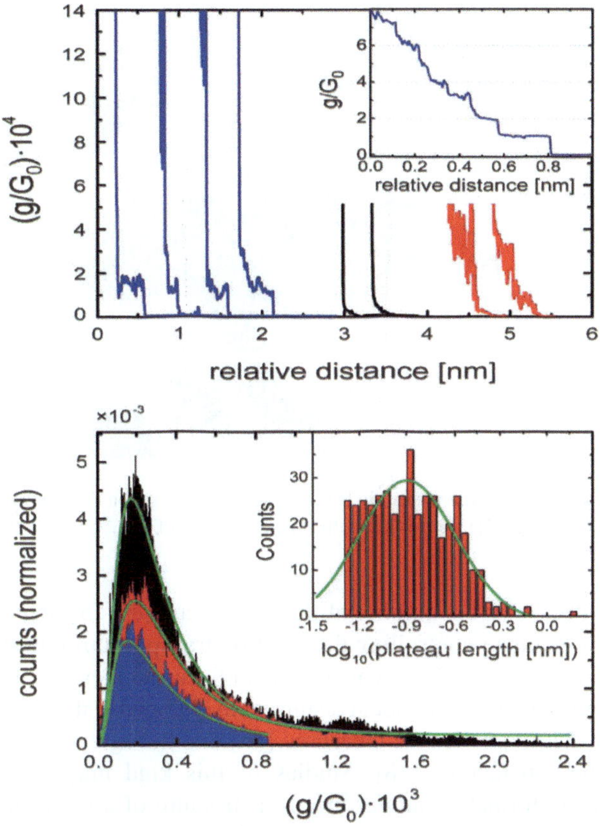

Fig. 1.15 The results of conductance measurements carried on for sulfur end-functionalized tercyclohexylidene molecules linking gold electrodes. Conductance histograms are shown on the *bottom* of the figure. The results demonstrate a crossover from electron tunneling to thermally activated transport (after [118]). Reprinted with the permission of the ACS

a common procedure in molecular electronics to describe transport properties of the junctions using conductance histograms obtained by repeating the junction formation and its consequent electrical characterization several times. An example of such histogram is shown in Fig. 1.15. The peaks on the histogram could be interpreted as signatures of the atomic-size contacts and linking molecules. Multiple peaks may appear, which correspond to multiple molecules wired in parallel or various molecular conformations.

Also, there are ways to demonstrate that indeed the desired molecules have made a contact with the leads and that one is measuring their characteristics and not those of some other objects. One possibility is to change chemical properties of the molecules and then to repeat the characterization, as was suggested in [14]. Applying this method to the molecules consisting of some repeated units (like the oligophenylene molecules), it may be shown that the molecular conductance

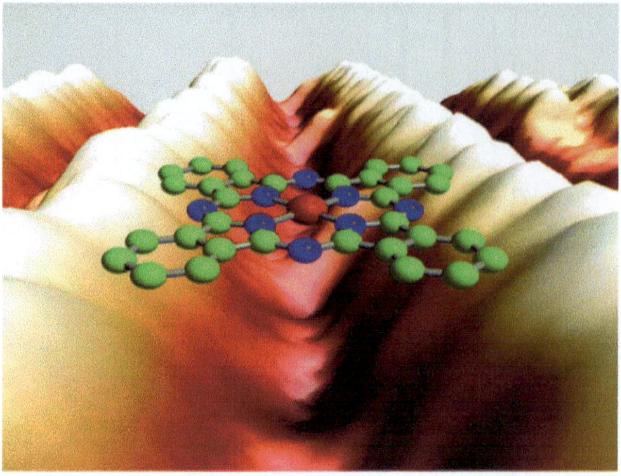

Fig. 1.16 A Cu-phthalocyanide molecule bridging adjacent crests on the corrugated Au-modified Ge(0 0 1) surface (after [193]). Reprinted with the permission of the ACS

exponentially decays as the number of the units increases [15], in consistence with the physical mechanism controlling the electron transport through short molecules which is the quantum tunneling. This result confirms the identity of the considered molecules. Similar length-dependent results were obtained in the experiments where the molecular thermopower was measured [16] and the agreement with the theory appeared to be satisfactory [18]. Studies of this kind may be carried out by modifying the conformation and/or electron structure of a molecule by means of functional side groups on the molecular backbone. These groups may serve as "dopants" donating electrons to the backbone or withdrawing them from the latter. Again, with knowledge of the performed modifications, one may specify the identity of the molecule. Also, by manipulating the electron structure of the molecule, one may to some extent tune the molecular orbitals alignments in the junction even without the gate electrode [192].

Recently, Berkelaar and coworkers suggested to study possible positions of the bridging molecule using a special way of building up contacts [193]. The contacts were built using well-ordered arrays of straight nanowires which form an Au-modified $Ge(0\,0\,1)$ surface. The adjacent wires were 1.6 nm wide, and they were separated by about 0.6 nm deep trenches. Cu-phthalocyanide molecules were brought near this warped surface, so they could link adjacent wires (see Fig. 1.16). Subsequent STM studies allowed to distinguish four possible configurations of the molecule bridges.

A very important diagnostic tool is based on the coupling of electrons to the molecular vibrations. When the differential conductance of the junction $g = dI/dV$ and the inelastic electron spectra d^2I/dV^2 are measured up to high enough values of the bias voltage, signatures originating from the inelastic transmission of electrons through the junction make their appearance. Inelastic tunneling signals can be very useful for identification of the molecules and specific features of their contact with the leads [26]. For instance, when the molecular contact is stretched, the energies of the vibration modes change in the predictable way, and the inelastic tunneling signals show these changes.

Another useful technique for identifying molecular junctions is Raman spectroscopy. This method is widely used in chemistry for fingerprinting molecules. In the case of MMM junctions, the structure of molecular linker may be sensitive to the presence of photo-excited surface plasmons on the electrode's surfaces. These excitations augment local electric fields near the contacts of the molecule with the electrodes and lead to significant (up to orders of magnitude) enhancement of Raman signals. This surface-enhanced Raman spectroscopy has been applied to study several MMM junctions [165]. It was demonstrated that electromagnetic field enhancement necessary for single-molecule detection can be achieved when gold electrodes are used. Combined measurements of molecular conductance and Raman spectrum were found to be especially fruitful. Comparing results of these combined measurements, one may explore correlations between variations in the molecular bridge geometry and its conductance. Furthermore, Raman spectroscopy can be employed to analyze inelastic electron–phonon interactions and heat dissipation in current-carrying MMM junctions.

Recently, some experimental results were obtained which suggest that molecule-vibration-induced noise can be used for single-molecule identifications [165]. In these experiments, current–voltage characteristics of Au-hexanedithiol-Al junctions were measured. It was observed that the current noise (defined as the standard deviation of current fluctuations at fixed voltage) exhibits a stepwise increase synchronous to excitations of vibrational modes involved into active electron–phonon interactions. This gives grounds to expect that vibrational spectra of molecules may be explored through examining noise spectra. Also, one may conjecture that information relevant to the local heating of MMM junctions could be extracted from the voltage-dependent noise in molecular junctions. This issue was theoretically analyzed [26], and it was suggested to study frequency dependencies of noise spectra in order to separate out contributions from the local heating which are embedded in the experimentally measured noise signals. So, inelastic noise spectroscopy may serve as a complimentary technique to IETS. Also, it may appear to be a unique method for investigating local heating in single-molecule MMM junctions.

Presently, experimental studies of MMM junctions are often carried out using electrochemical break junctions [68] which are among the newest and most effective ways to get statistically significant information concerning the MMMs transport properties [14, 15, 183, 192, 194–196]. Electrochemical gating in such structures [184] provides means for observation of different kinds of transport. Rich statistical information obtained from electrochemical break-junction characterization improves understanding of the transport spectroscopy. However, all these experiments are carried out for solvated junctions, and this creates some difficulties. First, the solvent presence modifies transport due to the effects of solvent polarization, and it may bring changes in the coupling of the molecule to the leads. Secondly, the solvated junctions can be studied only at rather high temperatures, which are above the solvent freezing point. Therefore, some interesting features which appear in the low-temperature transport cannot be observed.

These exist a significant interest in studies of large-area junctions where the electrodes are linked by self-assembled molecular monolayers. An example of such junction is schematically presented in Fig. 1.13d. Characterization methods include cross-wire test beds [179], suspended nanodots test beds [197], nanopores [174], in-wire junctions [198], and some other. In the studies of transport properties of self-assembled monolayers, it is necessary to take into account intermolecular interactions. It is true that in some works experimental results were reasonably interpreted in terms of simple additivity omitting the influence of these interactions (see, e.g., [199,200]). However, it was theoretically predicted [200] that intermolecular interactions may result in noticeable changes in transport characteristics of self-assembled monolayers, and these changes must be taken into consideration in extracting information concerning transport properties of individual MMM junctions from the experimental data obtained using self-assembled monolayers.

To achieve better understanding of the transport properties of a certain self-assembled monolayer, one should carry on contact angle measurements which would give information concerning the monolayer thickness. This information indicates the monolayer quality. The information concerning molecular bonding to the contacts could be obtained by using IETS and/or in the course of photoemission studies [201]. For example, the appearance of gold–sulfur modes instead of those originating from sulfur–hydrogen bonds stretching would indicate that on both sides of the junction, sulfur–hydrogen bonds are replaced by gold–sulfur ones. This would mean that the molecules are chemisorbed at both electrodes.

In near future, technological progress involving MMM junctions is expected to be evolutionary, meaning that these systems would be combined with silicon-based elements in manufacturing novel nanoelectronic devices. This gives rise to additional (and considerable) challenges. The manufacturing of silicon-based devices involves processes running through rather high temperatures which are well above the degeneration temperatures of the most structures including molecules. To give a few examples, one may mention metallization and high-temperature annealing of self-aligned polysilicon gates. Therefore, molecular components must be manufactured in such a way that they should be able to withstand the heat. In principle, this could be achieved through proper functionalization. Some experiments indicate that molecules could be fastened to silicon electrodes through very strong bonds (about 5 eV) which may allow these bonds to survive at elevated temperatures. By selective sublimation of molecules, it is possible to deposit desired molecules with strong bonding and high yield. This is crucially important for development of efficient methods of building up molecular circuits.

Chapter 2
Transport Theory

2.1 Retarded and Advanced Green's Functions for Electrons Traveling Through a MMM Junction

We start our theoretical analysis of MMM junctions transport properties by briefly introducing the concept of Green's functions which are necessary for the development of the transport theory. A very thorough treatment of the Green's functions properties and their quantum mechanical applications is given in numerous works on the subject. Here, we minimize mathematical intricacies by considering a system whose Hamiltonian H is independent of time. In this case the time evolution of a Schrodinger state vector $\{\psi(t)\}$ may be described by means of the time evolution operator $U(t,t')$:

$$\{\psi(t)\} = U(t,t')\{\psi(t')\} \tag{2.1}$$

which satisfies the equation of motion:

$$i\hbar \frac{\partial U(t,t')}{\partial t} - HU(t,t') = 0. \tag{2.2}$$

The time evolution operator must satisfy the condition $U(t,t) = [1]$ where $[1]$ is the unit operator. Keeping this in mind one can find the appropriate solution of Eq. (2.2), namely,

$$U(t,t') = \exp\left[\frac{1}{i\hbar} H(t-t')\right]. \tag{2.3}$$

The retarded and advanced Green's functions for the considered system are closely related to the time evolution operator. These operators are defined as

$$G^{(r)}(t,t') = \tfrac{1}{i}\theta(t-t')U(t,t'), \tag{2.4}$$

N.A. Zimbovskaya, *Transport Properties of Molecular Junctions*, Springer Tracts in Modern Physics 254, DOI 10.1007/978-1-4614-8011-2_2,

$$G^{(a)}(t,t') = -\tfrac{1}{i}\theta(t'-t)U(t,t'), \tag{2.5}$$

where

$$\theta(x) = \begin{cases} 1 & x > 0 \\ 0 & x \leq 0 \end{cases} \tag{2.6}$$

is the Heaviside step function. As follows from definitions (2.4) and (2.5), the retarded Green's function $G^{(r)}(t,t')$ only propagates forward in time (and hence effects caused by sources are always "retarded" for they occur later in time, as follows from the causality principle). The advanced Green's function $G^{(a)}(t,t')$ propagates backward. The time evolution operator obeys the rule $U^{\dagger}(t,t') = U(t',t)$, so the retarded and advanced Green's functions are Hermitian conjugates of each other:

$$\left[G^{(r)}(t,t')\right]^{\dagger} = G^{(a)}(t',t). \tag{2.7}$$

Using definitions (2.4) and (2.5), the equation of motion for the time evolution operator (2.2) and the relation between the step function and the delta function $(d\theta/dx = \delta(x))$, one may show that the operators $G^{(r,a)}(t,t')$ satisfy the equation with the delta function source term:

$$i\hbar\frac{\partial}{\partial t}G^{(r,a)}(t,t') - HG^{(r,a)}(t,t') = \hbar\delta(t-t') \tag{2.8}$$

which justifies the commonly accepted recognition of these operators as Green's functions.

For the considered case of the time-independent Hamiltonian H, the time evolution operator depends on the difference between the time coordinates t and t', and the same is valid for the Green's functions. So we may write

$$G^{(r,a)}(t,t') = G^{(r.a)}(t-t',0). \tag{2.9}$$

Now, Eq. (2.8) may be solved by applying the Fourier transformation. In general case, the relations between an arbitrary function $f(t)$ and its Fourier transform $F(E)$ are given by

$$F(E) = \int_{-\infty}^{\infty}\frac{dt}{\hbar}e^{iEt/\hbar}f(t), \tag{2.10}$$

$$f(t) = \int_{-\infty}^{\infty}\frac{dE}{2\pi}e^{-iEt/\hbar}F(E). \tag{2.11}$$

While computing Fourier transforms for the Green's functions, one meets a technical difficulty arising because the corresponding integrals diverge. To circumvent this difficulty, one must introduce damping factors $\exp[\mp\eta(t-t')]$ assuming that η

is a positive infinitesimal quantity ($\eta \to 0^+$). Then one may set $t - t' \to t$ and write out the following expressions for the Fourier transforms:

$$G^{(r,a)}(E) = \int_{-\infty}^{\infty} \frac{dt}{\hbar} e^{i(E \pm i\eta)t/\hbar} G^{(r,a)}(t,0). \tag{2.12}$$

Here, the sign before $i\eta$ corresponds to the retarded ($+$) and advanced ($-$) Green's functions, respectively. In agreement with Eq. (2.11), these functions are expressed in terms of their Fourier transforms as follows:

$$e^{\mp\eta t} G^{(r,a)}(t,0) = \int_{-\infty}^{\infty} \frac{dE}{2\pi} e^{-iEt/\hbar} G^{(r,a)}(E). \tag{2.13}$$

Substituting expressions of the form (2.13) into Eq. (2.8), one may reduce these differential equations for the original Green's functions to the algebraic equations for their Fourier transforms. As a result, one obtains the Green's functions in the energy representation:

$$G^{(r)}(E) = \big[(E - i\eta)[1] - H\big]^{-1}, \tag{2.14}$$

$$G^{(a)}(E) = \big[G^{(r)}(E)\big]^{\dagger} = \big[(E + i\eta)[1] - H\big]^{-1}. \tag{2.15}$$

Here, as before, $[1]$ stands for the unit operator, and H is Hamiltonian of the considered system. Due to the simple symmetry relation between the retarded and advanced Green's functions one may choose one of them to analyze its properties and then apply the obtained results to describe the properties of the remaining Green's operator. In further analysis we choose the retarded Green's function $G^{(r)}(E)$ for this purpose.

The Green's functions are often employed to theoretically analyze various properties of mesoscopic systems, including their transport properties. The important advantage of Green's function-based formalisms is that they allow to include into consideration various interaction effects. Now, we illustrate this by considering the Green's function of a system consisting of two electron reservoirs (source and drain) linked by a channel. Obviously the obtained results could be applied to study transport properties of metal–molecular junctions. In the analysis presented below we follow the method suggested and described by Datta in his book [4]. This method allows to bypass numerous computational difficulties arising when one follows more formal and rigorous methods described in several books and research articles (see, e.g., [202–210] and the last section of this chapter). The price which one pays for this simplification is that one must treat interactions perturbatively.

We start from partitioning of the system Hamiltonian H in such a way that the terms corresponding to the source (H_L) and drain (H_R) are separated from the part corresponding to the channel (H_C):

$$H = \begin{pmatrix} H_L - i\eta & \tau_L^\dagger & 0 \\ \tau_L & H_C & \tau_R \\ 0 & \tau_R^\dagger & H_R - i\eta \end{pmatrix}. \tag{2.16}$$

Here, the operators τ_β ($\beta \in L, R$) describe the coupling of the source and drain to the channel. We consider the system in the stationary state with the energy E, so the Schrodinger equation has the form $H\Psi = E\Psi$ where the Hamiltonian H is given by Eq. (2.16) and the state vector (wave function) may be presented as follows:

$$\Psi = \begin{pmatrix} \Phi_L + \chi_L \\ \psi \\ \Phi_R + \chi_R \end{pmatrix} \tag{2.17}$$

where $\{\Phi_\beta\}$ ($\beta \in L, R$) are the incident waves functions in the contact reservoirs, $\{\psi\}$ is the wave function transmitted into the channel, and $\{\chi_\beta\}$ correspond to the parts reflected back into the reservoirs. Implying that the functions $\{\Phi_\beta\}$ are the solutions of the equations $(H_\beta - i\eta)\{\Phi_\beta\} = E\{\Phi_\beta\}$, one may obtain the following expressions for $\{\chi_\beta\}$:

$$\{\chi_\beta\} = g_\beta^r(E)\tau_\beta\{\psi\}. \tag{2.18}$$

Here

$$g_\beta^r(E) = \left((E - i\eta)[1] - H_\beta\right)^{-1} \tag{2.19}$$

are retarded Green's functions for the contacts (reservoirs).

Substituting these expressions into the original Schrodinger equation, one gets

$$\left(E[1] - H_C - \Sigma_L^r - \Sigma_R^r\right)\{\psi\} = \{S\} \tag{2.20}$$

where $\Sigma_\beta^r = \tau_\beta^\dagger g_\beta^r \tau_\beta$ are self-energy terms originating from the interactions between the channel and the contacts and $\{S\} = \tau_L\{\Phi_L\} + \tau_R\{\Phi_R\}$. As mentioned above, $\{\Phi_\beta\}$ are the wave functions of the contacts in the absence of the coupling between them and the channel, so the term $\{S\}$ on the right side of Eq. (2.20) describes the electrons flow into the channel (inflow). At the same time, the term $(\Sigma_L^r + \Sigma_R^r)\{\psi\}$ is associated with the electrons flow from the channel to the contacts (outflow).

As follows from Eq. (2.20) and the definitions of the Green's functions, the retarded Green's function of the considered system (in the energy representation) has the form

$$G^r(E) = \left(E[1] - H_C - \Sigma_L^r - \Sigma_R^r\right)^{-1} \tag{2.21}$$

and it is related to the corresponding Green's function for the isolated channel

$$G_0^r(E) = \left((E - i\eta)[1] - H_C\right)^{-1} \tag{2.22}$$

by means of the Dyson equation:

$$G^r(E)^{-1} = G_0^r(E)^{-1} - \Sigma^r(E) \tag{2.23}$$

where $\Sigma^r(E) = \Sigma_L^r(E) + \Sigma_R^r(E)$. The Dyson equation separates out the response G_0^r of an isolated channel grouping the corrections due to interactions in the self-energy term Σ^r. In deriving Eq. (2.23), we took into account only interactions of the channel with the contacts, and this has determined the particular form of self-energy term. In general case, the self-energy terms may include contributions from other kinds of interactions such as electron–phonon interactions.

Equation (2.23) may be converted to the commonly used form:

$$G^r(E) = G_0^r(E) + G^r(E)\Sigma^r(E)G_0^r(E) \tag{2.24}$$

which can be rewritten as a series:

$$\begin{aligned} G^r(E) =& G_0^r(E) + G_0^r(E)\Sigma^r(E)G_0^r(E) \\ &+ G_0^r(E)\Sigma^r(E)G_0^r(E)\Sigma^r(E)G_0^r(E) + \cdots \end{aligned} \tag{2.25}$$

that schematically represents the role of various interactions.

The retarded and advanced Green's functions determine the spectral function $A(E)$ defined by the expression

$$A(E) = i\left[G^r(E) - G^a(E)\right] \equiv i\left[G^r(E) - G^r(E)^\dagger\right]. \tag{2.26}$$

The physical meaning of this operator is that its trace gives the electron density of states at a certain energy, namely,

$$D(E) = \frac{1}{2\pi}\mathrm{Trace}[A(E)] \equiv \sum_\alpha \delta(E - E_\alpha) \tag{2.27}$$

where the summation is carried over all eigenstates of the considered system. So, the spectral function $A(E)$ represents the matrix version of the electron density of states, and it is closely related to the density matrix of the considered system. Actually, when the electron system described by the Hamiltonian H is in the equilibrium state, the corresponding density matrix may be presented as

$$\rho \equiv \{\Psi\}\{\Psi\}^\dagger = \int \frac{dE}{2\pi} f(E - \mu)A(E) \tag{2.28}$$

where $f(E - \mu)$ is the Fermi distribution function with the chemical potential μ. When the system departs from the equilibrium state, its density matrix is given by

$$\rho = \int \frac{dE}{2\pi i} G^{<}(E) \tag{2.29}$$

where the lesser Green's function $G^{<}(E)$ is introduced. It could be expressed in terms of the retarded and advanced Green's functions. Now, we find the expression for $G^{<}(E)$ for our system including the channel coupled to the two contacts. In such a system, before the contact is established between the source and drain electron reservoirs and the channel, the electrons on the contacts remain in equilibrium states described by the density matrices

$$\rho_\beta = \{\Phi_\beta\}\{\Phi_\beta\}^\dagger = \int \frac{dE}{2\pi} f_\beta(E - \mu_\beta) A_\beta(E); \qquad \beta \in L, R. \tag{2.30}$$

Here, μ_β are the chemical potentials for the reservoirs and spectral functions $A_\beta(E)$ are equal:

$$A_\beta(E) = i\big[g^r_\beta(E) - g^a_\beta(E)\big] \tag{2.31}$$

where the retarded Green's functions for the reservoirs are determined by Eq. (2.19). Assuming that the contact between the source and drain reservoirs and the channel is established and using Eq. (2.19), one obtains

$$\begin{aligned} \rho &= \{\psi\}\{\psi^\dagger\} = G^r\{S\}\{S\}^\dagger G^a \\ &= G^r\big(\tau_L\{\Phi_L\}\{\Phi_L\}^\dagger\tau_L^\dagger + \tau_R\{\Phi_R\}\{\Phi_R\}^\dagger\tau_R^\dagger\big)G^a. \end{aligned} \tag{2.32}$$

The cross terms which appear when one calculates the product $\{S\}\{S\}^\dagger$ are zero because the wave functions $\{\Phi_L\}$ and $\{\Phi_R\}$ are related to different electron reservoirs which are connected only by the channel. Using the expressions for the density matrices on the contacts (2.30) and introducing denotations

$$\Gamma_\beta \equiv i\big(\Sigma^r_\beta - \Sigma^a_\beta\big) = \tau_\beta A_\beta \tau_\beta^\dagger \tag{2.33}$$

we may write down the following expression for the electron density matrix of the channel:

$$\rho = \int \frac{dE}{2\pi} G^r(E)\big[f_L(E-\mu_L)\Gamma_L + f_R(E-\mu_R)\Gamma_R\big]G^a(E). \tag{2.34}$$

Comparing Eqs. (2.29) and (2.34), one concludes that the lesser Green's function satisfies the Keldysh equation:

$$G^{<}(E) = G^r(E)\Sigma^{<}(E)G^a(E) \tag{2.35}$$

where the self-energy term has the form

$$\frac{1}{i}\Sigma^{<}(E) = f_L(E-\mu_L)\Gamma_L + f_R(E-\mu_R)\Gamma_R. \qquad (2.36)$$

As follows from the general theory, the relation (2.35) remains valid when other interactions apart from the interactions between the channel and the contacts are taken into account. However, the form of the self-energy term $\Sigma^{<}$ varies depending on the considered interactions.

The lesser Green's function including the self-energy term of the form (2.36) may be presented as the sum of two terms associated with the contacts:

$$G^{<}(E) = \sum_{\beta} f_{\beta} G^r \Gamma_{\beta} G^a. \qquad (2.37)$$

This means that the nonequilibrium density matrix for the channel retains the form similar to that given by Eq. (2.28). Now, however, the spectral function $A(E)$ consists of two parts, each part associated with the certain (left/right) contact. It is natural to think that the part $A_L(E) = G^r(E)\Gamma_L G^a(E)$ appears due spilling-over states belonging to the left contact which is filled according to the Fermi function f_L. Similarly, the contribution $A_R(E) = G^r(E)\Gamma_R G^a(E)$ arises from the spillover of the states from the right contact filled according to the Fermi function f_R. We remark that in the case when the channel is characterized by a single energy level E_0 (a single-site bridge), Eq. (2.35) becomes reduced to expression (1.9) for the electron density on the bridge which was obtained using the simplified model for a MMM junction.

Introducing a suitable set of basis functions, one may compute matrix elements of the Green's functions. Thus, the retarded Green's function $G^r(E)$ given by Eq. (2.21) may be presented in the form

$$G^r(E) = \left(ES - F - \Sigma_L^r - \Sigma_R^r\right)^{-1} \qquad (2.38)$$

where S and F are the corresponding overlap and Fock matrices. If the basis functions are orthonormalized, $S_{ij} = \delta_{ij}$, δ_{ij} being the Kronecker symbol. As for the self-energy matrices Σ_{β}, they need to be carefully computed taking into account the chemical coupling of individual lead orbitals (for the corresponding lead) with the bridge orbitals, as well as the band structure of the leads. Also, one must accurately describe the dependencies of self-energy terms of energy E.

To elucidate the crucial importance of a proper modeling of the self-energy term, we cite an example considered by Datta in his book [4]. The energy eigenvalues of a two-site system with the site energies E_1 and E_2 and site-to-site coupling τ_0 may be found by solving the quadratic equation:

$$(E-E_1)(E-E_2) - \tau_0^2 = 0. \qquad (2.39)$$

However, one could treat this system as a combination of a contact and a channel coupled to the latter. For certainty, we assume that the site with the energy E_1 takes on the part of the contact, and the remaining site corresponds to the channel. For such a simple system both the Green's function given by Eq. (2.38) and the self-energy term Σ are scalar quantities. We have

$$G^r(E) = \left(E - E_2 - \Sigma^r(E)\right)^{-1}. \tag{2.40}$$

The eigenenergies of the considered system coincide with the poles of this Green's function. Comparing Eqs. (2.39) and (2.40), one may see that the agreement in the results for the eigenenergies is achieved provided that the self-energy term has the form

$$\Sigma^r(E) = \frac{\tau_0^2}{E - E_1 + i\eta} \tag{2.41}$$

where, as before, η is a small positive quantity which is introduced to describe the broadening of energy levels.

Also, one may treat the site with the energy E_1 as a channel and that with the energy E_2 as a contact. Then the energy E_2 is to be replaced by E_1 in expression (2.40) for the Green's function, and E_2 takes the place of E_1 in expression (2.41) for the self-energy term. Nevertheless, the poles of the Green's function remain undisturbed. It is worthwhile to remark that the real and imaginary parts of the self-energy given by (2.41) obey Kramers–Kronig relation:

$$\mathrm{Re}\left[\Sigma^r(E)\right] = \frac{1}{\pi}\int \frac{dE'[\mathrm{Im}\Sigma^r(E')]}{E - E'}. \tag{2.42}$$

Formally, this constraint maintains the causality of the retarded response functions by shifting their poles to the lower part of the complex energy plane. Physical meaning of this relation is that it guarantees the conservation of states upon establishing contacts between atoms. The considered example illustrates that it is possible to formally treat a system in several ways separating out a certain part as a channel (bridge) and identifying the remaining parts with the contacts (leads). However, after choosing the way to subdivide the original system, one must properly and accurately compute the self-energy terms taking into account the detailed interfacial chemistry and the eigenenergies associated with the possible states on the contacts. Moreover, in subdividing the system, one must bear in mind that an arbitrarily chosen contact may not act as a good one, that is, the one whose properties are nearly independent of the dynamics of the bridge.

Another important property of a good contact is manifested as an absence of correlations between the states on the left and right contact included into the system. Also, a good contact must remain in equilibrium while electrons travel through the system. These inherent properties of the good contacts provide justification for the expression (2.36) which presents the self-energy term $\Sigma^<(E)$ as the sum of independent contributions from the left and right leads. Realistic contacts may possess the described properties only when there is a profusion of scattering in their bodies. The scattering processes should include dephasing scattering eliminating coherences between the contacts as well as inelastic scattering which helps to keep the contacts in the state of thermal equilibrium. It is worthwhile to emphasize that as the leads shrink in size, the above requirements may cease to be valid, and the expression (2.36) may become unjustified.

2.2 The Green's Function for the Atomic Wire; an Analytical Example

Now, we apply relations presented in the previous section to compute the retarded Green's function G^r for a periodical chain of identical atoms which can be parametrized as hydrogens. We assume that N atoms are included in the chain. The chain ends are put in contact with the source and drain reservoirs which are also imitated by semi-infinite atomic chains. Obviously, this model imitates a MMM junction, and the results described in this section could be used to qualitatively analyze transport properties of such systems.

In the following calculations we start from a tight-binding Hamiltonian for the chain

$$H = H_0 + H_1 \tag{2.43}$$

whose matrix elements between states $|k>$ and $|l>$ corresponding to the kth and lth sites of the chain (the states are supposed to be orthogonal) are given by

$$\begin{aligned}(H_0)_{kl} &= \alpha_k \delta_{kl}, \\ (H_1)_{kl} &= \mathcal{V}_{kl},\end{aligned} \tag{2.44}$$

where $\mathcal{V}_{kl}= 0$ when $k = l$. The diagonal matrix elements α_k are ionization energies of electrons at sites k, while $\mathcal{V}_{kl} = \mathcal{V}_{lk}$ describes the coupling between kth and lth sites and includes both direct and exchange energy contributions. Also, we assume that the contacts are coupled only to the end atoms of the chain. Then the nonzero matrix elements of the self-energy matrices $\Sigma_\beta^r(E)$ accept the form

$$\left(\Sigma_\beta^r\right)_{ii} = \sum_m \frac{|\tau_{im}|^2}{E - \epsilon_{m\beta} - \sigma_{m\beta}}. \tag{2.45}$$

Here, the index i takes on the values 1 and N for the left and right contacts, respectively. Summation over "m" in Eq. (2.45) is carried out over the contact states with the energies $\epsilon_{m\beta}$, and $\sigma_{m\beta} = \frac{1}{2}\left\{\theta_{m\beta} - i\sqrt{4\gamma_{m\beta}^2 - \theta_{m\beta}^2}\right\}$ are the self-energy corrections of the semi-infinite chains attached to the bridge ends [67].

In the expression for $\sigma_{m\beta}$, $\theta_{m\beta} = E - \epsilon_{m\beta}$ and $\gamma_{m\beta}$ are the nearest-neighbor hopping integrals for the contacts. When $E - \epsilon_{m\beta}$ is smaller than $\gamma_{m\beta}$, we have $\sigma_{m\beta} \approx -i\gamma_{m\beta}$. In the opposite limit $| E - \epsilon_{m\beta} | \gg \gamma_{m\beta}$, this quantity takes on values close to zero.

Due to the presence of self-energy corrections Σ^r_β, the eigenvalues of the effective Hamiltonian (for the atomic wire bridge) include imaginary parts Γ_i which represent broadening of the bridge energy-level E_i originating from the coupling of the bridge to the leads. Assuming that basis functions are orthogonal, we may calculate matrix elements for the retarded Green's function solving the system of linear equations which follows from the expression (2.38):

$$\begin{aligned}&\{E - \alpha_k - \left(\Sigma^r_L\right)_{kk}\delta_{k1}\delta_{l1} - \left(\Sigma^r_R\right)_{kk}\delta_{kN}\delta_{lN}\}G^r_{kl}\\&-\sum_{n\neq k}\mathcal{V}_{kn}G^r_{nl} = \delta_{kl}\end{aligned} \tag{2.46}$$

where the indexes k, l take on values corresponding to the chain sites.

We obtain the solution of equations (2.46) by using a graphic technique similar to that described in detail for the nearest neighbors (NN) approximation in an earlier work [211]. However, this technique may be generalized to include also next nearest neighbors (NNN) couplings [51]. This generalization may be useful for it is known that even modest fluctuations of the bridge geometry arising due to its flexibility or some other reasons can result in noticeable changes of transport characteristics of the chain. The NNN couplings are very sensitive to any changes in the geometry of the bridge, and in many practical molecules, they are rather strong, though weaker than NN couplings. Further generalization is also possible, but it significantly enlarges the whole body of calculations. Besides, NNNN and succeeding couplings are mostly very weak and could be neglected without noticeable changes in the results.

First, we present our chain of sites as a graph Δ_{1N}, using the following rules:

1. Any site included in the chain corresponds to a graph vertex.
2. A term $E - \alpha_k$ corresponds to a loop attached to the vertex "k" when $k \neq 1, N$.
3. For $k = 1, N$ the terms attached to the corresponding vertices equal $E - \alpha_1 - \Sigma^r_L$ and $E - \alpha_N - \Sigma^r_R$, respectively.
4. A hopping integral $\mathcal{V}_{kl}$ corresponds to an edge originating at the vertex "k" and ending at the vertex "l".

Taking into account only nearest neighbors (NN) and next nearest neighbor (NNN) couplings, we can introduce a new notation for the nonzero hopping integrals, i.e., $\mathcal{V}_{k,k+1} = \mathcal{V}_{k-1,k} = \beta_k$ and $\mathcal{V}_{k,k+2} = \mathcal{V}_{k-2,k} = \gamma_k$. Within this approximation, Δ_{ij} has the form shown in Fig. 2.1.

Now, we define a cycle as a sequence of graph edges without self-intersections where a successive edge originates at the point where the previous edge ends and the last edge ends where the first one starts. Then we also define the value of the cycle **O** as the product of values of the edges included into the cycle multiplied by the

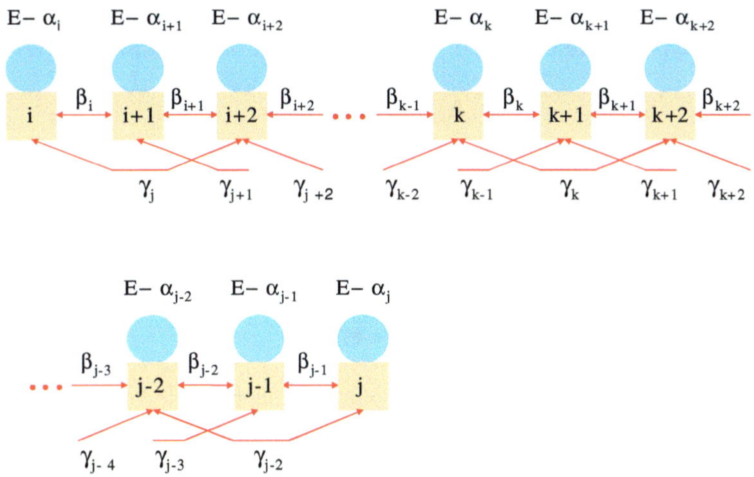

Fig. 2.1 Graph Δ_{ij} representing the chain segment starting at "i"th site and ending at "j"th site

Fig. 2.2 Schematics of cycles taken into account in computation of the Green's function matrix elements within the NNN approximation

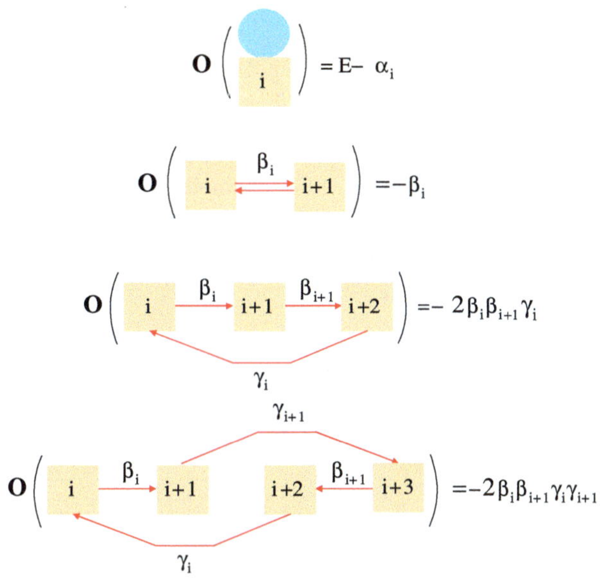

number of the edges connecting the first and last sites with a negative sign. When the numbers of the edges associated with the paths from the first site to the last one and back differ, we have to multiply by the maximum number of edges. The values of some simple cycles are given in Fig. 2.2.

We also introduce the definition of a cyclic term which is a set of cycles including all sites of the system. The value of the cyclic term is equal to the product of cyclic values of these cycles. According to [211], the sum of all possible cyclic terms $\mathbf{O}_r$ gives us the cyclic value $\Theta(\Delta_{ij})$ of the graph Δ_{ij} for our system, namely,

$$\Theta(\Delta_{ij}) = \sum_r (\mathbf{O}_r)_{ij}. \tag{2.47}$$

This quantity $\Theta(\Delta_{ij})$ is the determinant of the system of Dyson equations (2.46), and it is included in the expression for the matrix element G^r_{ij} which can be obtained in the form

$$G^r_{ij} = \sum_\sigma \frac{P^\sigma_{ij}}{\Theta(\Delta_{ij})}. \tag{2.48}$$

Here, the summation over σ is carried out over all possible pathways along the chosen chain, and the value of a pathway P^σ_{ij} has to be calculated as the product of all edges along the pathway. When the pathway includes all vertices of the graph Δ_{ij}, we have

$$\begin{aligned} P_{ij}(i \longrightarrow i+1 \longrightarrow \cdots \longrightarrow k \\ \longrightarrow k+1 \cdots \longrightarrow j-1 \longrightarrow j) = \prod_{k=i}^{j-1} \beta_k. \end{aligned} \tag{2.49}$$

When the pathway does not include some vertices, this produces different values for the corresponding P_{ij}. For example, when the electron is being transferred from the site l to the site $l+2$, passing the intermediate site $l+1$, the corresponding factor is equal to $(E - \alpha_{l+1})\gamma_l$. Consequently, we obtain for the pathway which excludes this vertex

$$\begin{aligned} &P_{ij}(i \longrightarrow i+1 \longrightarrow \cdots \longrightarrow l-1 \longrightarrow l \longrightarrow l+2 \\ &\longrightarrow \cdots \longrightarrow j-1 \longrightarrow j) = (E - \alpha_{l+1})\gamma_l \prod_{k=i, k\neq l, l+1}^{j-1} \beta_k. \end{aligned} \tag{2.50}$$

We can easily write the expressions for P_{ij} when the pathway passes more than one vertex. To obtain the corresponding result, we must bear in mind that, within the NNN approximation, omitted vertices cannot be adjacent ones. Suppose that we pass the sites $l+1$ and $m+1$, then

$$\begin{aligned} &P_{ij}(i \longrightarrow i+1 \longrightarrow \cdots \longrightarrow l-1 \longrightarrow l \longrightarrow l+2 \longrightarrow \cdots \\ &\longrightarrow m-1 \longrightarrow m \longrightarrow m+2 \longrightarrow \cdots \longrightarrow j-1 \longrightarrow j) \\ &=(E - \alpha_{l+1})(E - \alpha_{m+1})\gamma_l \gamma_m \prod_{k=i, k\neq l, l+1, k\neq m, m+1}^{j-1} \beta_k. \end{aligned} \tag{2.51}$$

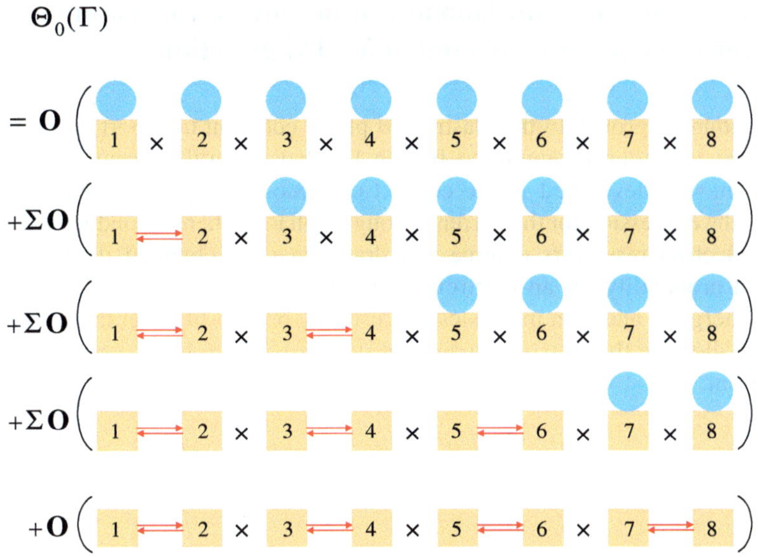

Fig. 2.3 Schematics of computation of $\Theta_0(\Delta)$ within NN approximation for the chain including eight sites

Within the NN approximation, the numerator in the expression for the Green's function G^r_{ij} is factorized as:

$$G^r_{ij} = \frac{1}{\Theta_0(\Delta_{ij})} \prod_{k=i}^{j-1} \beta_k. \tag{2.52}$$

Here, $\Theta_0(\Delta_{ij})$ is the cycle value of the graph calculated within the NN approximation. The method of computing of $\Theta_0(\Delta_{ij})$ is schematically shown in Fig. 2.3.

One may further simplify this result assuming that all hopping parameters β_i and ionization energies for all sites α_i take on the same values ($\beta_i = \beta$, $\alpha_i = E_0$, respectively) and omitting the real parts of the self-energy terms. Then, assuming that the chain includes an odd number of atoms ($N = 2m - 1$), one arrives at the following expression for the matrix element $G^r_{1,2m-1}$:

$$G^r_{1,2m-1}(E) = \frac{4^m \beta^{2(m-1)} \zeta}{(\lambda + \zeta)^{2(m-1)}(\lambda + \zeta + i\Gamma)^2 - (\lambda - \zeta)^{2(m-1)}(\lambda - \zeta + i\Gamma)^2}. \tag{2.53}$$

Here, $\lambda = E_0 - E$, $\zeta = \sqrt{\lambda^2 - 4\beta^2}$, $\Gamma = \Gamma_L + \Gamma_R$. This expression was first derived by Mujica et al. [45] using the method which differs from that described above.

2.3 Electron Transmission and Landauer Expression for the Current Through a MMM Junction

The standard approach to the elastic and phase-coherent transport through metal–molecular junctions is suggested by the Landauer–Buttiker formalism [1]. This formalism was developed in the context of mesoscopic physics. Within this approach, the expression for the current through a phase-coherent conductor connected to the electron reservoirs (source and drain) may be obtained if one calculates the total probability for an electron to travel between two electrodes at a certain tunnel energy E and then integrates the resulting probability over the whole energy range. Following this way one arrives at the well-known Landauer expression for the current:

$$I = \frac{e}{\pi\hbar}\int T(E)[f_L(E) - f_R(E)]dE \tag{2.54}$$

where

$$f_\beta(E) = \left\{1 + \exp\left[(E - \mu_\beta)/kT\right]\right\}^{-1} \tag{2.55}$$

are the Fermi distribution functions for the leads with the electrochemical potentials μ_β and $T(E)$ is the electron transmission function. To clarify the physical meaning of Eq. (2.54), one may rewrite it in the form

$$I = \frac{e}{\pi\hbar}\int T(E)\{f_L(E)[1 - f_R(E)] - f_R(E)[1 - f_L(E)]\}dE. \tag{2.56}$$

where the integrand is explicitly presented as the difference of two fluxes, one directed from the left to the right and another one flowing in the opposite direction.

The forms of these expressions coincides with Eqs. (1.11) and (1.13) which were derived within a simple model where the bridge was represented by a single orbital. This simplification resulted in the specific form of the electron transmission given by Eq. (1.12). Here, we derive the Landauer formula for the general case of a multilevel bridge. We use the formalism described in the first section. To obtain the expression for the current, one may compute the time rate of the probability density on the bridge. The latter is defined as $\psi^\dagger\psi$ where $\{\psi\}$ is the Schrodinger state vector for the bridge. For a system consisting of a sole electron reservoir and the attached channel, one may write down the following equation of motion [4]:

$$i\hbar\frac{d}{dt}\begin{Bmatrix}\psi\\ \Phi\end{Bmatrix} = \begin{pmatrix}H_C & \tau_R\\ \tau_R^\dagger & H_R - i\eta\end{pmatrix}\begin{Bmatrix}\psi\\ \Phi\end{Bmatrix}. \tag{2.57}$$

Here, we assume for certainty that the right end of the channel is put in contact with the reservoir. As before, the wave function $\{\Phi\}$ includes two terms ($\{\Phi\} =$

$\{\Phi_R + \chi_R\}$), where the second term is corresponding to the part which is reflected back to the channel. As follows from Eq. (2.57):

$$e\frac{d}{dt}(\psi^\dagger\psi) = \frac{e}{i\hbar}\text{Trace}\{\psi^\dagger\tau_R\Phi_R - \Phi_R^\dagger\tau_R^\dagger\psi\} \\ - \frac{e}{i\hbar}\text{Trace}\{\chi_R^\dagger\tau_R^\dagger\psi - \psi^\dagger\tau_R\chi_R\} \equiv I_{\text{in}} - I_{\text{out}}. \tag{2.58}$$

In this expression, the first term gives the electron charge flow into the channel from the reservoir (inflow), whereas the second term corresponds to the charge flow back to the reservoir (outflow). The same expression remains valid if one is considering the current through the contact between the bridge and the right reservoir in a system where two electron reservoirs are linked by the channel (bridge). We remark that all terms on the right side of Eq. (2.58) are matrices and/or vectors, whereas the product $\psi^\dagger\psi$ on the left side of this expression is a scalar quantity. Treating the system with two reservoirs, the inflow term in Eq. (2.58) may be reorganized in the following way:

$$I_{\text{in}} = \frac{e}{i\hbar}\text{Trace}\{\psi^\dagger\tau_R\Phi_R - \Phi_R^\dagger\tau_R^\dagger\psi\} = \frac{e}{\hbar}\text{Trace}\{S_R S_R^\dagger A\} \tag{2.59}$$

where $S_R = \tau_R\Phi_R$ and $A = i(G^r - G^a)$ is the spectral function.

Then one may use the expression (2.28) to obtain

$$I_{\text{in}} = \frac{e}{\hbar}\int \frac{dE}{2\pi} f_R \text{Trace}\{\Gamma_R A\}. \tag{2.60}$$

Similarity, the outflow term I_{out} may be rearranged as follows:

$$I_{\text{out}} = \frac{e}{i\hbar}\text{Trace}\{\psi^\dagger\tau_R g_R^\dagger\tau_R^\dagger\psi - \psi^\dagger\tau_R g_R\tau_R^\dagger\psi\} \\ = \frac{e}{\hbar}\text{Trace}\{\psi\psi^\dagger\Gamma_R\}. \tag{2.61}$$

Using the expression (2.29), one may present I_{out} in the form

$$I_{\text{out}} = \frac{e}{\hbar}\int \frac{dE}{2\pi}\text{Trace}[\Gamma_R G^<]. \tag{2.62}$$

Assuming that electrons flow from the left to the right and combining Eqs. (2.60) and (2.62), one gets

$$I = \frac{e}{\hbar}\int dE(\text{Trace}[\Gamma_R G^<] - f_R\text{Trace}[\Gamma_R A]). \tag{2.63}$$

Obviously, the electric charge flow through the considered system must remain the same at any part of the latter, so the expression may be treated as the general expression for the electron current. In the considered case, the energy of the electrons in the channel remains the same (it is equal to the energy E corresponding to the stationary state of the junction introduced in the previous section). Therefore, the electron transport is to be treated as elastic and coherent, and we may employ the expression (2.36) for the self-energy term $\Sigma^<$ in the Keldysh equation which determines the lesser Green's function $G^<$. Also, using the definition (2.37) and the relation

$$(G^a)^{-1} - (G^r)^{-1} = -i\Gamma_L - i\Gamma_R \tag{2.64}$$

we may rewrite the expression for the spectral function as follows:

$$A = G^r\Gamma_L G^a + G^r\Gamma_R G^a. \tag{2.65}$$

Substituting these relations into the expression for electron tunnel current given by Eq. (2.60), we may convert it to the Landauer form (2.54) where the electron transmission $T(E)$ is given by [82]

$$T(E) = \text{Trace}\big[\Gamma_L G^r \Gamma_R G^a\big]. \tag{2.66}$$

Imitating the MMM junction by a simple model where the bridge is represented by a single molecular orbital, one may significantly simplify the expression for the electron transmission. In this case it becomes reduced to the form given by Eq. (1.12).

The Landauer expression (2.54) confirms that a nonzero tunnel current may appear in the junction when the electrochemical potentials of the electrodes μ_L and μ_R differ due to the applied bias voltage and the bridge energy levels are situated between μ_L and μ_R. Otherwise, the difference of the Fermi functions becomes zero (assuming that the thermal broadening of the energy levels on the leads is neglected), and the current through the junction ceases to flow. As mentioned before, the Landauer expression for current (2.54), combined with the expression (2.66) for the electron transmission, correctly describes the coherent and elastic electron transport through MMM junctions. As an example, we present the results for the electron transmission and current calculated for the MMM including a porphyrin-nitrobenzene molecule as a linker (see Fig. 2.4). However, to adequately analyze the effects of electron–electron and electron–phonon interactions, one needs to go beyond the Landauer formula. A consistent approach to inelastic electron transport through molecular junctions and similar mesoscopic systems is provided by the nonequilibrium Green's function formalism [4, 202–204]. To compute the electron current through the junction within this approach, one must solve the system of interconnected equations for electron and phonon Green's functions, such as given in the works [26, 34]. To overcome extreme difficulties which one meets solving these equations, numerous approximate methods were developed. The validity of these approximations is determined by the relations between characteristic energies of the considered system, namely, the molecule-lead coupling parameters

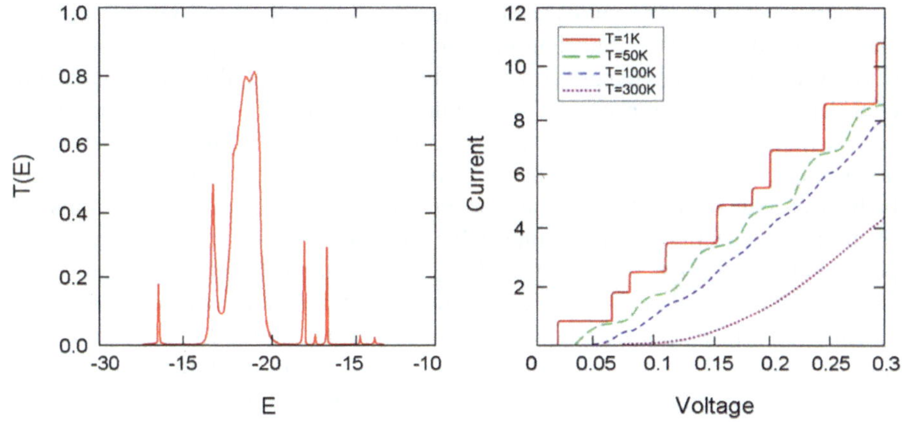

Fig. 2.4 *Left panel*: the calculated electron transmission function through the bridge $T(E)$ within a certain range of energies E (eV). *Right panel*: the calculated current (nA)–voltage (V) characteristic. Calculations are carried out for a porphyrin-nitrobenzene molecule using the Landauer formula [71]. Reprinted with the permission of the American Institute of Physics (AIP)

Γ_β, electron–phonon coupling strengths Λ_{km} and λ_{km} (see Eq. (1.41)), electron charging energy U_0, vibrational modes energies, and the thermal energy $k_B T$ which characterizes the intensity of stochastic fluctuations in the environment. Commonly used approximate computational techniques are described and thoroughly analyzed in the review article [26] and references therein. Nevertheless, these approximations insufficiently simplify the basic NEGF equations for practical molecular junctions which limit the usefulness of the powerful NEGF formalism to relatively simple models for MMMs.

A simple and rather obvious way to generalize the Landauer formula and to make it appropriate for analysis of inelastic transport was suggested by Wingreen and coworkers [205, 206] and used in several later works [207–209]. The modified expression has the form

$$I = \frac{e}{\pi\hbar} \int dE_i dE_f T(E_i, E_f) \times \left\{ f_L(E_i)\left[1 - f_R(E_f)\right] - f_R(E_i)\left[1 - f_L(E_f)\right] \right\} \tag{2.67}$$

where E_i, E_f are the initial and final energies of the electron traveling through the junction and the factor $T(E_i, E_f)dE_f$ corresponds to the probability that an electron starting its travel with the energy E_i is transmitted through the system possessing the energy E_F. The expression (2.67) agrees with the Landauer formula (2.54), when the transport is elastic. In this case $T(E_i, E_f) = T(E_i)\delta(E_i - E_f)$ ($T(E_i)$ being the elastic electron transmission. Substituting this expression into Eq. (2.67), one easily converts it to the form given by Eq. (2.54).

There exist some difficulties with the justification of the approximation (2.67). Datta showed [210] that cancellation of some terms, which happens in the course of derivation of the Landauer formula for the elastic current, does not occur while considering inelastic transport. Also, employing the full NEGF formalism, it was demonstrated that approximation (2.67) could be derived if one disregards terms describing the effect of back reflection from the leads and assumes the phonon bath to remain in the state of thermal equilibrium [26]. However, the omission of the back reflection terms is impossible to validate basing on an expansion in powers of some small parameter, and Eq. (2.67) cannot be interpreted as a first term of such expansion. Therefore, it is hard to estimate the error in the transport calculations which appears when we employ approximation (2.67). Nevertheless, this approximation has some advantages for it can be used regardless of the electron–phonon interactions strength, as it was indicated in the works [207, 208] or up to high-order perturbation theory terms in the electron–phonon interactions [212, 213]. One may further improve approximation (2.67) by taking into account that the transmission function may depend on the applied bias voltage V. The effect of the bias voltage which is negligible at small values of V may become noticeable as V increases. The expression (2.67) is not the simplest approximation for the inelastic current. There exists an especially simple approximation, which preserves the Landauer form of expression for the current including inelastic effects into the expression for the electron transmission function. This approximation is based on the phenomenological Buttiker model for quantum transport. It will be considered in the next section.

2.4 Electron Transport Through a Junction as a Multichannel Scattering Problem

An important advantage of the phenomenological model for the incoherent/inelastic quantum transport proposed by Buttiker [72] is that this model could be easily adapted to analyze various inelastic effects in the electron transport through molecules (and some other mesoscopic systems) avoiding complicated and time-consuming advanced methods such as based on the nonequilibrium Green's functions formalism. First, we present the Buttiker model for a simple junction including two electrodes linked by a single-site molecular bridge. The bridge is attached to a dissipative electron reservoir, as shown in Fig. 2.5a. Electrons tunnel from the electrodes to the bridge and vice versa via the channels 1 and 2. While on the bridge, an electron could be scattered into the channels 3 and 4 with a certain probability ϵ. Such electron arrives at the reservoir where it undergoes inelastic scattering accompanied by phase breaking, and then, the reservoir reemits it back to the channels 3 and 4 with the same probability. So, within the Buttiker model, the electron transport through the junction is treated as the combination of tunnelings through the barriers separating the molecule from the electrodes and interaction

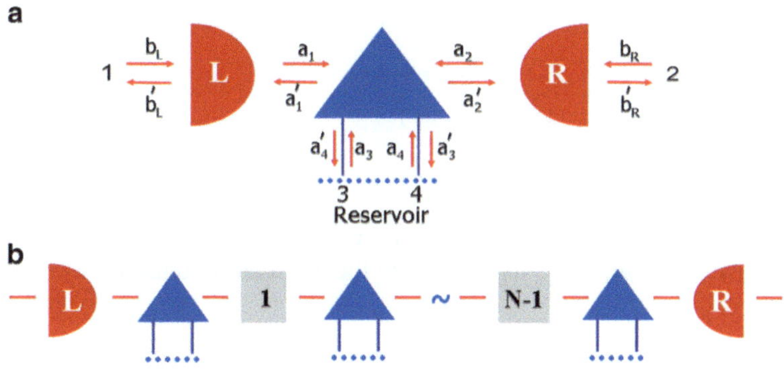

Fig. 2.5 Schematics of a junction including a single-site (**a**) and multisite (**b**) bridge linking the left and right electrodes. The triangles represent scatterers coupling the bridge sites with the electron reservoirs, and the squares indicate potential barriers separating the adjacent sites on the bridge

with the reservoir coupled to the bridge site. The key parameter of the model is the probability ϵ which is closely related to the coupling strength between the bridge site and the reservoir. When $\epsilon = 0$ the reservoir is detached from the bridge, and the electron transport is completely coherent and elastic. Within the opposite limit ($\epsilon = 1$), electrons are necessarily scattered into the reservoir which results in the overall phase randomization and inelastic transport.

Within the Buttiker model, the particle fluxes outgoing from the junctions J_i' could be presented as linear combinations of the incoming fluxes J_k where the indexes i, k label the channels for the transport $1 \leq i \leq k \leq 4$.

$$J_i' = \sum_k T_{ik} J_k. \tag{2.68}$$

The coefficients T_{ik} in these linear equations are matrix elements of the transmission matrix which are related to the elements of the scattering matrix S, namely, $T_{ik} = |S_{ik}|^2$. The matrix S expresses outgoing wave amplitudes b_L', b_R', a_3', a_4' in terms of the incident ones b_L, b_R, a_3, a_4. To provide the charge conservation in the system, the net current in the channels 3 and 4 linking the system with the dephasing reservoir must be zero, so we may write

$$J_3 + J_4 - J_3' - J_4' = 0. \tag{2.69}$$

The transmission for the quantum transport could be defined as the ratio of the particle flux outgoing from the system and that one incoming to the latter. Solving Eqs. (2.68) and (2.69), we obtain

$$T(E) = \frac{J_2'}{J_1} = T_{21} + \frac{K_1 \cdot K_2}{2R} \tag{2.70}$$

where:

$$K_1 = T_{31} + T_{41};$$
$$K_2 = T_{23} + T_{24};$$
$$R = T_{33} + T_{44} + T_{43} + T_{34}. \tag{2.71}$$

The scatterer coupled to the bridge is described by the 4×4 scattering matrix s relating wave amplitudes a_1', a_2', a_3', a_4' to the wave amplitudes a_1, a_2, a_3, a_4 :

$$s = \begin{pmatrix} 0 & \sqrt{1-\epsilon} & \sqrt{\epsilon} & 0 \\ \sqrt{1-\epsilon} & 0 & 0 & \sqrt{\epsilon} \\ \sqrt{\epsilon} & 0 & 0 & -\sqrt{1-\epsilon} \\ 0 & \sqrt{\epsilon} & -\sqrt{1-\epsilon} & 0 \end{pmatrix}. \tag{2.72}$$

Also, the electron tunneling through a single barrier is described by the transmission and reflection amplitudes (t and r, respectively). These are the matrix elements of a 2×2 matrix:

$$s_{\text{el}} = \begin{pmatrix} r & t \\ t & r \end{pmatrix}. \tag{2.73}$$

Combining Eqs. (2.72) and (2.73), we arrive at the expression for the scattering matrix S.

For the junction including the single-site bridge, the matrix S has the form [70,72]

$$S = \frac{1}{Z} \begin{pmatrix} r_L + \alpha^2 r_R & \alpha t_L t_R & \beta t_L & \alpha\beta t_L r_R \\ \alpha t_L t_R & r_R + \alpha^2 r_L & \alpha\beta r_L t_R & \beta t_R \\ \beta t_L & \alpha\beta r_L t_R & \beta^2 r_L & \alpha r_L r_R - \alpha \\ \alpha\beta t_L r_R & \beta t_R & \alpha r_L r_R - \alpha & \beta^2 r_R \end{pmatrix}. \tag{2.74}$$

Here, $Z = 1 - \alpha^2 r_L r_R$, $\alpha = \sqrt{1-\epsilon}$, $\beta = \sqrt{\epsilon}$ and $r_{L,R}$ and $t_{L,R}$ are the reflection and transmission coefficients for the barriers separating the bridge from the left and right electrode, respectively. The model may be generalized to include an arbitrary number of the bridge sites that is to describe a practical molecule presented as a set of energy levels. Here, however, we follow [45,70,109], and we mimic a molecular bridge linking two electrodes as a periodical chain of identical sites. The schematics of the adopted model is shown in Fig. 2.5b. The first and the last sites in the chain are attached to the leads and remain fixed. All sites are put in contact with electron reservoirs. We assume that the parameter ϵ may accept different values for scatterers attached to different bridge sites. So, we may characterize the system including N bridge sites with the set of parameters ϵ_j $(j = 1, \ldots, N)$.

To simplify the following calculations of the electron transmission we conjecture that an electron could be injected into the system shown in Fig. 2.5b and/or leave from there solely via $2N+2$ channels indicated in the figure. Then the scattering matrix $S^{(N-1)}$ corresponding to the system including $N-1$ ($N>1$) bridge sites has dimensions $2N \times 2N$. This matrix relates wave amplitudes outgoing from the system $b'_L,\ b'_R,\ a'_3,\ \ldots, a'_{2N-2}$ to the incoming amplitudes $b_L,\ b_R,\ a_3,\ \ldots, a_{2N-2}$. Here, the wave amplitudes $a_k,'\ a_k$ ($k=3,4,\ldots,2N-2$) are associated with the channels connecting the sites with the dissipative reservoirs.

Now, we may derive recursive expressions bringing into correlation matrix elements of the scattering matrices $S^{(N-1)}$ and $S^{(N)}$. For certainty, we put an extra site at the right end of the bridge chain. Then we split the whole system shown in Fig. 2.5b in two subsystems. The first subsystem consists of the left lead and the bridge including $N-1$ sites, and the second one includes the extra site and the right lead. To proceed we introduce the matrix $\tilde{S}^{(N-1)}$ closely related to the matrix $S^{(N-1)}$. One may obtain expressions for matrix elements of $\tilde{S}^{(N-1)}$ replacing the reflection and transmission coefficients for the right lead r_R, t_R in the matrix elements of $S^{(N-1)}$ by the corresponding characteristics for the barrier separating $(N-1)$th and Nth sites. The second subsystem could be considered as a junction with a single-site bridge and an absolutely transparent left lead ($t_L=1,\ r_L=0$). So, the scattering matrix for this subsystem has the 4×4 dimensions, and it could be written in the form

$$\theta^{(N)} = \begin{pmatrix} \alpha_N^2 r_R & \alpha_N t_R & \beta_N & \alpha_N \beta_N r_R \\ \alpha_N t_R & r_R & 0 & \beta_N t_R \\ \beta_N & 0 & 0 & -\alpha_N \\ \alpha_N \beta_N r_R & \beta_N & -\alpha_N & \beta_N^2 r_R \end{pmatrix}. \tag{2.75}$$

Here, the parameters α_N, β_N characterize the coupling of the Nth site to the electron reservoir attached to the latter. The matrix $\tilde{S}^{(N-1)}$ relates the wave amplitudes $b'_L,\ b'_{N-1},\ a'_3, \ldots,\ a'_{2N-2}$ to the wave amplitudes $b_L,\ b_{N-1},\ a_3, \ldots,$ a_{2N-2} where the outgoing (b'_{N-1}) and incoming (b_{N-1}) wave amplitudes at the right end of the first subsystem appear. At the same time, b'_{N-1} corresponds to the wave entering the second subsystem from the left, whereas b_{N-1} is associated with the wave outgoing from there. Accordingly, the matrix $\theta^{(N)}$ relates $b_{N-1},\ b'_R,\ a'_{2N-1},\ a'_{2N}$ to $b'_{N-1},\ b_R,\ a_{2N-1},\ a_{2N}$. Excluding the wave amplitudes $b'_{N-1},\ b_{N-1}$ associated with the electron transport inside the system, we arrive at the desired recurrence formulas [109]

$$S_{in}^{(N)} = \theta_{in}^{(N)} + \theta_{i1}^{(N)} \tilde{S}_{22}^{(N-1)} \theta_{1n}^{(N)} / Z_N, \qquad (i,n = 2,\ 2N+1,\ 2N+2), \tag{2.76}$$

$$S_{kl}^{(N)} = \tilde{S}_{kl}^{(N-1)} + \theta_{11}^{(N)} \tilde{S}_{k2}^{(N-1)} \tilde{S}_{2l}^{(N-1)} / Z_N, \qquad (k,l = 1,3,4,\ldots,2N-1,2N) \tag{2.77}$$

$$S_{ik}^{(N)} = \theta_{i1}^{(N)} \tilde{S}_{2k}^{(N-1)} / Z_N, \tag{2.78}$$

where $Z_N = 1 - \theta_{11}^{(N)} \tilde{S}_{22}^{(N-1)}$. Similar expressions were derived in the earlier work [70] to analyze long-range electron transfer reactions in macromolecules.

As before, the incoming particle fluxes are related to outgoing ones by the system of linear equations of the form (2.68), which now includes $2N + 2$ equations ($1 \leq i \leq k \leq 2N + 2$. Also, the charge conservation is required in the channels connecting every single site on the bridge with the corresponding dissipative reservoir. Solving this system we obtain the following result for the electron transmission [70]:

$$T(E) = T_{21} + \sum_{i,j=1}^{N-1} K_i^{(2)} (W^{-1})_{ij} K_j^{(1)} \tag{2.79}$$

where $1 \leq i, j \leq N - 1$;

$$\begin{aligned} K_i^{(1)} &= T_{2i+1,1} + T_{2i+2,1}; \\ K_i^{(2)} &= T_{2;2i+1} + T_{2,2i+2}; \end{aligned} \tag{2.80}$$

and W^{-1} is the matrix inversed with respect to the $(N - 1) \times (N - 1)$ matrix W, whose matrix elements are given by

$$W_{ij} = (2 - R_{ii})\delta_{ij} - \tilde{T}_{ij}(1 - \delta_{ij}), \tag{2.81}$$

In this expression we used the following denotations:

$$\begin{aligned} R_{ii} &= T_{2i+1,2i+1} + T_{2i+2,2i+2} + T_{2i+2,2i+1} + T_{2i+1,2i+2}, \\ \tilde{T}_{ij} &= T_{2i+1,2j+1} + T_{2i+1,2j+2} + T_{2i+2,2j+1} + T_{2i+2,2j+2}. \end{aligned} \tag{2.82}$$

Returning to the case of a single-site bridge and assuming for certainty that the charge flows from the left to the right, one may obtain

$$T(E) = \frac{t_L(E)t_R(E)(1 + \alpha^2)\big[1 - \alpha^2\big(1 - t_L(E)t_R(E)\big)\big]}{2\big[1 + \alpha^2\big(1 - t_L(E)t_R(E)\big)\big]^2}. \tag{2.83}$$

For elastic transport ($\alpha = 1$) the electron transmission accepts the form

$$T(E) = \frac{t_L^2 t_R^2}{(2 - t_L t_R)^2}. \tag{2.84}$$

At the same time, the elastic transmission is described by Eq. (1.12). Equating these two expressions, one finds

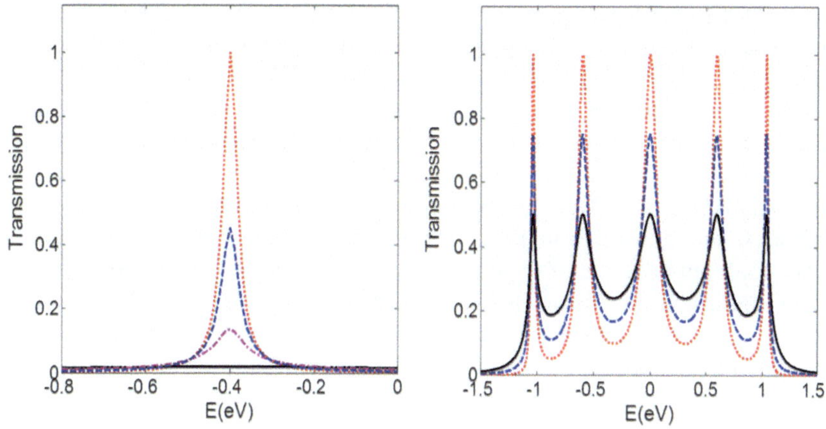

Fig. 2.6 Electron transmission through a junction including a single-site (*left panel*) and multisite (*right panel*) bridge. The curves are plotted assuming $\Gamma_L = \Gamma_R = 0.1\,\text{eV}$, $T = 30\,\text{K}$, $E_0 = -0.5\,\text{eV}$, (*left panel*) and $E_0 = 0$ (*right panel*). The parameter ϵ takes on values $0.0,\ 0.1,\ 0.3,\ 0.7$ (*left panel*). In plotting the curves in the right panel it was assumed that there are five electronic states on the bridge, $\epsilon_i = 0$ $(i \neq 3)$, and $\epsilon_3 = 0$ (*dash-dotted line*) 0.5 (*dashed line*) and 1.0 (*solid line*) (after [52, 109]). Reprinted with the permission of the AIP

$$t_L t_R = \frac{2g}{1+g} \tag{2.85}$$

where

$$g = \sqrt{\frac{\Gamma_L \Gamma_R}{(E-E_0)^2 + (\Gamma/2)^2}}. \tag{2.86}$$

So, the expression (2.84) may be converted to the form [51]:

$$T(E) = \frac{g(E)(1+\alpha^2)[g(E)(1+\alpha^2)+1-\alpha^2]}{2\big[g(E)(1-\alpha^2)+1+\alpha^2\big]^2}. \tag{2.87}$$

Now, the electron transmission strongly depends on the scattering probability ϵ. As shown in Fig. 2.6, in the case of elastic transport ($\epsilon = 0$), the transmission exhibits a sharp peak at $E = E_0$ which gives a step-like shape to the volt-ampere curve, as was discussed in the previous chapter. In the presence of dissipation the peak gets eroded. When the value ϵ approaches 1, the I–V curve becomes linear, corroborating the Ohmic law for the inelastic transport. For a multisite bridge, the electron transmission depends on the whole set of the dissipative parameters ϵ_j, as well as on the characteristics of the tunneling barriers separating the sites. For instance, within the model adopted in [109], the bridge chain included an odd number of sites ($N = 2m - 1$), and only the single site in the middle was assumed to be attached to a dissipative reservoir ($\epsilon_j = 0$ for $j \neq m$). The energy dependence of the electron transmission calculated using this model is shown in the right panel of Fig. 2.6.

As expected, instead of a single maximum we now get a set of peaks. The number of peaks equals to the number of sites in the chain. These peaks correspond to the energy levels of the molecular bridge within the adopted model. The coupling of the chain ends to the leads affects the minimum values of transmission. As the coupling strengthens, the transmission minimum values increase. When the reservoir is attached to the middle site of the chain ($\epsilon_m > 0$), it causes a decrease in the range of the transmission variations. However, even within the strong dephasing/inelastic limit ($\epsilon_m = 1$), the transmission peaks remain well distinguishable for multisite molecular bridges, whereas the peak in the transmission through the single-site bridge appears to be completely washed out under such conditions. This discrepancy could be explained as an effect of elastic and coherent electron tunneling through the left and right halves of the bridge chain, which affects the transport characteristics even in the presence of a strong dissipation at the middle site. The single dissipative reservoir is not sufficient to completely suppress these effects. To fully wash out the peaks in the transmission one needs to attach dissipative reservoirs to all bridge sites and to assume $\epsilon_j = 1$ for $j = 1, \ldots, 2m - 1$.

Despite of its extreme simplicity, Eq. (2.87) for the electron transmission through a single-site molecular bridge may be of practical use. For example, it was successfully employed to explain some transport characteristics of conducting polymer (doped polyaniline–polyethylene (PANi) oxides) nanofibers, observed in experiments [214]. Starting from their discovery, conducting polymers have undergone intense research. These materials are significant mostly due to various possible applications in fabrication of nanodevices for polymer-based devices should have advantages of low cost and flexible, controlled chemistry [215]. Also, there are some unsolved problems concerning the physical nature of charge-transfer mechanisms in conducting polymers, making them interesting subjects for fundamental research [95, 96]. Conducting chemically doped polymers are known to be very inhomogeneous. In some regions polymer chains are disordered, forming an amorphous poorly conducting substance. In other regions the polymer chains are ordered and densely packed [216, 217]. These regions behave as metallic-like grains embedded in a disordered amorphous environment. This gives grounds to apply the model of a granular metal [218] to describe such materials. Prigodin and Epstein [219] suggested that the grain-to-grain transport in conducting polymers mostly occurs due to the resonance tunneling of electrons. The latter is provided by intermediate states on the chains connecting the grains. Applying this approach to the charge transport in conducting polymers, it was shown that the results are consistent with those obtained in transport experiments on individual PANi nanofibers.

The transmission coefficient for the electron resonance tunneling is determined with the probability of finding the resonance state. This probability T is estimated as $T \sim \exp(-L/\xi)$, where L is the average distance between the interacting grains

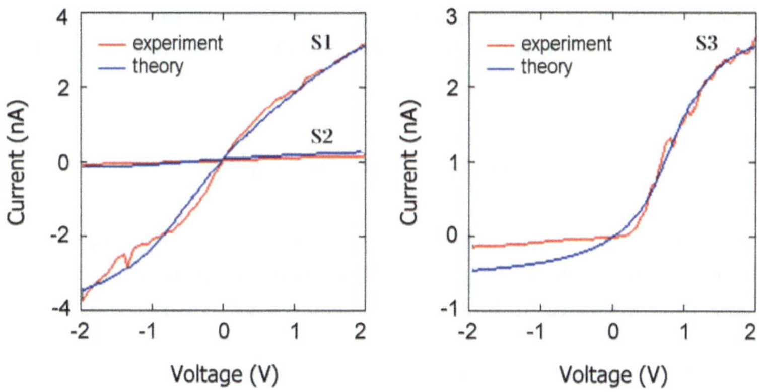

Fig. 2.7 Calculated and experimental current (nA)–voltage (V) characteristics for PANi-PEO nanofibers $T = 300\,\mathrm{K}$ (after [221]). Reprinted with the permission of the American Physical Society (APS)

and ξ is the localization length for electrons [219]. The tunneling probability takes on values which are small compared to unity but much greater than the transmission probability for sequential hoppings along the chains, $T_h \sim \exp(-2L/\xi)$ [220]. The probability for existence of a resonance state at a certain chain is rather low, so only a few out of the whole set of the chains connecting two grains are participating in the process of intergrain electron transport. Therefore, one could assume that any two adjacent metallic-like domains are connected by a single chain providing an intermediate state for the resonance tunneling. So, one may treat the adjacent islands as leads linked by single-site bridges. Realistic PANi nanofibers have diameters in the range $\sim$20–100 nm and lengths of the order of several micrometers. This is significantly greater than typical for conducting polymers sizes of metal-like grains and intergrain distances, which take on values $\sim$5 nm [215]. Therefore, one may treat a nanofiber as a set of parallel channels for the tunneling current, any channel being a sequence of metallic domains connected by resonance chains. The total current in the fiber is the sum of the contributions from all these channels. Also, the voltage V applied across the whole fiber is a sum of contributions δV from sequential pairs of grains along a single channel. One may assume as the first approximation that $\delta V \sim VL/L_f$, where L_f is the fiber length. Under these assumptions, the Landauer expression for the current (2.54) combined with the expression for the electron transmission (2.87) may be used to compute the current in the fibers. Such computations were carried out in [221], and a good agreement with the experiments [214] was obtained, as illustrated in Fig. 2.7.

Within the Buttiker's model, ϵ is introduced as a phenomenological parameter whose relation to the microscopic characteristics of dephasing/dissipative processes affecting electron transport through molecular junctions remains uncertain. To further advance this model one should find out how to express ϵ in terms of the relevant microscopic characteristics for various transport mechanisms. This should open the way to make a link between the phenomenological Buttiker model and

NEGF. Such attempt was carried out in [52, 222] where the effect of stochastic nuclear motion on the electron transport through molecules was analyzed.

2.5 Nonequilibrium Green's Functions Formalism

In the present section, a brief description of the nonequilibrium Green's function formalism is given which generalizes and extends the simplified approach presented in the previous sections of this chapter. The NEGF provides a powerful and systematic framework for computing optical and transport properties of various open many-body systems including metal–molecule junctions. Within the more frequently used NEGF version, the primary objects employed to describe and analyze transport properties of MMM junctions and other mesoscopic systems are Green's functions on the Keldysh contour [4, 93, 202–205], their projections to the real-time axis, and the corresponding self-energy parts. The Green's functions matrix elements may be expressed in terms of field operators. For instance, the electron Green's function matrix elements $G_{ij}(\tau,\tau')$ are defined as

$$G_{ij}(\tau,\tau') = -i\big\langle T_c c_i(\tau) c_j^\dagger(\tau')\big\rangle. \tag{2.88}$$

where $\big\langle \ldots \big\rangle$ stands for averaging, τ is the Keldysh time starting and ending at $-\infty$, T_c is the contour time-ordering operator, and $c_i(\tau), c_j^\dagger(\tau)$ are the annihilation and creation operators for the electrons taken at the respective times τ and τ'. The crucial impact of electron–phonon interactions on the characteristics of electron transport makes it necessary to analyze the combined electron–phonon system coherently evolving according to the corresponding many-body Hamiltonian. The latter includes terms describing the electronic subsystem, the vibrational (phonon) subsystem, and the electron–phonon interactions. Consequently, one may introduce the phonon Green's functions of the form

$$D_{lk}^{(1)}(\tau,\tau') = -i\big\langle T_c A_l(\tau) A_k^\dagger(\tau')\big\rangle, \tag{2.89}$$

$$D_{mk}^{(2)}(\tau,\tau') = -i\big\langle T_c B_m(\tau) B_k^\dagger(\tau')\big\rangle, \tag{2.90}$$

which correspond to the vibrational and thermal phonons, respectively. Different versions of NEGF were developed basing on another time-ordering prescriptions known as "the single time" and the "physical" representations [223–225]. A Liouville space formulation of NEGF was presented in [226–228].

Now, we consider a metal–molecule junction assuming that the molecule is coupled to a set of vibrational phonons. For this system, the Green's functions defined by Eqs. (2.88) and (2.89) satisfy the integral equations given below [25]:

$$G_{ij}(\tau,\tau') = G^0_{ij}(\tau,\tau') + \sum_{k,l}\int d\tau_1 \int d\tau_2 G^0_{ik}(\tau,\tau_1)\Sigma_{kl}(\tau_1,\tau_2)G_{lj}(\tau_2,\tau'), \quad (2.91)$$

$$D^{(1)}_{lk}(\tau,\tau') = D^{(10)}_{lk}(\tau,\tau') + \sum_{m,n}\int d\tau_1 \int d\tau_2 D^{(10)}_{lm}(\tau,\tau_1)\Pi_{mn}(\tau_1,\tau_2)D^{(1)}_{nk}(\tau_2,\tau'). \quad (2.92)$$

In these equations, the integration over τ_1, τ_2 is carried out over the Keldysh contour; the Green's functions $G^0(\tau,\tau')$ and $D^{(10)}(\tau,\tau')$ are associated with the molecule detached from the leads and phonons and to the system of vibrational phonons decoupled from the electron system, respectively. The terms $\sum(\tau_1,\tau_2)$ and $\Pi(\tau_1,\tau_2)$ represent the electron and phonon self-energies. If necessary, a Dyson-type equation similar to Eqs. (2.91) and (2.92) may be written out for the Green's function $D^{(2)}(\tau,\tau')$ as well.

Restricting oneself with the consideration of steady-state processes, one may project the Green's functions onto the real-time axis. Then one may use the fact that under steady-state conditions, all two-time quantities actually depend only on differences in times, so one may turn to Fourier transformed functions in the energy space ($t - t' \to E$). The relevant projections are retarded (r), advanced (a), less ($<$), and greater ($>$) Green's functions discussed in the first section of this chapter. Now, the retarded electron Green's function Fourier transform is introduced by the relation

$$G^r_{ij}(E) \equiv \langle\langle c_i, c_j^\dagger\rangle\rangle = -i\int_0^\infty \langle\{c_i(t), c_j^\dagger\}\rangle \exp[i(E+i\eta)t]dt \quad (2.93)$$

where curly brackets denote the anticommutator and η is a positive infinitesimal parameter. The advanced electron Green's function is a Hermitian conjugate of the retarded one: $G^a(E) = [G^r(E)]^\dagger$. As for the functions $G^>, G^<$, they are determined by Keldysh equations. In matrix form the lesser electron Green's function $G^<$ satisfies equation (2.35). Similar equation could be written for the greater Green's function $G^>$, namely,

$$G^>(E) = G^r(E)\Sigma^>(E)G^a(E). \quad (2.94)$$

Also, Eq. (2.91) could be reduced to the standard Dyson equation for the retarded Green's function and the equivalent equation for its advanced counterpart. Similar equations may be written for the phonon Green's functions.

As discussed above, the significance of Green's functions and self-energies for the transport theory is based on the fact that they could be used to calculate important observable transport characteristics. For example, the relaxation rates for electrons $\Gamma(E)$ and phonons $\gamma(E)$ have the form

$$\Gamma(E) = i[\Sigma^r(E) - \Sigma^a(E)];$$
$$\gamma(E) = i[\Pi^r(E) - \Pi^a(E)] \tag{2.95}$$

where $\Pi^{r,a}(E)$ are the retarded and advanced phonon self-energies. Also, the molecular spectral function describing the electron density of states is proportional to the difference of the retarded and advanced Green's functions, as given by Eq. (2.26), and the current flowing through the junction has the form [93, 229, 230]

$$I = \frac{e}{\hbar} \sum_{\beta} \int \frac{dE}{2\pi} \text{Trace} \left[\Sigma_{\beta}^{<}(E) G^{>}(E) - \Sigma_{\beta}^{>}(E) G^{<}(E) \right]. \tag{2.96}$$

Here, as before, the relevant Green's functions and self-energy terms are matrices and must be treated as such. In general, the expressions for self-energies include contributions from all kinds of interactions occurring in the system. To simplify the forms of these terms the non-crossing approximation [231] is often used. Within this approximation, it is assumed that the molecule independently interacts with the leads and the environment (in the considered case with phonons). Correspondingly, the self-energies are presented as sums including terms associated with various kinds of interactions (see, e.g., [25, 26]). For instance, one may write

$$\Sigma^{r,a}(E) = \Sigma_L^{r,a}(E) + \Sigma_R^{r,a}(E) + \Sigma_{\text{ph}}^{(1)r,a}(E) + \Sigma_{\text{ph}}^{(2)r,a}(E) \tag{2.97}$$

where two first terms correspond to the couplings of the molecule to the leads and two last terms represent the effects of electron–phonon interactions.

The terms $\Sigma_{\beta}^{r,a}(E)$ may be obtained basing on the expressions (2.19) for the retarded Green's functions for the contacts. They have the form

$$\Sigma_{\beta}^r = \Sigma_{\beta}^a \equiv \Sigma_{0\beta}, \tag{2.98}$$

$$\Sigma_{0\beta}^{ij}(E) = \sum_{r,\sigma} \frac{\tau_{ir\beta\sigma}^* \tau_{jr\beta\sigma}}{E - \epsilon_{r\beta\sigma} + i\eta}. \tag{2.99}$$

Imaginary parts of the $\Sigma_{0\beta}^{ij}$ determine matrix elements of the matrices Γ_{β} which describe broadening of the molecular levels due to the coupling of the molecule to the leads:

$$\Gamma_{\beta}^{ij}(E) = \pi \sum_{r,\sigma} \tau_{ir\beta\sigma}^* \tau_{jr\beta\sigma} \delta(E - \epsilon_{r\beta\sigma}). \tag{2.100}$$

Also, assuming that electron–electron interactions on the molecule do not affect its coupling to the leads, we may write out the following expressions for the matrices $\Sigma^<$ and $\Sigma^>$ [4, 25]:

$$(\Sigma_\beta^<)^{ij} = i f_\beta(E)\Gamma_\beta^{ij}(E), \tag{2.101}$$

$$(\Sigma_\beta^>)^{ij} = -i\{1 - f_\beta(E)\}\Gamma_\beta^{ij}(E) \tag{2.102}$$

where $f_\beta(E)$ is the Fermi distribution function for the lead β and the matrix elements $\Gamma_\beta^{ij}(E)$ are given by Eq. (2.100). Within the often used wide-band approximation for the leads, Γ_β^{ij} become independent of energy E. We remark that the expression (2.99) for $\Sigma^<$ agrees with the previous result (2.36) which was obtained within the simplified approach.

Assuming that the transport through the junction is elastic and using Eqs. (2.35) and (2.96), one may rearrange the expression (2.96) for the current through the junction as follows:

$$\begin{aligned} I = \frac{e}{\pi\hbar}\int dE\,\mathrm{Trace}\,\{&\Sigma_L^<(E)G^r(E)\left[\Sigma_L^>(E) + \Sigma_R^>(E)\right]G^a(E) \\ &- \Sigma_R^>(E)G^r(E)\left[\Sigma_L^<(E) + \Sigma_R^<(E)\right]G^a(E)\}\,. \end{aligned} \tag{2.103}$$

Then, substituting into this formula the expressions (2.101) and (2.102) for the self-energy terms $\Sigma_\beta^<$ and $\Sigma_\beta^>$, one may reduce the expression (2.103) to the Landauer form given by Eqs. (2.54) and (2.66).

Putting aside the effect of electron–phonon interactions on the electron transport through molecular junctions (which is justified at low temperatures and in the absence of molecular vibrations), one may analytically calculate the electron Green's functions and then compute the current through the junction. This approach is often used in studies of diverse phenomena occurring due to electron–electron interactions on the molecular bridges such as the Coulomb blockade and/or the Kondo maximum in the conduction [34, 35, 86–90]. Here, we illustrate this by deriving the expression for the retarded electron Green's function by means of the commonly used technique based on the Heisenberg equations of motion for the electron creation and annihilation operators [86]. We mimic the molecular bridge in the considered junction by a single spin-degenerated level, so the bridge Hamiltonian H_M takes on the form given by Eq. (1.31), and the terms describing the molecule-lead couplings have the form

$$H_T = \sum_{r,\beta,\sigma} \tau_{r\beta\sigma}^* c_{r\beta\sigma}^\dagger c_\sigma + H.C. \tag{2.104}$$

The equation of motion for the retarded Green's function for the bridge $G_\sigma^r(E) \equiv \langle\langle c_\sigma; c_\sigma^\dagger\rangle\rangle$ reads

$$
\begin{aligned}
&(E - E_0 + i\eta)\langle\langle c_\sigma; c_\sigma^\dagger \rangle\rangle \\
&= 1 + \sum_{r,\beta} \tau_{r\beta,\sigma}\langle\langle c_{r\beta\sigma}; c_\sigma^\dagger \rangle\rangle \\
&\quad + U_0\langle\langle n_{-\sigma}c_\sigma; c_\sigma^\dagger \rangle\rangle.
\end{aligned}
\tag{2.105}
$$

Here, $n_{-\sigma} \equiv c_{-\sigma}^\dagger c_{-\sigma}$, and $\langle\langle c_{r\beta\sigma}; c_\sigma^\dagger \rangle\rangle$ obeys the equation

$$
(E - \epsilon_{r\beta\sigma} + i\eta)\langle\langle c_{r\beta\sigma}; c_\sigma^\dagger \rangle\rangle = \tau_{r\beta\sigma}^*\langle\langle c_\sigma; c_\sigma^\dagger \rangle\rangle. \tag{2.106}
$$

Substituting this expression into (2.105), we get

$$
(E - E_0 - \Sigma_{0\sigma})\langle\langle c_\sigma; c_\sigma^\dagger \rangle\rangle = 1 + U_0\langle\langle n_{-\sigma}c_\sigma; c_\sigma^\dagger \rangle\rangle. \tag{2.107}
$$

The self-energy part $\Sigma_{0\sigma}$ is given by Eq. (2.99) where the indices i, j are omitted in accordance with the chosen model for the bridge. The equation of motion for the four-operator function $\langle\langle n_{-\sigma}c_\sigma; c_\sigma^\dagger \rangle\rangle$ includes higher-order averages. Omitting them we write out

$$
\begin{aligned}
&(E - E_0 - U_0 + i\eta)\langle\langle n_{-\sigma}c_\sigma; c_\sigma^\dagger \rangle\rangle \\
&= \langle n_{-\sigma} \rangle \\
&\quad + \sum_{r,\beta} \big(\tau_{r\beta\sigma}\langle\langle c_{r\beta\sigma}n_{-\sigma}; c_\sigma^\dagger \rangle\rangle \\
&\quad + \tau_{r\beta;-\sigma}\langle\langle c_{-\sigma}^\dagger c_{r\beta;-\sigma}c_\sigma; c_\sigma^\dagger \rangle\rangle \\
&\quad - \tau_{r\beta;-\sigma}^*\langle\langle c_{r\beta;-\sigma}c_{-\sigma}c_\sigma; c_\sigma^\dagger \rangle\rangle\big).
\end{aligned}
\tag{2.108}
$$

To proceed we must write equations for the averages inserted in the right side of Eq. (2.108). We get

$$
\begin{aligned}
&(E - \epsilon_{r\beta\sigma} + i\eta)\langle\langle c_{r\beta\sigma}n_{-\sigma}; c_\sigma^\dagger \rangle\rangle \\
&= \tau_{r\beta\sigma}^*\langle\langle n_{-\sigma}c_\sigma; c_\sigma^\dagger \rangle\rangle \\
&\quad + \sum_{r'\beta} \big(\tau_{r'\beta;-\sigma}\langle\langle c_{r\beta\sigma}c_{-\sigma}^\dagger c_{r'\beta;-\sigma}; c_\sigma^\dagger \rangle\rangle \\
&\quad - \tau_{r'\beta;-\sigma}^*\langle\langle c_{r'\beta-\sigma}c_{-\sigma}c_{r\beta;\sigma}; c_\sigma^\dagger \rangle\rangle\big),
\end{aligned}
\tag{2.109}
$$

$$
\begin{aligned}
&(E - \epsilon_{r\beta;-\sigma} + i\eta)\langle\langle c_{-\sigma}^\dagger c_{r\beta;-\sigma}c_\sigma; c_\sigma^\dagger \rangle\rangle \\
&= \tau_{r\beta;-\sigma}^*\langle\langle n_{-\sigma}c_\sigma; c_\sigma^\dagger \rangle\rangle
\end{aligned}
$$

$$- \sum_{r'\beta} \left(\tau_{r'\beta-\sigma} \langle\langle c^{\dagger}_{r'\beta;-\sigma} c_{r\beta;-\sigma} c_{\sigma}; c^{\dagger}_{\sigma} \rangle\rangle \right.$$

$$\left. + \tau_{r'\beta\sigma} \langle\langle c^{\dagger}_{-\sigma} c_{r\beta;-\sigma} c_{r'\beta\sigma}; c^{\dagger}_{\sigma} \rangle\rangle \right), \tag{2.110}$$

$$(E - 2E_0 - U_0 + \epsilon_{r\beta;-\sigma} + i\eta) \langle\langle c^{\dagger}_{r\beta;-\sigma} c_{-\sigma} c_{\sigma}; c^{\dagger}_{\sigma} \rangle\rangle$$

$$= - \tau_{r\beta;-\sigma} \langle\langle n_{-\sigma} c_{\sigma}; c^{\dagger}_{\sigma} \rangle\rangle$$

$$+ \sum_{r'\beta} \left(\tau_{r'\beta\sigma} \langle\langle c^{\dagger}_{r'\beta;-\sigma} c_{-\sigma} c_{r\beta\sigma}; c^{\dagger}_{\sigma} \rangle\rangle \right.$$

$$\left. - \tau_{r'\beta;-\sigma} \langle\langle c^{\dagger}_{r\beta;-\sigma} c_{\sigma} c_{r'\beta;-\sigma}; c^{\dagger}_{\sigma} \rangle\rangle \right). \tag{2.111}$$

Writing out these equations (2.109)–(2.111), we neglected averages like $\langle c^{\dagger}_{\sigma} c_{r\beta\sigma} \rangle$ including a creation/annihilation operator for the molecule combined with the annihilation/creation operator for the left/right leads. Such averages are omitted in further calculations as well. Now, we decouple four-operator averages included in the sums over "r'" in these equations following the scheme [4, 91]:

$$\langle\langle A^{\dagger} BC; D^{\dagger} \rangle\rangle \approx \langle A^{\dagger} B \rangle \langle\langle C; D^{\dagger} \rangle\rangle - \langle A^{\dagger} C \rangle \langle\langle B; D^{\dagger} \rangle\rangle. \tag{2.112}$$

Also, we use the approximation

$$\langle c^{\dagger}_{r\beta\sigma} c_{r'\beta\sigma} \rangle = \langle c_{r'\beta\sigma} c^{\dagger}_{r\beta\sigma} \rangle = \delta_{rr'} f^{\beta}_{r\sigma} \tag{2.113}$$

where $f^{\beta}_{r\sigma}$ is the Fermi distribution function corresponding to the energy $\epsilon_{r\beta\sigma}$. As a result, the Green's functions included in the right-hand side of Eq. (2.108) get expressed in terms of $\langle\langle n_{-\sigma} c_{\sigma}; c^{+}_{\sigma} \rangle\rangle$ and $\langle\langle c_{\sigma}; c^{+}_{\sigma} \rangle\rangle$. Substituting these expressions into Eq. (2.108), we obtain

$$\langle\langle n_{-\sigma} c_{\sigma}; c^{+}_{\sigma} \rangle\rangle = \frac{\langle n_{-\sigma} \rangle + \Sigma_{1\sigma} \langle\langle c_{\sigma}; c^{\dagger}_{\sigma} \rangle\rangle}{E - E_0 - U_0 - \Sigma^{\sigma}_{02}}. \tag{2.114}$$

Here, self-energy parts $\Sigma_{1\sigma}, \Sigma_{2\sigma}, \Sigma^{\sigma}_{02}$ are given by

$$\Sigma_{1\sigma} = \sum_{r\beta} |\tau_{r\beta,-\sigma}|^2 f^{\beta}_{r,-\sigma}$$

$$\times \left\{ \frac{1}{E - 2E_0 - U_0 + \epsilon_{r\beta,-\sigma} + i\eta} + \frac{1}{E - \epsilon_{r\beta,-\sigma} + i\eta} \right\}, \tag{2.115}$$

$$\Sigma_{2\sigma} = \sum_{r\beta} |\tau_{r\beta,-\sigma}|^2$$

$$\times \left\{ \frac{1}{E - 2E_0 - U_0 + \epsilon_{r\beta,-\sigma} + i\eta} + \frac{1}{E - \epsilon_{r\beta,-\sigma} + i\eta} \right\}, \tag{2.116}$$

$$\Sigma_{02}^{\sigma} = \Sigma_{0\sigma} + \Sigma_{2\sigma}. \tag{2.117}$$

To arrive at the resulting expression for the electron Green's function $G_{\sigma}^{r} \equiv \langle\langle c_{\sigma}; c_{\sigma}^{\dagger}\rangle\rangle$, we substitute Eq. (2.114) into Eq. (2.107). We have

$$G_{\sigma}^{r}(E) = \frac{E - E_0 - \Sigma_{02}^{\sigma} - U_0(1 - \langle n_{-\sigma}\rangle)}{(E - E_0 - \Sigma_{0\sigma})(E - E_0 - U_0 - \Sigma_{02}^{\sigma}) + U_0\Sigma_{1\sigma}}. \tag{2.118}$$

This expression was first derived by Meir et al. [86], and it was repeatedly used in theoretical studies of the Coulomb blockade and Kondo peak ever since.

The above-mentioned non-crossing approximation does not provide the decoupling of the Keldysh and Dyson equations for the electron and phonon subsystems. The equations remain interconnected, which creates significant computational difficulties, calling for development of suitable approximate schemes for evaluating the relevant Green's functions. The commonly used computational techniques are described and analyzed in detail in the review paper [26], and we do not discuss them in the present work. Nevertheless, we consider it necessary to introduce a certain expression for the self-energy terms $\Sigma_{\mathrm{ph}}(E)$ which appear in the expressions for the electrons Green's functions due to electron–phonon interactions. Accordingly, we derive approximate expressions for these terms following [4, 5]. We start from the Hamiltonian describing the electron–phonon system which is written in the form

$$H = \sum_{k} E_k c_k^{\dagger} c_k + \sum_{l} \hbar\Omega_l a_l^{\dagger} a_l + \sum_{k,l} \Lambda_{kl} A_l c_k^{\dagger} c_k. \tag{2.119}$$

Here, as in the previous expressions (1.40) and (1.41), Ω_l and Λ_{kl} are the frequencies of the phonon modes and electron–phonon coupling constants, respectively, and E_k correspond to the eigenenergies of the electron system. We choose basis functions in a multiparticle electron–phonon Hilbert space as direct products of state vectors for the electron and phonon subsystems. Employing this basis we may write the Hamiltonian matrix in the following way:

$$\begin{bmatrix} \cdots & \epsilon_{N-1}^{k} & \Lambda_{kl}\sqrt{N_l} & \cdots & \cdots \\ \cdots & \Lambda_{kl}\sqrt{N_l} & \epsilon_{N}^{k} & \Lambda_{kl}\sqrt{N_l+1} & \cdots \\ \cdots & \cdots & \Lambda_{kl}\sqrt{N_l+1} & \epsilon_{N+1}^{k} & \cdots \end{bmatrix} \tag{2.120}$$

where $\epsilon_N^k = E_k + N_l\hbar\Omega_l$ and N_l is the number of phonons with such energy which is determined by the Bose–Einstein distribution function for the phonons with the energy $\hbar\Omega_l : \; N_l = \{\exp[\hbar\Omega_l/k_{\mathrm{B}}T] - 1\}^{-1}$. Comparing the explicitly presented part of the Hamiltonian matrix (2.119) with the Hamiltonian matrix (2.16), one may conclude that the two resemble each other. On these grounds one may consider the subspace containing N_l phonons as an open system which is put in contact with the $(N_l - 1)$ and $(N_l + 1)$ phonon subspaces just like the channel is connected with the source and drain contacts [4]. As was discussed before, the effect of the source and/or drain contact is represented by the self-energy matrix of the form $\tau^\dagger g \tau$ (or its Hermitian conjugate) whose imaginary part represents the broadening of the energy levels in the channel (electron relaxation rates). As follows from comparison of matrices (2.16) and (2.120), the analogs of the matrices τ_β and $\tau_\beta^\dagger$ are expressed in terms of the sandwiching terms $\Lambda_{kl}\sqrt{N_l}$ and $\Lambda_{kl}\sqrt{N_l+1}$. These terms could be identified with τ_β within a simplest model for a junction with a single-site bridge where both leads are represented by single sites as well. In general, one may treat Λ as a nondiagonal matrix with matrix elements Λ_{km}. Combining Λ terms in the expressions for τ matrices, one may generate the deformation potential L, which is a fourth-rank tensor [5]:

$$L_{ijkl} = \Lambda_{ij}\Lambda_{kl}^\dagger. \tag{2.121}$$

Assuming that the electron interacts with a sole vibrational mode with the frequency Ω, one could reduce the deformation potential to a second-rank tensor: $\Lambda_{ij} = \Lambda_i\Lambda_j^\dagger$. Holding on this assumption and following [5], one may write

$$G^<(E) = G^r(E)\Sigma_{\rm ph}^<(E)G^a(E). \tag{2.122}$$

Here, the self-energy term $\Sigma_{\rm ph}^<(E)G^a(E)$ satisfies the equation

$$\begin{aligned}\Sigma_{\rm ph}^<(E) = L(\Omega) \otimes \big[&N(\Omega)G^<(E-\hbar\Omega)\\ &+(N(\Omega)+1)G^>(E+\hbar\Omega)\big].\end{aligned} \tag{2.123}$$

In Eq. (2.123), $N(\Omega)$ is the Bose–Einstein distribution function for the phonons with the energy $\hbar\Omega$, and the sign $\otimes$ indicates element by element multiplication as opposed to the ordinary matrix multiplication $[(A \otimes B)_{ij} = a_{ij}b_{ij}]$. As usual, Eqs. (2.122) and (2.123) may be complemented by the similar equations including the greater self-energy term $\Sigma_{\rm ph}^>(E)$. In these equations the self-energy terms arising due to phonons $\Sigma_{\rm ph}^<(E)$, $\Sigma_{\rm ph}^>(E)$ are interconnected with the electrons Green's functions. Therefore, Eqs. (2.122) and (2.123) are to be solved together employing a self-consistent computational procedure. Also, these equations may be generalized to cover the case when the phonons with diverse energies participate in the interactions with electrons. This could be carried out by multiplying the right side of Eq. (2.122) by the phonon density of states and integrating the obtained expression over the phonon frequencies.

The broadening of the bridge energy levels originating from the electron–phonon interaction is given by

$$\Gamma_{\text{ph}}(E) = i\left[\Sigma_{\text{ph}}^r(E) - \Sigma_{\text{ph}}^a(E)\right] \equiv \Sigma_{\text{ph}}^<(E) + \Sigma_{\text{ph}}^>(E). \tag{2.124}$$

Starting form this expression, one may find the imaginary part of $\Sigma_{\text{ph}}^r(E)$ and then compute its real part using the Kramers–Kronig relations.

Electron-phonon interactions contribute to the current flowing through the junction. Here, we present the expression for the relevant contribution ΔI originating from the electron–phonon interactions. This expression was derived employing the model of a single-site bridge for the molecular linker in the junction [24, 25]. It has the form

$$\begin{aligned}\Delta I =& \frac{2e}{\hbar}\int dE \frac{\Gamma_L\Gamma_R}{\Gamma^2} D(E)\Gamma_{\text{ph}}(E) \\ &\times \frac{(E-\tilde{E}_0)^2 - (\Gamma/2)^2}{(E-\tilde{E}_0)^2 + (\Gamma/2)^2}\left[f_L(E) - f_R(E)\right].\end{aligned} \tag{2.125}$$

Here, $\tilde{E}_0$ is the energy of the bridge level renormalized due to the electron–phonon interactions, $\Gamma = \Gamma_L + \Gamma_R + \Gamma_{\text{ph}}(E)$; $D(E)$ is the electron density of states on the molecule, which (in the case of a single-level bridge) is given by Eq. (1.1), and the imaginary part of the phonon self-energy $\Sigma_{\text{ph}}(E)$ is given by

$$\begin{aligned}\Gamma_{\text{ph}}(E) = 2\pi\Lambda^2 \int_0^\infty & d\omega \rho_{\text{ph}}(\omega) \\ &\times \Big\{\left[1 + N(\omega)\right] D(E - \hbar\omega)\left[1 - n(E - \hbar\omega)\right] \\ &+ N(\omega) D(E + \hbar\omega)\left[1 - n(E + \hbar\omega)\right] \\ &+ N(\omega) D(E - \hbar\omega) n(E - \hbar\omega) \\ &+ \left[1 + N(\omega)\right] D(E + \hbar\omega) n(E + \hbar\omega)\Big\}\end{aligned} \tag{2.126}$$

where the constant Λ describes the electron–phonon coupling strength, $\rho_{\text{ph}}(\omega)$ is the phonon density of states,

$$n(E) = \frac{1}{2\pi}\text{Im}G^<(E) \tag{2.127}$$

is the electron occupation number, and $N(\omega) = \left\{\exp[\hbar\omega/kT] - 1\right\}^{-1}$. The expressions (2.125) and (2.126) will be used in further analysis of the inelastic transport through molecular junctions.

2.6 Rate Equations

Presently, there exist numerous works where the electron transport through a MMM junction is analyzed using the system of kinetic equations for occupation probabilities of the molecular bridge. This approach was first suggested by Beenakker [74] and Averin and coworkers [75] to analyze the effects of electron–electron interactions on the electron transport through molecular junctions. The proposed approach is justified when the junction is weakly coupled, that is, the quantities Γ_β are significantly smaller than the electron charging energy U_0 and the spacings between the molecular orbitals involved in the transport. In general, the occupation probabilities for various states of the system determined by the relevant set of quantum numbers ν are given by diagonal elements of the density matrix of this system ρ. Time evolution of the density matrix is described by the Liouville equation:

$$i\hbar\frac{\partial\rho}{\partial t} = [H, \rho]. \tag{2.128}$$

Considering a weakly coupled junction, this equation may be transformed to a set of equations for the diagonal elements $\rho_{\nu\nu} \equiv P_\nu$:

$$\frac{dP_\nu}{dt} = -\sum_{\nu'}(W_{\nu\nu'}P_\nu - W_{\nu'\nu}P_{\nu'}). \tag{2.129}$$

Here, $W_{\nu\nu'}$ is the transition rate between the $|\nu\rangle$ and $|\nu'\rangle$ states of the system. The first term on the right-hand side of Eqs. (2.129), which are also known as "master equations," describes the transitions from the state $|\nu\rangle$ to other states of the system, whereas the second term corresponds to the transitions to the state $|\nu\rangle$ from the remaining states. The set of rate equations (2.129) is to be taken along with the normalization equation for the occupation probabilities

$$\sum_\nu P_\nu = 1. \tag{2.130}$$

In analyzing the electron transport through the molecular junctions, it is useful to consider the states $|\nu\rangle$ as many-body states $|N, i\rangle$ with N electrons on the molecule and the total energy $E_i(N)$. It is the commonly used assumption that the states involved in the transport process differ by a single electron. Then one may keep in Eq. (2.129) only those transition rates which correspond to the relevant transitions between the leads and the molecule. As a result, Eq. (2.129) accepts the form

$$\frac{dP_{N,i}}{dt} = -\sum_j \left(W_{N,i;N\pm1,j}P_{N,i} - W_{N\pm1,j;N,i}P_{N\pm1,j}\right) \tag{2.131}$$

where the rates $W_{N,i;N\pm1,j}$ and $W_{N\pm1,j;N,i}$ include contributions associated with the transport between the left/right lead and the molecule. For instance,

$$W_{N,i;N\pm1,j} = \sum_{\beta} W^{\beta}_{N,i;N\pm1,j} \qquad (\beta \in L, R). \tag{2.132}$$

Tunnel currents through both (left and right) barriers separating the leads from the molecular bridge must remain equal to provide the charge conservation. Therefore, the expression for the current through the junction may be written including only the transition rates associated with the chosen lead. Choosing the left lead, one obtains [83]

$$I_L = I = \pm e \sum_{ij} \left(W^{L}_{N,i;N\pm1,j} P_{N,i} - W^{L}_{N\pm1,j;N,i} P_{N\pm1,j} \right) \tag{2.133}$$

where the transitions corresponding to removing of an electron by the lead contribute with the negative sign. The transition rates included in Eq. (2.133) are usually presented as products of "unconditional" rates $\Gamma_\beta(N)$ determined by the coupling of the molecule to the leads and the occupancy factors for the electron states on the leads. We may write [81]

$$\begin{aligned} W^{\beta}_{N-1,i;N,j} &= \frac{1}{\hbar}\Gamma^{ij}_{\beta}(N)\left[1 - f\left(\epsilon_{ij}(N) - \mu_\beta\right)\right], \\ W^{\beta}_{N,i;N-1,j} &= \frac{1}{\hbar}\Gamma^{ij}_{\beta}(N) f\left(\epsilon_{ij}(N) - \mu_\beta\right). \end{aligned} \tag{2.134}$$

Here, $\epsilon_{ij}(N) = E_i(N) - E_j(N-1)$, and the leads occupancy factors are expressed in terms of the Fermi distribution functions $f(\epsilon - \mu_\beta)$.

The sets of rate equations, similar to those presented above, were derived and used in numerous works where various aspects of the electron transport through molecular junctions were analyzed, setting aside the effects of electron–phonon interactions (see, e.g., [81, 83, 85, 147, 232]). For instance, such equations were successfully used to describe the effects of Coulomb and spin blockade observed in molecular junctions and some other similar systems. However, as already mentioned, these equations could be applied to analyze electron transport only in weakly coupled junctions. When the coupling of the molecule to the leads is strong, so that the coupling energies Γ_β have the same order of magnitude as some other relevant energies (e.g., the charging energy U_0), the rate equations fail to provide an adequate description of the junction transport characteristics. For instance, occurrence of the Kondo peak in the zero bias voltage conductance cannot be described following this way.

The rate ("master") equations were and still are employed to analyze the effects of the electron–vibron interactions on the electron transport through molecular junctions in the situations when these interactions are sufficiently strong [22, 103, 233–245]. Actually, the approach based on the kinetic equations is shown to be justified when the typical time spent by the traveling electron on the molecule is significantly longer than characteristic times of the decoherence processes arising due to electron–electron interactions and/or stochastic nuclear fluctuations in the environment. Usually, these conditions are satisfied for weakly coupled junctions.

One of the early attempts to analyze the effect of the electron–vibron interactions on the electron transport basing on the rate equations was undertaken by Gorelik and coworkers [234]. They discussed a specific bridge-shuttle mechanism for the electron conduction through the molecular bridge considering the latter as a classical harmonic oscillator. It was assumed that the center of mass of the vibrating bridge is moving forth and back in between the leads. As a result, the widths of tunneling barriers separating the leads from the bridge vary in accordance with the variations of the bridge center of mass position x. The effect of the electronic transport on the bridge vibrations was expressed through the bridge charging, so that the bridge motion was described by the equation

$$m\frac{d^2x}{dt^2} = -kx - \gamma\frac{dx}{dt} + \alpha q \tag{2.135}$$

where m and k are the bridge mass and force constant, γ is the damping coefficient, and the parameter α characterizes the bridge charging. The charge q is determined by the probability $P_n(t)$ to have the excess of n electrons on the bridge at a certain instant t. Within this model, the rate equations accept the form

$$\begin{aligned}\frac{dP_n}{dt} =& e^{-x/\lambda}W(n-1,n)P_{n-1} + e^{x/\lambda}W(n+1,n)P_{n+1} \\ &- \left[e^{-x/\lambda}W(n,n+1) + e^{x/\lambda}W(n,n-1)\right]P_n. \end{aligned} \tag{2.136}$$

These equations were used to find the probabilities $P_n(t)$ which depend on the bridge position in the junction. Similar models were employed in several other works (see, e.g., [235–237]). However, in the majority of presently existing research articles, the molecular vibrations are treated within the quantum mechanical approach.

When electron–phonon interactions are taken into consideration, the transition rates in the kinetic equations are associated with the transitions between different states of the many-body electron–phonon system. These states are characterized by certain numbers of electrons and phonons. Assuming that the molecule is represented by a single spin-degenerated level coupled to a vibrational mode with the frequency Ω, one may write down the equations for probabilities P_m^N of the molecular bridge being in a state with N electrons and m phonons as follows [22]:

$$\frac{dP_m^N}{dt} = -\sum_{\beta,m'}\left(W^\beta_{N,m;N\pm1,m'}P_m^N + W^\beta_{N\pm1,m';N,m}P_{m'}^{N\pm1}\right) \tag{2.137}$$

where β denotes the left/right lead in the junction, and the expressions for the rates have the form

$$
\begin{aligned}
W^{\beta}_{N,m;N-1,m'} &= \frac{\Gamma_\beta}{\hbar}\gamma_{mm'} f\left[\hbar\Omega(m'-m)+U_0(N-1)-\mu_\beta\right], \\
W^{\beta}_{N,m;N+1,m'} &= \frac{\Gamma_\beta}{\hbar}\gamma_{mm'}\left\{1-f\left[\hbar\Omega(m'-m)+U_0N-\mu_\beta\right]\right\}, \\
W^{\beta}_{N-1,m';N,m} &= \frac{\Gamma_\beta}{\hbar}\gamma_{m'm}\left\{1-f\left[\hbar\Omega(m-m')+U_0(N-1)-\mu_\beta\right]\right\}, \\
W^{\beta}_{N+1,m';N,m} &= \frac{\Gamma_\beta}{\hbar}\gamma_{m'm} f\left[\hbar\Omega(m-m')+U_0N-\mu_\beta\right].
\end{aligned}
\tag{2.138}
$$

Here, $\frac{1}{\hbar}\Gamma_\beta\gamma_{mm'}$ represents the transition rate involving hopping an electron from the bridge to the lead β and changing the phonon occupancy from m (measured with respect to the ground state of the bridge with the electron occupancy N) to m' (measured relative to the bridge ground state with the electron occupancy $N-1$). In Eq. (2.138) the terms proportional to the charging energy U_0 appear due to electron–electron interactions on the bridge. The matrix elements $\gamma_{mm'}$ introduced in Eq. (2.138) are given by the expressions [22, 103]

$$
\gamma_{mm'} = \begin{cases}
\exp\left(-\dfrac{\tilde{\Lambda}^2}{2}\right)(-\tilde{\Lambda})^{m'-m}\sqrt{\dfrac{m'!}{m!}}\,L_m^{m'-m}(\tilde{\Lambda}^2) & \text{for} \quad m'>m \\[2ex]
\exp\left(-\dfrac{\tilde{\Lambda}^2}{2}\right)(-\tilde{\Lambda})^{m-m'}\sqrt{\dfrac{m!}{m'!}}\,L_{m'}^{m-m'}(\tilde{\Lambda}^2) & \text{for} \quad m'<m
\end{cases}
\tag{2.139}
$$

where $L_q^p(x)$ are the associated Laguerre polynomials and the dimensionless constant $\tilde{\Lambda} = \Lambda/\hbar\Omega$ determines the electron–phonon coupling strength. The model adopted in [22] may be generalized to describe a multilevel molecular bridge. Such generalization was carried out by Siddique and coworkers [103].

The rate equations (2.137) include the rates for transitions where the number of electrons on the bridge changes. The electrons enter or leave the molecular bridge by emitting and/or absorbing phonons associated with the vibrational mode. Also, the bridge vibrations produce electrostatic potential fluctuations in the environment, which result in the environment-mediated transitions between the vibrational states. These transitions are not accompanied with changes in the electron occupation. The relation between the transition rates corresponding to the phonon emission and absorption processes mediated by the thermal environment is given by the Boltzmann's ratio [103, 238]:

$$
\begin{aligned}
W^{\beta}_{N,m;N,m+1} &= \exp\left[-\frac{\hbar\Omega}{k_\mathrm{B}T}\right] W^{\beta}_{N,m+1;N,m} \\
&= K(m)\frac{\Omega}{\exp\left[\hbar\Omega/k_\mathrm{B}T\right]-1}.
\end{aligned}
\tag{2.140}
$$

Here, $K(m)$ is the dimensionless coefficient. As follows from this result, extra terms associated with nonzero transition rates described by Eq. (2.140) are to be included into the kinetic equations (2.137).

A different form of "master" equations invoking the transitions between the states of a many-body electron–phonon system was suggested and used in the works of Koch, von Oppen, and coworkers [239–242]. They presented the kinetic equations for the joint probabilities P_m^N as follows:

$$\frac{dP_m^N}{dt} = - \sum_{N' \neq N, m', \beta} \left(W^{\beta}_{N,m;N',m'} P_m^N - W^{\beta}_{N',m';N,m} P_{m'}^{N'} \right) - \frac{1}{\tau} \left(P_m^N - P_m^{\text{eq}} \sum_{m'} P_{m'}^N \right). \quad (2.141)$$

Here, the last term on the right describes the direct vibrational relaxation towards the equilibrium phonon distribution P_m^{eq}. It is given by the Bose–Einstein distribution function $N(\Omega)$ where again Ω is the frequency of the vibrational mode coupled to electrons. The relaxation process is characterized by a single relaxation time τ.

To understand the effect of the direct vibrational relaxation on the phonon distribution, one must keep in mind that typical speeds of diffusion and drift processes in the phonon space strongly depend on the electron–phonon coupling strength. These processes slow down when the parameter $\tilde{\Lambda}$ characterizing the electron–phonon coupling strength decreases. It was shown [239] that as $\tilde{\Lambda}$ decreases, it reaches a crossover value $\tilde{\Lambda}_0$. When $\tilde{\Lambda} \gg \tilde{\Lambda}_0$, the vibrational diffusion is limited by the phonon drift which originates from the asymmetry between vibrational excitations and de-excitations. In the opposite limit $\tilde{\Lambda} \ll \tilde{\Lambda}_0$, the predominating limiting process is the direct vibrational relaxation leading to a decrease in the width of the phonon distribution. The crossover coupling parameter $\tilde{\Lambda}_0$ depends on the relaxation time τ but only moderately. The $\tilde{\Lambda}_0$ versus τ dependence is close to the power law with an exponent accepting values 1/4–1/3 [239]. When the electron–phonon coupling is strong ($\tilde{\Lambda} > 1$), this leads to exponentially small transition rates provided that the temperature is low ($k_B T \ll \hbar\Omega$) and the bias voltage is small ($|e|V < \tilde{\Lambda}^2 \hbar\Omega$). This results in the suppression of the current through the junction at low bias voltage arising due to electron–phonon interactions and termed Franck–Condon blockade [240]. We further discuss this issue in Chap. 4. Also, in the limit of strong electron–electron interactions ($U_0 \to \infty$) and at low temperatures, the expressions for the transition rates may be factorized as [239, 241]

$$W^{\beta}_{N,m;N',m'} = (1 + \delta_{N',1})\, \delta_{|N-N'|,1} w_{mm'} \quad (2.142)$$

where $\delta_{nn'}$ is the Kronecker symbol and $w_{mm'}$ is the phonon factor. Using this result one may factorize the expressions for the occupational probabilities $P_m^N = P^N P_m$ and derive the rate equations for the phonons, as was shown in [239].

Finally, we note that the rate (master) equations were proven to be very useful, especially while analyzing transport through the molecular junctions with strong electron–electron and electron–phonon interactions and weak coupling of the linker to the leads. The general NEGF formalism should yield the rate equations as an appropriate limiting case thus establishing the agreement between these two approaches. Promising advances in this direction are offered (see, e.g., [229]), but the issue still needs further consideration.

Chapter 3
Ballistic Transport

3.1 Charge Transfer and Electrostatic Potential Distribution in Unbiased Junctions

As it was briefly mentioned above, the contact of a molecule with the leads significantly changes some important properties of the molecule. The perturbation induced by the metal–molecule coupling initiates the response from the molecule. The response is determined by two most important electronic processes occurring in the MMM junction, namely, the charge transfer between the leads and the molecular linker and the changes in the spatial distribution of the electrostatic potential. These processes strongly contribute to the forming of electrical characteristics of molecular junctions.

Significant effort was applied to study the charge transfer and voltage drop inside MMM junctions and their effect on the transport characteristics. These studies were based on first-principles electronic structure calculations combined with the NEGF [53–63]. In the present section, we give a brief overview of the obtained results concentrating on unbiased junctions. Following [56, 58] we put aside all effects originating from ionic dynamics in the considered molecular junctions. Also, we assume that electron–electron interactions are accounted for within the mean field approximation, which prevents manifestations of such effects as Coulomb blockade. Under these conditions, the sole scattering mechanism permitted to act in the junction is that originating from the contacts between the molecule and the leads, and the only temperature effect originates from the Fermi–Dirac distributions of electrons in the leads. These distributions are characterized by the electronic temperature in the electrodes T. In principle, a thermoionic component appears in the current through the junction at nonzero temperature T, but it is usually much smaller than the predominating tunneling current and may be neglected. Therefore, within the adopted model the electron transport through the junction may be treated as coherent and elastic. Correspondingly, the electron current is described

N.A. Zimbovskaya, *Transport Properties of Molecular Junctions*, Springer Tracts in Modern Physics 254, DOI 10.1007/978-1-4614-8011-2_3,

by Eqs. (2.54) and (2.66) where the matrix form for the retarded Green's function for the extended molecule could be written as follows [56]:

$$G^r(E) = \left[ES - F - V_{\text{ext}} - \Sigma_L^r - \Sigma_R^r\right]^{-1}. \tag{3.1}$$

In this expression, the Fock matrix F is contributed by both ionic and electron charge distributions in the extended molecule, and the term $V_{\text{ext}}(\mathbf{r})$ represents the long-range Coulomb potential which arises due to the charge distribution over the contact regions. When the bias is applied across the junction, the corresponding voltage drop must be included in the term $V_{\text{ext}}(\mathbf{r})$, as well.

The appearance of the term $V_{\text{ext}}(\mathbf{r})$ in the expression for the Green's function requires a self-consistent procedure to be used in calculations of the Fock matrix. To compute this matrix, one must simultaneously and self-consistently solve the Schrodinger equation for the extended molecule and the Poisson equation for the electrostatic potential. The charge density en included in the Poisson equation may be expressed in terms of the relevant Green's function, namely, $en = e\text{Trace}\{\rho S\}$, where the density matrix ρ is given by its expression (2.29) and S represents the overlap matrix. For an unbiased junction, boundary conditions for the Poisson equation could be specified by the equality of potentials at the metallic contacts. In general case, when a bias voltage is applied across the junction, the boundary conditions are determined by its value and polarity. The self-consistent calculation proceeds by computing the input density matrix for the next iteration from the lesser Green's function obtained in the previous iteration. Once the self-consistent calculation converges, the resulting Green's functions could be applied to explore transport properties of the considered MMM junction. To successfully analyze electric charge and/or potential distribution over the real space occupied by the molecule, one needs to turn to the physical picture of electron transport through a molecule on atomic scale. This means that the molecule is to be viewed as comprised of chemically defined atomic groups or even single atoms. Correspondingly, electrical characteristics of the molecule may be interpreted as the collective response of these atomic groups/atoms to the perturbation, which originates from the molecule coupling to the leads and from the applied bias voltage. This physical picture has a significant advantage, for it gives means to treat electron transport through the molecule as a process developing in the real space. As discussed above, within this approach, the electron transport is a sequence of tunnelings between potential wells representing atomic or atomic group orbitals belonging to atoms/atomic groups situated at different parts of the molecule. In general, a traveling electron may follow several routes (pathways) as it moves along the molecule. The number of possible pathways rapidly increases as the molecule acquires complicated chemical structure, which makes transport computations very time-consuming and cumbersome. These difficulties first emerged in building up the theory of long-range electron transfer reactions in macromolecules such as proteins and DNA. A number of semiempirical models have been proposed to estimate the electron transmission coefficient in large molecules basing on the above-described visualization of electron traveling through the molecule, and some of them still remain in use. The most commonly employed is the primary pathways model suggested by Beratan and Onuchic [73, 246].

Within this model, every single pathway is considered as a combination of "through bond" and "through space" steps, and the probability for electron to follow the pathway is assumed to be proportional to the product [246]:

$$\left(\prod_i \epsilon_{\text{bond}}(i)\right)^2 \left(\prod_j \epsilon_{\text{H-bond}}(j)\right)^2 \left(\prod_k \epsilon_{\text{space}}(k)\right)^2 \tag{3.2}$$

where the factors ϵ_{bond}, $\epsilon_{\text{H-bond}}$, and ϵ_{space} are related to covalent and hydrogen bonded steps and to through space tunneling of an electron following the pathway, respectively. Due to the differences in decay rates for "through bond" and "through space" steps, the products in Eq. (3.2) significantly differ for different routes, so one may search for the routes maximizing the net product (3.2). These routes are called primary pathways for the electron propagation through the molecule. After the primary pathways are detected one may simplify the structure of the considered molecule putting aside the portions which do not contain the preferable routes, for these portions scarcely participate in the electron transport. The concept of primary pathways originally developed to analyze long-range electron transfer reactions may be employed to study electron transport through molecular bridges in MMM junctions.

Also, we may use the physical picture of a molecule arranged in the real space to explore how inhomogeneities in the electric charge and potential distributions over the MMM junction affect its transport properties. In accordance with the atomic-scale approach to the electron transport, matrix elements of the relevant Green's functions may be computed using local orbital functions $\Phi_i(\mathbf{r})$ associated with the states of single atoms/atomic groups, as basis functions. Using such a basis one may find space distributions of the important transport characteristics. For instance, the local density of states is given by [59]:

$$A(\mathbf{r}, E) = -\frac{1}{\pi} \sum_{i,j} \text{Im}[G^r(E)] \Phi_i(\mathbf{r}) \Phi_j^*(\mathbf{r}). \tag{3.3}$$

The above-described or similar self-consistent computational procedures were used to study charge transfer and electrostatic potential distribution in MMM junctions, which consisted of gold leads connected by various linkers such as carbon [53] and gold [56] chains and benzene-based molecular radicals [59–61]. The obtained results strongly varied for different linkers. Nevertheless, some important common features were revealed.

When the considered junction remains unbiased ($V = 0$), and the contact between the leads and the molecular bridge is established, the charge transfer between the leads and the molecule immediately starts. Electric charge flows into the system due to the difference between the chemical potential of the metal electrodes and the charge neutrality level of the free molecule [247]. The self-consistent

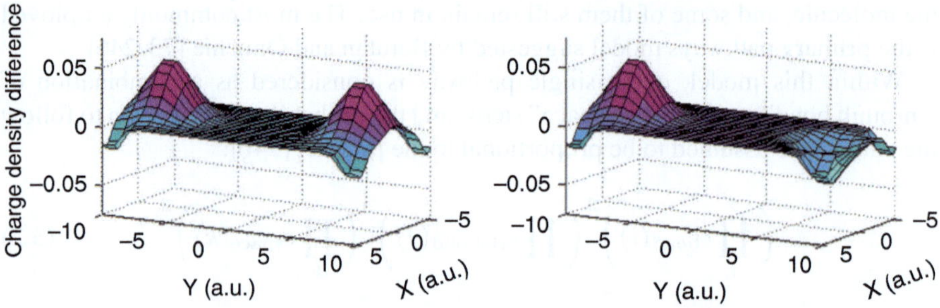

Fig. 3.1 Schematics of spatial distribution of the electric charge transferred to a molecule upon the formation of its contacts with the leads. The molecule is supposed to be located in the "X–Y" plane within the shown region. The dimensions of the region are given in atomic units, and they match the size of simple molecules (such as phenylene dithiol), which are used in transport experiments. Symmetric (*left panel*) and asymmetric (*right panel*) charge density distributions are shown

process of the charge transfer continues until the molecule and the electrodes reach an equilibrium state. As a result, the originally neutral molecule becomes charged. The transferred charge is mostly accumulated at the ends of the molecular bridge, as schematically shown in Fig. 3.1. The amount of the transferred charge and its space distribution depend on the multitude of factors such as the chemical composition of the molecule in general and its end groups directly providing the metal–molecule coupling in particular, the atomic structure at the metallic interface, the molecular absorption geometry, and the closeness of the molecule to the metal leads. The interplay of all these factors determines the resulting distribution of the induced electric charge as well as the conduction through the junction in whole. Thorough microscopic studies of the effect of the aforesaid factors on the transport properties of metal–molecular junctions were carried out. The obtained results are presented in [59–61] and some other works. Here, we schematically illustrate the effect of the molecule position amidst the leads, which is commonly manifested for various kinds of molecular linkers. If the molecule with the identical end groups is equidistantly placed in between the identical leads, one may expect the transferred charge to be equally distributed between the molecule ends, as presented in the left panel of the figure. When the molecule is so shifted that its distance from one lead is significantly smaller than from the other, the transferred charge division becomes nonequal, and it could happen that the sign of this charge is different for different ends of the molecule, as shown in the right panel of Fig. 3.1.

The charge transfer process is accompanied by changes in the profile of the electrostatic potential over the molecular bridge. Due to the charge accumulation/depletion at the ends of the molecule, the potential barriers/wells could be formed at the bridge ends, as schematically shown in Fig. 3.2. The electrostatic potential profiles shown in this figure were obtained by deducting the sum of potentials computed for isolated leads and the molecular linker from the net potential computed for the whole system. The potential barriers at the ends of the molecular

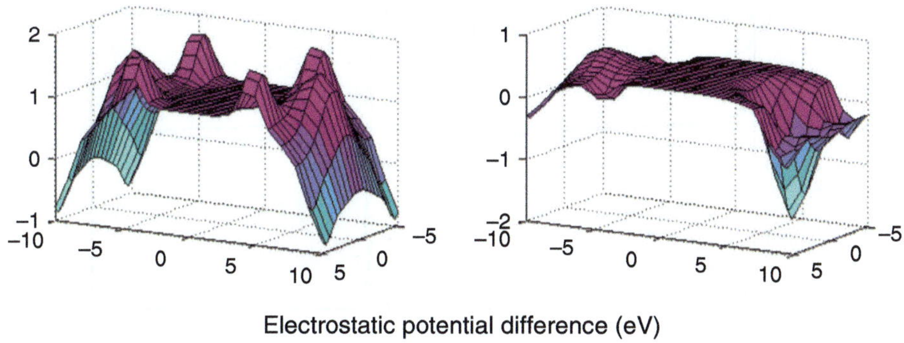

Fig. 3.2 Schematics of the cross-sectional view of the electrostatic potential spatial distribution over a molecule upon the formation of the molecule contacts with the leads. The molecule is assumed to be arranged in the "X–Y" plane. Symmetric (*left panel*) and asymmetric (*right panel*) spatial distributions of the electrostatic potential correspond to the transferred charge density distribution presented in Fig. 3.1

linker may act as Schottky-like barriers in semiconducting devices, and these barriers can significantly influence the electron transmission through the junction.

The charge-transfer-induced changes in the electrostatic potentials cause noticeable shifts in the energy levels of the molecular bridges in the metal–molecule junction with respect to the molecular orbitals corresponding to an isolated molecule of the same kind. To analyze these effects, one may use the local density of states given by Eq. (3.3), the electron transmission $T(E)$ given by Eq. (2.66), and their projections onto the individual molecular orbitals. The shifts of the molecule states which originate from the coupling of the molecule to the leads strongly vary for various MMM junctions, so we choose a particular system for illustration which was already used in the first chapter. The chosen junction consists of copper leads and a hydrocarbon chain including eight carbon atoms. The band lineup for this junction was computed using the NRLMOL code [248] based on the density-functional theory. The electron transmission corresponding to four frontier molecular states closest to the equilibrium Fermi energy of the system is shown in Fig. 3.3 along with the projected density of states. The states of the isolated molecule have energies -0.282, -3.071, -3.167, and 4.644 eV. So, one sees that there are significant shifts in the positions of the molecular orbitals originating from the coupling between the molecule and the leads.

Studying electron transport through MMM junctions, one often assumes that the leads are large enough to have a featureless electron density of states below their Fermi energies. However, if the transport through the molecule is explored in STM experiments, this approach is not justified. It would not be appropriate to mimic the STM tip by a large (practically semi-infinite) crystal, for atomic-scale structure may exist at the tip apex. Similar situation occurs in break-junction experiments when both contacts in the junction may include atomic-scale structures on their surfaces and in studies of transport properties of molecular networks consisting of metal

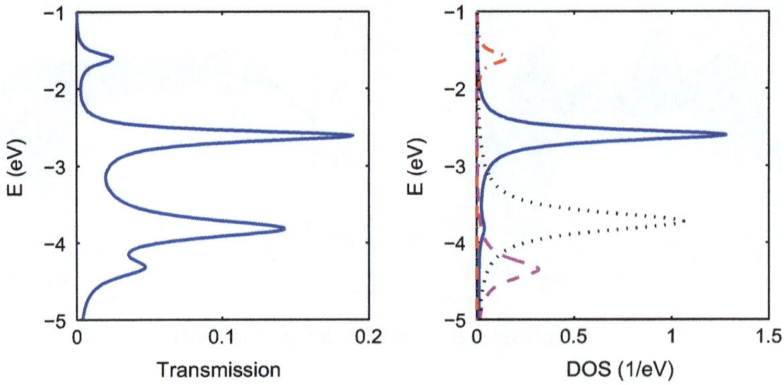

Fig. 3.3 Energy dependencies of the electron transmission (*left panel*) and of the electron density of states projected on four frontier molecule states (*right panel*) plotted for the MMM junction, which consists of copper leads linked by a hydrocarbon chain including eight atoms. The energy levels of the unshifted molecular states equal -0.282, -3.071, -3.167, and $-4.644\,$eV, respectively

nanoparticles linked by molecules. Due to the presence of atomic-scale structures on the surface of the leads, their electron density of states may reveal some sharp and distinct features. In these cases the conductance of a considered MMM junction should depend on the relative positions of these features in the electron density of states on the leads and the molecular bridge states [60, 249]. We further discuss this issue in the fifth section of the present chapter.

3.2 Electric Charge and Potential Distribution Over a Biased MMM Junction: Current Rectification

When the bias voltage is applied across a MMM junction, this brings significant changes in both charge and electrostatic potential distributions over the system. The charge response of the molecule to the bias voltage includes redistribution of the electronic charge on the molecule as well as the charge injection/extraction. The injection of the electric charge occurs at the source-molecule contact, and its extraction from the molecule takes place at the drain-molecule contact. The net charge on the molecule in the presence of the bias is determined by the balance between these two processes. In many cases, especially when symmetrically coupled junctions are considered, the extra charge appearing on the molecule due to the applied electric field is small and may be neglected [60]. However, one may expect a noticeable accumulation of the net injected charge on the molecular linkers of some junctions with strongly asymmetric coupling of the bridge to the leads. Although the net charge injection on the molecule due to the applied voltage is often negligible, significant redistribution of electric charge over the molecule is likely to occur.

Usually, charge transport in the molecular junctions is called "electron transport" if the conduction through molecule is maintained by the unoccupied molecular orbitals. When the conduction is mediated by occupied molecular states, the transport is characterized as "hole transport." For certainty, we consider the case of electron transport to describe the redistribution of electric charge over the molecule subjected to the external electric field. As the bias voltage increases, the electrons move along the molecule from its source-connected end to the drain-connected end. However, this flow of electrons is hindered by a potential barrier at the molecule–drain contact, and this barrier prevents the electron extraction from the molecule. This leads to the accumulation of electrons near the molecule–drain interface and their depletion on the molecule–source contact. Due to their negative charge, the electrons move towards the electrode connected to the positive terminal of the bias-voltage source, and their accumulation nearby creates a resistivity dipole at the corresponding metal–molecule interface. Simultaneously, another dipole appears at the opposite interface due to the electron depletion. Therefore, the charge redistribution over the molecule leads to partial screening of the applied electric field. Similar consideration applies to the case of hole transport. Naturally, the tunneling transport through molecules has a quantum mechanical nature; therefore, one cannot visualize tunneling electrons as semiclassical quasiparticles being physically injected into the molecule or extracted from the latter. Nevertheless, the simple physical picture presented above helps to understand the origin of charge redistribution in the biased molecular junctions.

The concept of a resistivity dipole is well known in mesoscopic electron transport [250]. Resistivity dipoles play an important part in transport processes in inhomogeneous media where their appearance near scattering centers helps to overcome the barriers for transport. Typically, strong spatial variations in local charge density and electrostatic field occur in mesoscopic semiconductor heterostructures, and this leads to nonlinear transport characteristics [251]. MMM junctions resemble mesoscopic heterostructures since the charge and potential distributions over the molecular bridge may be largely inhomogeneous. Also, one may control the charge and potential distributions in these junctions by varying their structural components, which opens opportunities for engineering of molecular nanodevices.

Spatial variations of the current-induced charge could affect the structural stability of the molecular bridge. As the current increases, the electron accumulation at one end of the molecule and its depletion on another end become more pronounced, and this leads to the changes in the bond lengths at the ends of the molecule [60]. At certain value of the applied bias voltage, the lengthened bonds could break, leading to the conformational changes of the molecule as illustrated in Fig. 3.4. The problem of current-induced changes in the geometry of molecular linkers in MMM junctions is similar to the electromigration problem in metals [252]. The possibility of current-induced conformational changes in the molecules was first considered a decade ago [253, 254], and these changes were observed in the experiments where electron current-induced isomerization of azobenzene molecules occurring above the threshold bias voltage was reported [255]. The corresponding changes in the

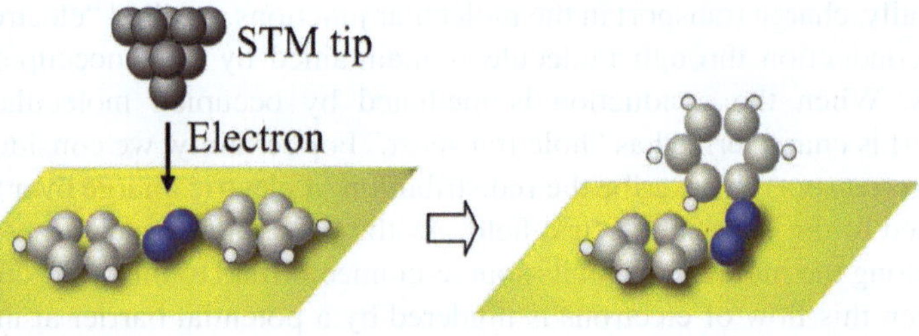

Fig. 3.4 A concept of current-induced conformational change of an azobenzene molecule adsorbed onto a metal substrate with a scanning tunneling microscope [255]. Reprinted with the permission of the APS

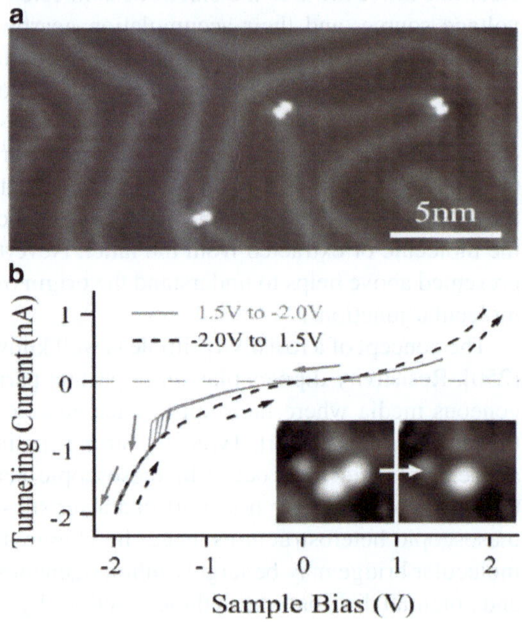

Fig. 3.5 (**a**) A large-scale image of three trans-azobenzene molecules on Au(1 1 1). (**b**) A current–voltage curve measured at the center of trans-azobenzene (*left inset*) and a molecule which underwent a conformational change (*right inset*) by sweeping the voltage twice; from 1.5 to −2.0 V on trans-azobenzene (*solid line*) and from −2.0 to 2 V on the new product (*dashed line*). *Insets*: topographic images (2.2 nm × 2.2 nm) before the I–V measurement (*left*) and after (*right*) the current jump taken at $V = 1$ V and $I = 70$ pA [255]. Reprinted with the permission of the APS

molecule geometry and transport characteristics are illustrated in Fig. 3.5. Research concerning current-induced conformational changes in molecules is still going on.

Due to both charge transfer occurring upon establishment of the contact between the electrodes and the linking molecule and current-induced charge redistribution, the spatial distribution of the electrostatic potential in the biased molecular junctions significantly differs from the profile of the bias voltage itself. Again, we emphasize that the potential distribution is determined by the cooperation of numerous factors originating from the characteristics of considered junction. Nevertheless, a few general features could be indicated. First, the screening of the applied electric field by the charge distributed on the molecule results in the greater voltage drop near the molecule-lead contact region than over the central part of the molecule. Secondly, for symmetrically coupled junctions, one may expect the electric potential profile to

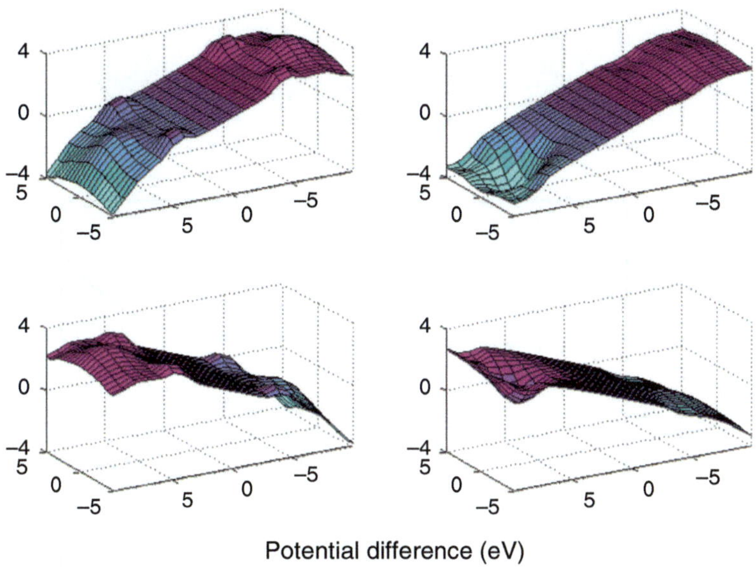

Fig. 3.6 Schematics of the potential drop across a MMM junction. Spatial distributions of electrostatic potential over biased MMM junctions for a nonrectifying (*left panels*) and rectifying (*right panels*) MMM junction

be more or less symmetric, as illustrated in the left panels of Fig. 3.6. However, if the coupling of the junction for some reason is asymmetric, the potential profile is likely to acquire an asymmetric shape. Examples of such profiles are shown in the right panels of Fig. 3.6.

As discussed in the previous section, the energy levels of the molecule included in the MMM junction become shifted as a result of charge transfer between the molecule and the leads which gives rise to the spatial redistribution of the electrostatic potential. The applied bias voltage brings further modifications in the molecule energy levels. These modifications remain insignificant when the bias voltage is low, but they gain importance as the voltage grows. Correspondingly, in transport computations one must calculate the transmission $T(E,V)$ given by Eqs. (2.66) and (3.1) to account for the changes in the positions and heights of the peaks in the transmission induced by the applied bias [59]. The equilibrium transmission $T(E,0)$ may be used in transport calculations only at low values of the bias voltage.

One of the first applications suggested for MMM junctions was to use them as molecular diodes [1], for these systems are capable of providing current rectification. In general, the rectification properties of metal–molecular junctions are caused by the cooperation of the same factors which are responsible for asymmetries in the spatial distribution of the electrostatic potential over the junction. However, it is worthwhile to discuss the mechanisms bringing current rectification in more detail. For a strongly coupled system where a molecule is chemisorbed at both ends, the asymmetry of current–voltage characteristics is mostly determined by the

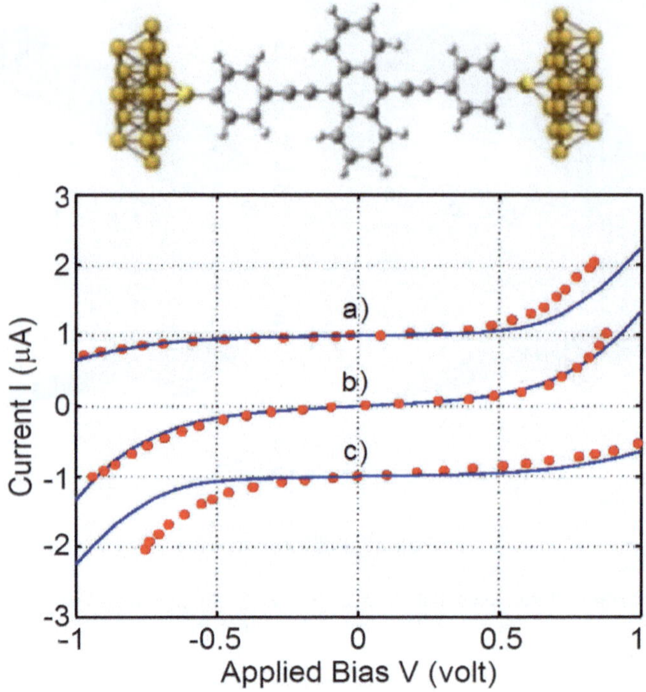

Fig. 3.7 Current-voltage characteristics experimentally observed (dots) and theoretically computed for a metal-molecule-metal junction with gold electrodes and the linker shown above. The upper curve (**a**) and lower curve (**c**) shown current rectification originating from stretching the sulphur-gold bound at the left (**a**) and right c) electrode. The middle curve (**b**) corresponds to the symmetrically coupled system (after [84]). Reprinted with the permission of the APS

respective coupling strengths of the molecule with the source and drain reservoirs. As discussed before, the current starts to flow through a certain molecular orbital when the corresponding energy level appears in the conducting window between μ_L and μ_R. A strongly asymmetric coupling (say $\Gamma_L \gg \Gamma_R$) makes the level available for quick and easy occupation for a certain bias direction and likewise emptied for the opposite bias direction. Depending on either the level was empty or full when the bias was switched on, a net charge appears on the level for the appropriate bias polarity.

The Coulomb cost of setting this charge postpones the onset of conduction, thus bringing asymmetry to the current–voltage curves. Specific features of these curves are directly correlated with the identities of conducting levels. For instance, in the case of conduction through HOMO, an otherwise symmetric molecule should exhibit a retarded rise of the current for positive bias on the stronger coupled contact. The situation is precisely reversed when the conduction occurs via LUMO. In this case, the current rise in magnitude is slowered down for negative bias on the stronger coupled lead [30, 84]. An example of a current rectification in a strongly coupled junction is shown in Fig. 3.7. Another example is discussed in

[5]. In this case, the system consisted of $GaAs$ electrodes linked by an alkanethiol chain of varying length. The observed current–voltage characteristics exhibited a distinguishable current rectification. Also, the kinks indicating the crossover from a low-bias thermoionic emission over the depletion (Schottky) barrier in the $GaAs$ electrodes to the tunneling through the potential barriers at the ends of the molecular bridge occurring at higher bias are clearly visible in the presented current–voltage curves.

To analyze current rectification in strongly coupled (chemisorbed) metal–molecular junctions in the simplest way, one may introduce the division parameter η into the expressions for the electrochemical potentials of the electrodes [82]:

$$\mu_L = E_{\rm F} + \eta|e|V,$$
$$\mu_R = E_{\rm F} - (1-\eta)|e|V \tag{3.4}$$

where $\eta = \Gamma_R/(\Gamma_L + \Gamma_R)$. For symmetrically coupled junctions ($\Gamma_L = \Gamma_R$) the division parameter equals 0.5, and Eq. (3.4) are reduced to the form given by Eq. (1.16). In this case, the current–voltage curves described by Eqs. (2.54), (2.66), and (3.4) are symmetrical with respect to the bias-voltage polarity. When the electrostatic potential profile indicates that the bias voltage is nonequally divided between the molecule–leads contacts, the current–voltage curves reveal asymmetric shapes. This means that the junction to some extent acquires rectifying properties. To provide a workable current rectification, the junction must be coupled with a well-pronounced asymmetry, so that the division parameter η should accept values close to zero/one.

We remark again that the previous description is justified for MMM junctions being in strongly chemisorbed regime when slight bond distortions upon contacts would stretch the current–voltage curves in either direction depending on the nature of the conducting levels and the bias polarities. However, when the junction is physisorbed at one end, the mechanism creating current rectification changes. In such cases, the molecule quickly reaches Coulomb blockade. The current through a Coulomb blockaded molecule is known to exhibit intermediate steps. The steps are associated with the so-called conduction through open shells. For an asymmetrically coupled junction, an electron quickly travels through the molecule when it enters from the weakly coupled contact and exits into the contact with strong coupling to the lead. This process is referred to as shell tunneling [5]. When the voltage polarity is reversed, and the electron enters the molecule from the strongly coupled end, it lingers inside the molecule, and it takes a relatively long time for a weaker coupled contact to extract it from there. Lingering on the bridge, the electron is exposed to strong Coulomb forces which can excite the electrons while in residence. As a result, current through such a junction which is physisorbed at least at one end should exhibit a pronounced asymmetry with bias. This mechanism leads to a significant difference in the saturation current levels for opposite bias-voltage polarities. On the contrary, within the above-discussed chemisorbed limit, the asymmetry is manifested rather in voltage range over which the current rises than in the different magnitudes of the saturation current for opposite bias polarities [84].

One more factor which could give rise to the current rectification is the dependence of coupling parameters $\Gamma_{L,R}$ of the bias voltage. The sensitivity of $\Gamma_{L,R}$ to the voltage applied across the junction often arises due to the specifics of the electrostatic potential distribution. If there exists a significant potential barrier at one side of the junction, this may lead to a distinct asymmetry of the current–voltage curves. The reason is that the bias voltage tilts the potential in opposite directions for the opposite polarities, thus enhancing the barrier height in one direction and reducing it in another one. This scenario resembles that observed in semiconductor p–n junctions with built-in potential. As for the nature of the potential barrier, it could be a vacuum barrier typical for STM experiments with MMM junctions or a Schottky barrier at the interface with a contact.

3.3 Fano Resonances

An important advantage of the NEGF approach is its ability to directly include some quantum effects while analyzing transport characteristics of the corresponding systems. These effects arise from quantum mechanical interference between different transport channels which may create distinct signatures in the output current. A well-known example of such an effect is Fano interference, leading to Fano resonance in the conductance. This type of resonance appears as a result of quantum interference between two coherent transmission paths, one involving a continuum of energy levels and another one including a set of discrete levels. A quantum dot or a molecule may take on the part of the second transmission path. However, the requirement of continuity of the energy spectrum of the first transmission channel is not an obligatory condition for the resonance to appear. In principle, Fano resonance may appear in the system where each of the two transmission paths is represented by a single level. For simplicity, in the following analysis, we adopt just this model. We assume that the bridge in the MMM junction is represented by a single site which is coupled to another site not directly attached to the leads. Such situations may occur in molecular junctions where the linking molecules include certain types of side groups attached to their backbones. In these cases, the sites associated with the side groups are coupled to the backbone sites, but their coupling to the leads may be weak enough to be omitted. The Hamiltonian of the bridge which includes two sites with the energies E_0 and E_s coupled to each other with the coupling strength τ may be written in the form

$$H = \begin{pmatrix} E_0 & \tau \\ \tau^\dagger & E_s \end{pmatrix}. \tag{3.5}$$

Here, the energy E_0 is corresponding to the site directly coupled to the leads, and E_s is the energy associated with the side site. These two levels represent two

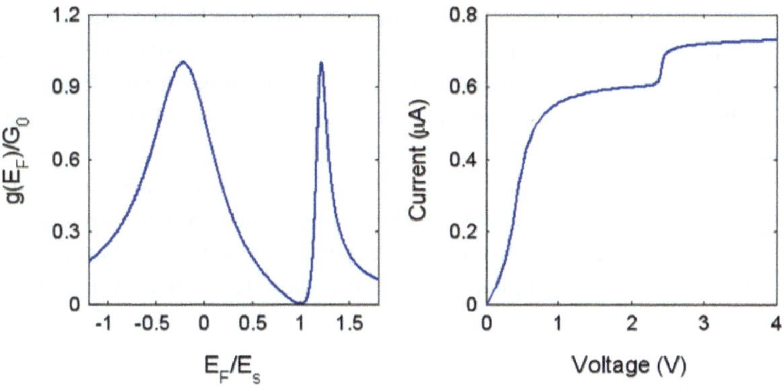

Fig. 3.8 Fano resonance in the molecular conductance (*left*) and the corresponding step-like feature in the current–voltage curve (*right*). The curves are plotted using the expression (3.8) for the electron transmission and assuming $\gamma_L = \gamma_R = 0.1E_S,\ \tau = 0.5E_s,\ E_0 = 0$

different transport channels. Within the adopted model, the matrices Γ_β describing the coupling of the bridge to the leads have the form

$$\Gamma_\beta = \begin{pmatrix} \gamma_\beta & 0 \\ 0 & 0 \end{pmatrix} \tag{3.6}$$

where $\beta \in L, R$, and γ_β are coupling constants. The retarded Green's function corresponding to the Hamiltonian (3.5) is given by the expression:

$$(G^r)^{-1} = \begin{pmatrix} E - E_0 + i\gamma/2 & \tau \\ \tau^\dagger & E - E_s \end{pmatrix}. \tag{3.7}$$

In this expression, $\gamma = \gamma_L + \gamma_R$. Substituting Eqs. (3.5)–(3.7) into the expression (2.66) for the electron transmission, one gets

$$T(E) = \frac{\gamma_L \gamma_R (E - E_s)^2}{\left[(E - E_0)(E - E_s) - \tau^2\right]^2 + \gamma^2 (E - E_s)^2/4}. \tag{3.8}$$

For the considered case of elastic and coherent transport, the differential conductance $g = dI/dV$ at small bias voltage is given by the expression:

$$g = -\frac{e^2}{\pi\hbar} \int T(E) \frac{\partial f}{\partial E} dE. \tag{3.9}$$

At low temperatures, the derivative of the Fermi function behaves as the Dirac delta function, and the conductance becomes proportional to the transmission at the Fermi energy. Therefore, the resonance features in the transmission should be reproduced in the dependence of conductance of the position of the Fermi level, as well. This is illustrated in Fig. 3.8. One sees that there is the conduction maximum at $E_0 = E_{\mathrm{F}}$, which corresponds to the situation when the bridge channel is opening up for the transport. However, along with this maximum, one observes an extra maximum in the conductance at $E_{\mathrm{F}} = E_s$ which corresponds to the Fano resonance. Accordingly, the current–voltage curve exhibits an additional step associated with the Fano peak in the conductance, as shown in the right panel of Fig. 3.8.

The shape of Fano resonances varies depending on the relations between the relevant energies. Actually, analyzing local extrema of the electron transmission given by Eq. (3.8), one concludes that the extra peak in the transmission/conductance at $E = E_s$ or $E_{\mathrm{F}} = E_s$ appears when the parameter τ describing the coupling between the bridge site and the side site takes on values considerably smaller than the separation between the corresponding energy levels ($|E_0 - E_s| \gg \tau$). At stronger coupling between these sites the peak may be replaced by a resonance–antiresonance pair. To further explore the matter one may assume that both transport channels are coupled to the leads. Then one may introduce the retarded Green's functions for the isolated bridge (g_b) and for the isolated side site (g_s), namely,

$$g_b(E) = \frac{1}{E - E_0 + i\gamma/2},$$
$$g_s(E) = \frac{1}{E - E_s + i\gamma_s/2} \tag{3.10}$$

where $\gamma_s = \gamma_{sL} + \gamma_{sR}$ describes the coupling of the side site to the contacts. The retarded Green's function for the bridge coupled to the side site satisfies the Dyson equation. In the considered case it has the form

$$G^{(r)}(E)^{-1} = g_b(E)^{-1} - \tau^2 g_s(E). \tag{3.11}$$

Now, one may proceed by comparing the spectral functions for the bridge in the absence and in the presence of the side site. As shown in [5], the coupling of the bridge site to the side site brings the following correction to the spectral function:

$$\Delta A = a_0 \left[1 - \frac{(q_0 + \xi)^2}{\xi^2 + 1}\right]. \tag{3.12}$$

Various spectral functions, level broadenings, and other parameters introduced in this expression are defined as follows:

$$a_0(E) = a_b(E)\frac{\gamma_0(E)}{\gamma^*(E)}, \tag{3.13}$$

$$a_b(E) = -\frac{1}{\pi}\mathrm{Im}(g_b(E)), \tag{3.14}$$

$$\gamma_0(E) = \pi\tau^2 a_b(E), \tag{3.15}$$

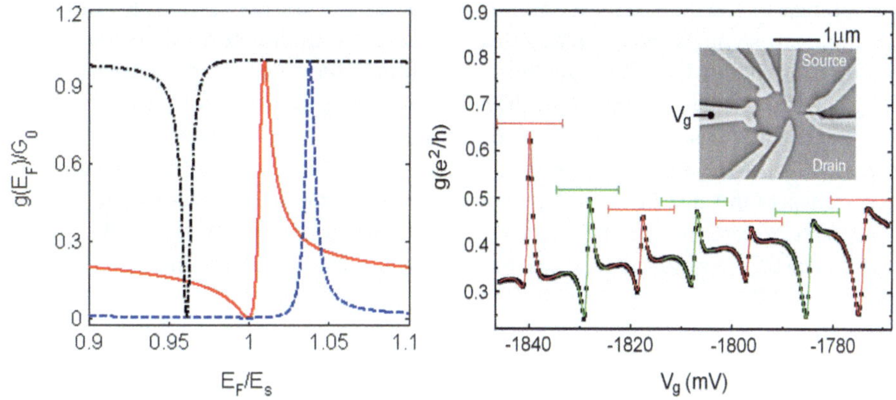

Fig. 3.9 The variety of shapes of the Fano resonance features in the molecular conductance (*left*) and Fano resonance observed in the experiments on a one lead quantum dot [256]

$$\gamma^*(E) = \gamma_L(E) + \gamma_R(E) + \gamma_0(E),$$

$$q_0(E) = \frac{E - E_s}{\gamma_s(E)},$$

$$\xi = \frac{E - E_0 - \gamma_0(E) q_0(E)}{\gamma^*(E)}.$$

The expression (3.12) is describing the resonant–nonresonant superposition of two transport pathways for the electrons tunneling through the system. Depending on the value of the parameter ξ, the spectral function reveals a peak, a dip, or a resonance–antiresonance pair near $E = E_s$, and similar features should appear in the energy dependencies of the electron transmission and conductance, as shown in the left panel of Fig. 3.9. We remark that the Fano resonance could appear as a minimum in the conductance only provided that both transport channels are coupled to the electrodes. The expression (3.8), which was derived assuming that there is no direct contact between the side site and the electrodes, does not describe the Fano minimum in the electron transmission/conductance. Fano resonances were observed in transport experiments involving molecules and quantum dots. Some experimental results reported in [256] are reproduced in the right panel of Fig. 3.9.

3.4 Simmons Model for Electron Transport Through Molecular Insulators

Here, we briefly discuss a model which was first suggested by Simmons to describe electron tunneling through insulating films sandwiched in between two electrodes [257]. Such systems are referred to as tunnel junctions. In the last two decades the interest of the research community to studies of transport properties of tunnel junctions arose anew, when it was discovered that ultrathin perovskite films (only as thick as a few lattice parameters) may retain ferroelectric properties. It was suggested to use thin ferroelectric films as insulating layers in ferroelectric tunnel junctions. Such systems are expected to serve as essential components in various nanodevices. Also, the Simmons model could be applied to describe coherent electron transport through large-area metal–molecule–metal junctions consisting of two electrodes of certain area A linked by a molecular monolayer. Actually, this model was repeatedly employed to describe measured current–voltage characteristics in large-area MMM junctions, especially those using alkanethiol series as linkers. Typically, these molecules have a very low conductance, so a monolayer made out of them may be treated as a molecular insulator.

Within the Simmons model, it is assumed that the barrier electrostatic potential varies only in the direction perpendicular to the planes of the electrodes. Accordingly, one may decompose the wave vector $\mathbf{k}$ of the tunneling electron in two components. The component $\mathbf{k}_{||}$ is a projection of the wave vector to the electrode's plane, and $\mathbf{k}_{\perp}$ is directed perpendicularly to this plane. Also, we assume that the energy of a tunneling electron may be presented as the sum of two terms which depend on $\mathbf{k}_{||}$ and $\mathbf{k}_{\perp}$, respectively. Employing a simple isotropic form for the energy-momentum relation, we may write

$$E = E_{||} + E_{\perp} \equiv \frac{\hbar^2 k_{||}^2}{2m} + \frac{\hbar^2 k_{\perp}^2}{2m} \tag{3.16}$$

where m is the electron effective mass. Only $E_{\perp}$ matters for the transmission over or through the barrier.

There are various approximations for the tunneling transmission. Most often, a Wentzel–Kramers–Brillouin expression is employed, namely,

$$T(E_{\perp}) \approx t_L(E_{\perp}) t_R(E_{\perp}) \exp\{-2dQ(E_{\perp})\}. \tag{3.17}$$

Here, $t_L(E_{\perp})$ and $t_R(E_{\perp})$ are the interface transmission functions for the left and right electrodes, respectively, d is the thickness of the film or molecular monolayer, and $Q(E_{\perp})$ is determined by the height and shape of the potential barrier. Generally speaking, it also depends on the bias voltage V but one may disregard this at low bias and write the expression for $Q(E_{\perp})$ in the form

$$Q(E_{\perp}) = \sqrt{\frac{2m}{\hbar^2}(E_{\mathrm{F}} + \langle \Phi \rangle - E_{\perp})}. \tag{3.18}$$

Here, $\langle \Phi \rangle$ stands for the averaged barrier potential. Choosing the coordinate system in such a way that "z" axis is directed perpendicularly to the electrodes planes, $\langle \Phi \rangle$ may be defined as follows:

$$\langle \Phi \rangle = \frac{1}{d} \int_0^d \Phi(z) dz \tag{3.19}$$

where $\Phi(z)$ describes the potential barrier profile. The approximation for the electron transmission (3.17) and (3.18) works only for electron energies below the barrier, and it is to be modified when electrons travel above the latter.

To compute a tunnel current, we may start from the Landauer formula (2.54) where the transmission is approximated by Eq. (3.17). We observe that in the considered case, the transmission depends only of $E_\perp$. Therefore, the expression (2.54) could be converted to the form

$$\begin{aligned} I =& \frac{e}{h} \sum_{\mathbf{k}_{||}} \int dE_\perp T(E_\perp) \left[f_L(E_\perp) - f_R(E_\perp) \right] \\ =& \frac{e}{h} \int dE_\perp T(E_\perp) \left[F_L(E_\perp) - F_R(E_\perp) \right] \end{aligned} \tag{3.20}$$

where

$$\begin{aligned} F_\beta(E_\perp) =& \sum_{\mathbf{k}_{||}} f_\beta \left(E_\perp + \frac{\hbar^2 k_{||}^2}{2m} \right) \\ =& \frac{mAk_\mathrm{B}T}{2\pi\hbar^2} \ln \left\{ 1 + \exp \left[\frac{E_\perp - \mu_\beta}{k_\mathrm{B}T} \right] \right\}, \end{aligned} \tag{3.21}$$

and μ_β ($\beta \in L, R$) are the chemical potentials for the electrodes. After some algebra, the expression (3.20) may be reduced to the standard Simmons expression for the current density $J = I/A$:

$$\begin{aligned} J =& J_0 \left\{ \langle \Phi \rangle \exp \left[-2d \sqrt{\frac{2m \langle \Phi \rangle}{\hbar^2}} \right] \right. \\ & \left. - \left(\langle \Phi \rangle + eV \right) \exp \left[-2d \sqrt{\frac{2m}{\hbar^2} \left(\langle \Phi \rangle + eV \right)} \right] \right\}. \end{aligned} \tag{3.22}$$

Here:

$$J_0 \approx \frac{e t_L(E_\mathrm{F}) t_R(E_\mathrm{F})}{2\pi h d^2}. \tag{3.23}$$

As mentioned before, the Simmons model brings satisfactory results when applied to tunnel junctions where the electrodes are separated by an insulating film. As an example, in Fig. 3.10, we present the current–voltage characteristics computed for a model imitating a practical $Pt/BaTiO_3/Pt$ system where the ferroelectric

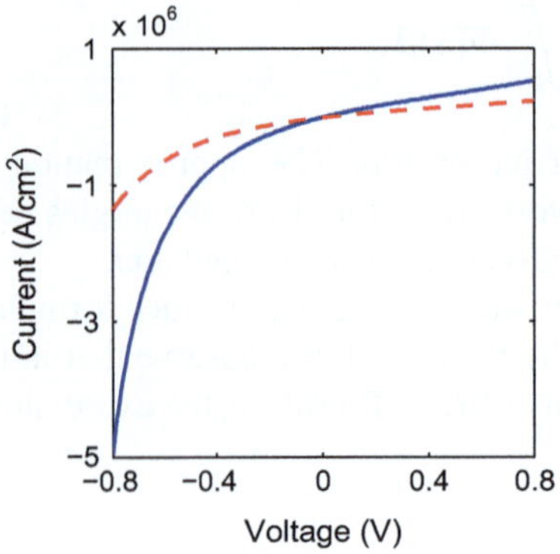

Fig. 3.10 Current–voltage characteristics computed for a ferroelectric tunnel junction using the Simmons model [258]. Reprinted with the permission of AIP

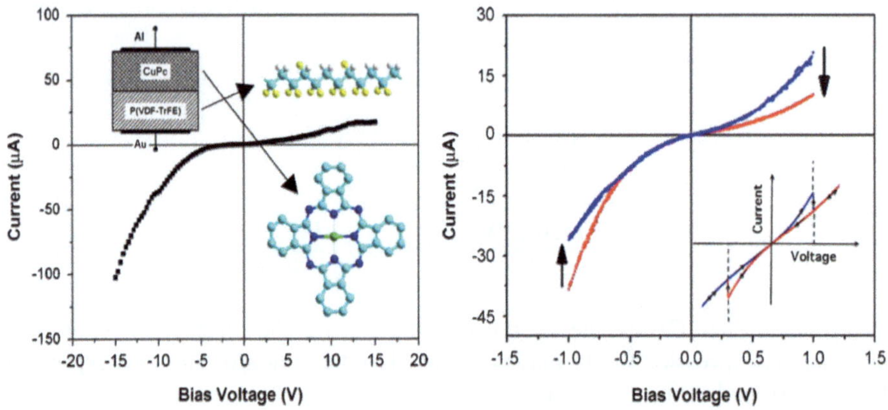

Fig. 3.11 *Bottom*: current–voltage characteristics measured for a heterojunction fabricated using 4 nm CuPc deposited on 3 nm of $P(VDF - TrFE)$. Schematics of the relevant molecules are presented in the *left panel* where *blue* stands for carbon, *yellow* for fluorine, *navy blue* for nitrogen, and *green* for copper (to simplify the schematics for CuPc, the hydrogens are left off) [259]. Reprinted with the permission of ELSEVIER

$BaTiO_3$ film is 1.8 nm thick [258]. The computed curves are consistent with the measured ones reported in [259] (see Fig. 3.11).

When we turn from the above-discussed tunnel junctions to metal–molecular junctions, serious difficulties arise. These difficulties originate from simplifications accepted within the Simmons model. First, there is a nontrivial task of defining the electron effective mass which enters the expression (3.21). It was introduced assuming a parabolic energy-momentum relation (3.16) which is valid near the band-edge dispersions for an infinite solid. However, for a small object where the effect of boundaries cannot be neglected, the energy-momentum relation of the form (3.16) becomes questionable. Moreover, the Simmons model was developed to analyze electron tunneling through an isotropic insulator where the current flow should be equally effective in all directions. This gives grounds to introduce the same effective mass into both terms of the expression (3.16). Unlike insulating films, molecular monolayers cannot be treated as isotropic media transmitting the electron current. Practical molecular monolayers always exhibit a large anisotropy in their in-plane and out-of-plane (along backbone) conductances. This means that one must introduce two different effective masses: a large one for the transport between the molecules (characterized by a low conductance) and a small one for the transport along the backbones with a significantly higher conductance.

Introducing two different effective masses helps to circumvent the difficulty arising due to the well-established anisotropy of transport properties of molecular monolayers. Nevertheless, there remains a problem of applicability of the very concept of effective mass in the case of electron transport through molecular bridges. The "effective masses" introduced in the Simmons model when the latter is employed to describe the electron transport through molecular bridges are not related to the effective masses of electrons in solid. They are no more than parameters characterizing the relative ease of electron tunneling along the molecular backbones and between the adjacent molecules and should be treated as such.

Another difficulty arises when one tries to define the current density in the case of electron transport through the molecular monolayer. It makes a perfect sense to introduce an electric current per unit area when one is considering a well-packed isotropic system such as an insulating film. However, for a molecular monolayer, the current per unit area is determined by the number of molecules attached to the substrate, and this may vary even in the case of a well-ordered molecule assembly. In this case the relevant metric is the current per molecule.

Introducing two "masses" ($m_\perp$ and $m_{||}$), respectively, we may rewrite Eq. (3.16) as follows:

$$E = \frac{\hbar^2 k_{||}^2}{2m_{||}} + \frac{\hbar^2 k_\perp^2}{2m_\perp}. \tag{3.24}$$

It is a lot more difficult for the electron to tunnel between molecules than along them, so $m_{||} \gg m_\perp$. This makes the first term in (3.24) much smaller than the second one: $E_{||} \ll E_\perp$. Therefore, we may omit the second term in the argument of the Fermi function in Eq. (3.21). As a result, the summation over $\mathbf{k}_{||}$ simply yields the prefactor which is equal to the number of molecules N_{mol}. So, the expression for the current (3.20) takes on the form

$$I = \frac{e}{\pi\hbar} N_{\text{mol}} \int dE_\perp T(E_\perp) \left[f_L(E_\perp) - f_R(E_\perp) \right]. \tag{3.25}$$

Comparing this expression with that given by the previous formula (3.20), we see the difference between the two, which originates from different physical properties of an isotropic insulating film and the molecular monolayer. As remarked by Ghosh [5], in some experimental works where electron transport through molecular monolayers

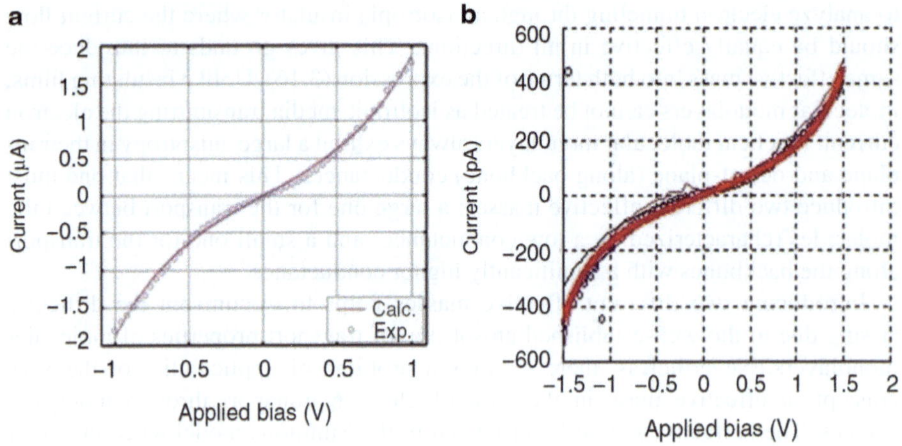

Fig. 3.12 Measured current–voltage curves and theoretical fits obtained using the modified Simmons equation (3.26) for alkanethiol chains on gold [260]. Reprinted with the permissions of AIP

was studied, the current–voltage curves were analyzed using Eq. (3.21) instead of the modified expression (3.25) which resulted in discrepancies between computed and measured current values.

To derive the correct Simmons equation for the current through a MMM junction, we substitute the Wentzel–Kramers–Brillouin expression for the electron transmission (3.17) into Eq. (3.25). Then we get

$$I = I_0 \left\{ \sqrt{\langle \Phi \rangle - \frac{eV}{2}} \exp\left[-2d\sqrt{\frac{2m_\perp}{\hbar^2}\left(\langle \Phi \rangle - \frac{eV}{2}\right)} \right] \right.$$
$$\left. - \sqrt{\langle \Phi \rangle + \frac{eV}{2}} \exp\left[-2d\sqrt{\frac{2m_\perp}{\hbar^2}\left(\langle \Phi \rangle + \frac{eV}{2}\right)} \right] \right\} \tag{3.26}$$

where $I_0 = eN_{\mathrm{mol}} t_L(E_{\mathrm{F}}) t_R(E_{\mathrm{F}}) / d\sqrt{m_\perp}$, d is the length of the molecular bridge, the "mass" $m_\perp$ characterizes the electron transport along the molecules, and $\langle \Phi \rangle$ is the average barrier potential. As discussed before, the modified Simmons equation which is appropriate for describing electron transport through a molecular insulator yields the current instead of the current density. The current is proportional to the number of molecules included in the assembly.

The modified Simmons equation (3.26) could provide a good match with the measured current–voltage curves. Two examples of a fit between the theory based on the Simmons equation and the experiment are shown in Fig. 3.12. Comparing theoretical and measured curves one may extract information concerning the number of well-contacted molecules in the considered system, the averaged barrier height, and the "effective mass" of a tunneling electron. The Simmons model does not provide means to extract more detailed information concerning the electrostatic potential profile in the transport channel. This deficiency originates from of the Wentzel–Kramers–Brillouin expression for the electron transmission.

The exponential dependence of the zero-bias resistance of the length of a molecule is a characteristic feature of coherent electron tunneling which was repeatedly observed in experiments. The expression for the resistance $R = (dI/dV)^{-1}$ derived from the modified Simmons formula (3.26) is consistent with these observations. Within the considered model, the zero-bias resistance is proportional to $\exp(q_0 d)$ where d could be treated as the molecular length, and the coefficient q_0 depends on the barrier height: $q_0 = 2\sqrt{2m_\perp\langle\Phi\rangle}\big/\hbar$. The experiments carried out for alkanethiol chains of varying length confirm this theoretical estimation. However, one must keep in mind that this exponential dependence of the resistance of the molecule length could be exhibited only in sufficiently short molecules where the scattering free transport regime may be maintained. When the molecule length exceeds a certain limit, backscattering processes come into play, and this changes the exponential dependence into a linear Ohmic dependence through multiple reflections. The critical length corresponding to the crossover from exponential to linear dependence of the molecular resistance varies for different kinds of molecules. For instance, a series of conjugated oligophenyleneimine molecules showed the crossover near the molecule length of 4 nm [261], whereas alkanethiol chains demonstrated the exponential behavior at significantly greater lengths.

3.5 Conformational Gating of Molecular Bridges

An important inherent property of the majority of molecules is their conformation flexibility. It creates possibilities for conformational changes in the molecular bridges. An example of current-induced changes in the molecules geometry was previously discussed. The conformational changes may significantly (and sometimes dramatically) affect transport properties of metal–molecule–metal junctions, showing potentialities for design and manufacturing of nano-electromechanical devices such as molecule-based switches. These potentialities and existing devices are discussed in more detail in the last chapter of the present book. Conformational changes in molecules could be triggered by visible light shining at the molecule and/or by ultraviolet photons. Also, they could occur at certain values of the bias voltage applied across the junction, as was briefly discussed above. Here, we discuss how the gating of MMM junctions may activate conformational changes in molecular bridges. Technically, the gating of MMM junctions meets very serious difficulties which occur due to the small length of molecular bridges which impede positioning of the gate electrode in a close proximity of the bridge. However, sometimes a molecule could preferentially respond to the gate voltage rather than to the bias voltage even if the gate sits farther from it than the source and drain. This happens when the direction of the electric field created by the gate is chosen in such a way that it couples with the molecular dipole more efficiently and exercises a bigger torque on an oriented molecular dipole than the field corresponding to the bias voltage. The torque of the gate field may change the direction of the molecular dipole which could be achieved by changing the molecular geometry. In this way, the gate field may cause an alteration of the molecular geometry.

To describe the effect of the gate-induced conformational changes in the molecular linker on the electron current, one starts from the Landauer formula (2.54) written out in a slightly different way [5, 262]:

$$I = \frac{e}{\pi\hbar}\int_{\mu_L}^{\mu_R} \langle T(E)\rangle \otimes F_T(E)dE, \tag{3.27}$$

where

$$\langle T(E)\rangle = \frac{\prod_i \int d\theta_i T(E,\theta_i)\omega(\theta_i)}{\prod_i \int d\theta_i \omega(\theta_i)} \tag{3.28}$$

and $F_T(E)$ is the thermal broadening function. Equations (3.27) and (3.28) are general equations which can take into account the effect of several factors on the electron transport at low bias voltage. Each factor is represented by a variable θ_i and the weighting function $\omega(\theta_i)$. For instance, for spintronic applications the averaging of the electron transmission specified by Eq. (3.28) involves a partial trace over the spin degrees of freedom weighted by the distribution function for spins implemented by an external bath. Similarly, when the phonons affect electron transport, $\langle T(E)\rangle$ is to be computed as a thermal average over the phonon distribution function. In the present case, one may accept an approximation:

$$\omega(\theta) \equiv \omega(U) = \exp\left[-\frac{U}{k_B T}\right] \tag{3.29}$$

where U is representing the conformational potential which depends on the voltage V_G applied to the gates.

Starting from Eqs. (3.28) and (3.29) and employing some algebra, one gets an expression for the derivative of the average transmission with respect to the gate voltage [262]:

$$\frac{\partial\langle T\rangle}{\partial V_g} = \left\langle\frac{\partial T}{\partial V_g}\right\rangle - \frac{1}{k_B T}\left\{\left\langle T\frac{\partial U}{\partial V_G}\right\rangle - \langle T\rangle\left\langle\frac{\partial U}{\partial V_G}\right\rangle\right\}. \tag{3.30}$$

In this expression, the first term describes the electrostatic contribution arising from the direct modulation of the transmission by the gate voltage. The second term takes on a nonzero value only when the considered molecule possesses a conformational potential which reasonably strongly varies as the gate potential changes. This term describes the relation between the transmission and $\partial U/\partial V_{\mathrm{G}}$. For instance, in the case when the stable configuration of a molecule which minimizes the potential U is corresponding to a maximum in the electron transmission, the gate modulation of this configuration should bring further increase of the transmission as well as of the transconductance $\partial I/\partial V_{\mathrm{G}}$.

It seems that molecules with softer bonds which are more responsive to the gate voltage variations are more likely to show the effects induced by conformational changes in their transport characteristics. However, such molecules are more responsive to thermal fluctuations, as well. Usually, a gate voltage drives a molecule between some states within the conformational potential manifold whose energies only slightly differ. Then, in the absence of a barrier between an initial and a final conformation of the molecule, thermal fluctuations may cause backflow returning the molecule to the original state and thus destroying the stability of the effect. To stabilize the molecular states corresponding to different geometries, one needs to create a barrier preventing the backflow. In principle, a barrier separating local stability wells may be built and operated by the gate voltage, but this costs energy and contributes to dissipation. A better way is to build the barrier using a separate gate-independent mechanism such as a sticking potential which appears due to van der Waals forces or hydrogen bonding between the molecule and the neighboring electrodes, as schematically shown in Fig. 3.13.

The considered issue of conformational gating in molecules highlights some points which are important for further development of the whole molecular electronics. Future electronics should use unique properties of the electric field on the gate (including its vector directionality and tensor coupling in piezomaterials) as an additional handle to modify the transport properties of certain metal–molecule junctions. The effect of the gate field should be maximized when the molecular dipole moment is directed along the electric field associated with the bias voltage and perpendicularly to the gate field. Within such geometry, the torque due to the former electric field would be eliminated (at least partially). The predominating torque due to the gate field may physically rotate certain parts of the molecule to put its dipole in parallel with the gate field, thus causing conformational changes and corresponding alterations in the molecular conductance. This may happen if the considered molecule possesses a significantly large dipole moment to robustly couple to the gate field withstanding thermal fluctuations. At the same time, the molecule must have sufficiently soft bounds providing for fairly strong response of the conformational potential to the variations of the gate voltage.

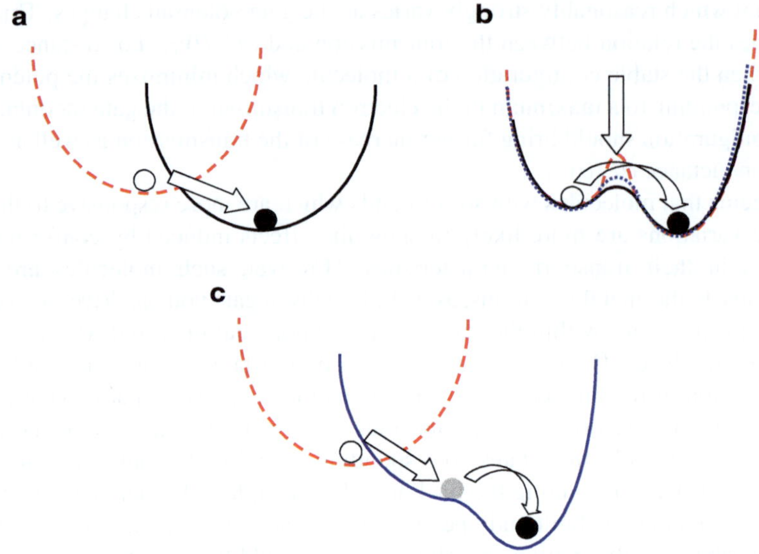

Fig. 3.13 (**a**) Gate-driven barrierless transition between the molecule configurations which could be backflowed by thermal fluctuations. (**b**) Transition within the same potential landscape where the energy minima corresponding to the molecule configurations are separated with a barrier operated using the gate potential. (**c**) The gate-independent mechanism (e.g., sticking potential arising due to van der Waals forces or hydrogen bonding) shapes an additional potential well preventing the backflow

3.6 Effect of the Electronic Structure of the Leads on the Electron Transport Through a Junction: Negative Differential Resistance

Among various important properties of the electron transport through metal–molecule junctions, one may separate out the negative differential resistance (NDR), that is, the decrease of the current I while the bias voltage V across the molecule increases. The NDR effect was originally observed in tunneling semiconducting diodes [263]. Later, the NDR was viewed in quantum dots [264] and metal–molecule–metal junctions (see, e.g., [187, 265–270]). Several possible scenarios are proposed to explain the NDR occurrence in the electron transport through molecules. The NDR could appear if the conducting system includes weakly linked segments which have rather narrow and sharp features in their electron density of states. Then, as bias voltage V varies, the electron energy levels on these segments could move with respect to each other due to the electrostatic potential distribution across the whole system. At some values of the applied voltage, the energy levels of the adjacent segments could be matching, whereas at another values of V, they mismatch [267, 271]. One may expect the coupling between the segments to strengthen when the energy levels belonging to them match and weaken when the levels mismatch. This may noticeably modify the coupling of the molecule to the leads. The variations in the coupling strengths could serve as an immediate reason for the NDR peaks to appear. Also, such peaks could occur as a Coulomb blockade-induced effect [81, 264], and/or they could originate from conformational changes in the molecule [272] and electron interaction with the molecule vibrational modes [19, 187]. It is likely that different mechanisms could play a major part in the NDR appearance in different molecular junctions where it was observed so far. However, reviewing the available experimental data, one may conclude that the most distinguished NDR features in the current–voltage curves (sharp and narrow peaks separated by intervals of extremely low conductivity, like those reported in [265]) are usually attributed to the matching-mismatching of the molecule energy levels with those of the leads. Keeping in mind that precisely such NDR characteristics are potentially valuable for molecular electronics applications, we further concentrate on this mechanism in the present section. Some other scenarios will be analyzed below.

Correspondingly, we consider the NDR effect in the electron transport through a molecular junction where the leads are small metal clusters whose electron density of states reveals sharp and distinct features [249]. This system provides better opportunities to study the effects of matching-mismatching of the molecule energy levels with those of the "nanoleads" in the electron transport characteristics. Under certain conditions, such junctions may show very distinguished multiple NDR features in the I–V characteristics. To maintain a steady supply of electrons tunneling through the junction, we assume that the metal clusters ("nanoleads") keep contact with large source and drain electron reservoirs. The latter, however, are separated from the molecule in such a way that electrons cannot directly tunnel from these reservoirs to the molecule. In the absence of the applied bias voltage the whole system is supposed to be in the equilibrium state characterized with the equilibrium Fermi energy E_{F}.

To describe the transport through this junction we employ Eqs. (2.54), (2.66), and (3.1). The self-energy terms $\Sigma_{L,R}$ now include corrections describing the coupling of the nanoleads with the large electron reservoirs:

$$\Sigma_{\beta}^{ij}(E) = \sum_{r,\sigma} \frac{\tau_{ir\beta\sigma}^{*}\tau_{jr\beta\sigma}}{E - \epsilon_{r\beta\sigma} - \sigma_{r\beta\sigma}} \tag{3.31}$$

where $\epsilon_{r\beta\sigma}$ are energy levels of the corresponding leads and the parameters $\sigma_{r\beta\sigma}$ are the self-energy corrections, which originate from the coupling of the metal clusters to the large electron reservoirs. Their imaginary parts characterize the width of the clusters energy levels. When the bias voltage V is applied across the system, this causes charge redistribution and subsequent changes in the molecular energy states E_i and in the energy levels of the leads. In consequence, the Green's function matrix elements G_{ij} and self-energy terms Σ_{β}^{ij} vary as V changes. This affects the electron transmission function T given by the expression (2.66).

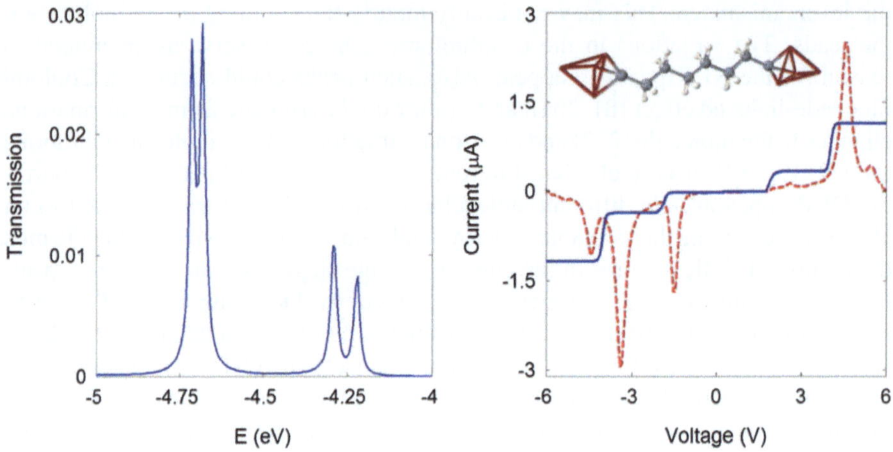

Fig. 3.14 *Left panel*: the electron transmission function through the hydrocarbon chain $T(E, 0)$ within a certain range of energies E. *Right panel*: The current–voltage characteristics for the electron tunneling through the junction including the hydrocarbon chain and two copper clusters. The curves are plotted assuming $T = 50\,\mathrm{K}$, and $V_{\mathrm{mol}} = 0.1\,\mathrm{V}$ (*solid line*) and $V_{\mathrm{mol}} = 0.9\,\mathrm{V}$ (*dashed line*) [249]. Reprinted with the permission of the APS

The current–voltage characteristics shapes crucially depend on the electrostatic potential profile in the considered system. If the voltage mostly drops between the large electron reservoirs and the small metal nanoparticles included in the junction, so that the bias applied across the MMM junction V_{mol} is much smaller than that applied across the whole system ($V_{\mathrm{mol}} \ll V$), one may approximate $T(E, V)$ as $T(E, 0)$. Then the applied bias voltage does not change relative positions of the energy levels $\epsilon_{r\beta\sigma}$ and E_i, and low-temperature current–voltage characteristics should display step-like shapes. These are typical for electron tunneling through molecules. Current increases as the voltage V increases, and the NDR does not appear. On the contrary, when no significant voltage drop occurs between the electron reservoirs and the metal nanoleads ($V_{\mathrm{mol}} \sim V$), the effect of the bias voltage on the energy levels of the clusters $\epsilon_{r\beta\sigma}$ could considerably differ from its effect on the molecular energy E_i. Due to these differences in the voltage-induced shifts, the relative positions of the energy levels on the clusters and these on the molecule vary as the voltage changes. This creates opportunities for alignment/disalignment of the molecule orbitals with those associated with the metal clusters. Therefore, in such a case, one may expect the NDR to occur. The above suppositions were confirmed by calculations of the tunneling electron current through a junction which consisted of two copper nanoclusters linked by an aliphatic-saturated hydrocarbon chain, assuming that the junction was placed in between two large metal electrodes [249].

The resulting I–V curves are presented in Fig. 3.14. One can see that when the electrostatic potential mostly drops between the source/drain reservoirs and the nearby copper clusters ($V_{\mathrm{mol}} = 0.1V$), the corresponding I–V curve reveals a stepwise profile without NDR features. On the contrary, distinct NDR peaks emerge provided that there exists a significant drop in the electrostatic potential between the copper clusters and the molecule tips ($V_{mol} = 0.9V$). Also, one may notice that the NDR peaks are rather sharp and narrow, and the current peak values are much greater than in the valleys between the peaks. This is consistent with the experimental data reported in the [265]. The analysis of [249] was carried out using a simplified model for the electrostatic potential distribution over the considered system. To quantitatively study the effects of the lead electron structure on the transport properties of MMM junctions, one must properly compute the spatial profile of the electrostatic potential employing a self-consistent computational procedure discussed before. However, it seems likely that such analysis will not bring qualitative changes in the above-described results.

3.7 Coulomb Blockade and Charge Oscillations: NEGF Approach

As well known in mesoscopic physics, the conduction of a tunnel junction including a small-sized conducting island situated between two tunnel barriers may be suppressed at low values of the bias voltage due to the Coulomb repulsion of electrons on the island [273, 274]. This effect is called the Coulomb blockade, and it may be manifested regardless of the character of the energy spectra of electrons. The Coulomb blockade can be removed by capacitive charging of the island by means of a gate electrode. As the gate voltage or chemical potential of the leads varies, the conduction shows oscillations as a function of the gate voltage. The theory of these Coulomb blockade oscillations appropriate for junctions where the discreteness of the electron energy spectrum on the island may be ignored was first suggested by Kulik and Shekhter [275]. This classical theory could serve as a good approximation in metals, where the energy level separation is typically much smaller than both charging and thermal energies. Also, the Coulomb blockade oscillations are manifested in the tunnel junctions including a quantum dot or another entity with the discrete electron spectrum. The role of the spectrum discreteness in the conduction oscillations was addressed in the works [74–76, 86, 276, 277] and some other. The Coulomb blockade oscillations were repeatedly observed in experiments on molecular and carbon nanotube junctions [27–33]. The effect could be manifested when the junction is weakly coupled, that is, the charging energy associated with the Coulomb repulsion of electrons on the bridge U_0 significantly exceeds the energies characterizing the coupling of the leads to the molecule.

To analyze the electron transport within Coulomb blockade regime one may start from the simple case when the bridge is represented by a single spin-degenerated level with the energy E_0. In this case the current through the junction given by the general expression (2.103) takes on the form

$$I = \frac{ie}{2\hbar}\sum_{\sigma}\int\frac{dE}{2\pi}$$
$$\times\left\{\left[\Gamma_{L\sigma}f_{L\sigma}(E) - \Gamma_{R\sigma}f_{R\sigma}(E)\right]\right.$$
$$\left.\times\left[G_{\sigma}^{r}(E) - G_{\sigma}^{a}(E)\right] + (\Gamma_{L\sigma} - \Gamma_{R\sigma})G_{\sigma}^{<}(E)\right\} \tag{3.32}$$

where the retarded Green's function $G_{\sigma}^{r}(E)$ has the form (2.118). As follows from Eq. (2.27), the occupation numbers $\langle n_{\sigma}\rangle$ are given by

$$\langle n_{\sigma}\rangle = \frac{1}{2\pi}\int dE\,\mathrm{Im}\left[G_{\sigma}^{<}(E)\right]. \tag{3.33}$$

The lesser Green's function obeys the Keldysh equation (2.35) where the self-energy terms $\Sigma_{\beta}^{<}$ may be approximated by Eq. (2.101). Employing this approximation for $\Sigma_{\beta}^{<}$, we assume that the Coulomb interactions on the molecule do not affect the coupling of the latter to the leads. When the coupling to the leads differs only by a constant factor $\left(\Gamma_{L\sigma}(E) = \lambda\Gamma_{R\sigma}(E)\right)$, the expression for the current may be significantly simplified and presented in the form [228]

$$I = \frac{e}{2\hbar}\sum_{\sigma}\int\frac{dE}{2\pi}$$
$$\times\, G_{\sigma}^{a}\Gamma_{R\sigma}G_{\sigma}^{r}\Gamma_{L\sigma}\Sigma_{0}^{-1}\Sigma\left[f_{L\sigma}(E) - f_{R\sigma}(E)\right]. \tag{3.34}$$

Here, Σ_0 and Σ are self-energy terms for the cases of noninteracting and interacting electrons, respectively. It is obvious that the difference between the self-energies Σ_0 and Σ determines the diversity in the form between Eq. (3.34) and the Landauer formula [see Eqs. (2.54) and (2.66)]. Also, the expressions for the relevant Green's functions are certainly different due to the Coulomb interactions between the electrons on the bridge. In the limit of noninteracting electrons ($U_0 = 0$), Eq. (3.34) is reduced to the form (2.54) where the electron transmission is given by Eq. (2.66).

Simultaneously solving Eqs. (3.32) and (3.33) we may explore the dependencies of the occupation numbers and current on the applied bias voltage V. We assume the symmetrical voltage division between the leads, and we set the electrochemical potentials for the unbiased junction $\mu_L = \mu_R = E_F = 0$. Also, we introduce a nonzero gate voltage V_g to shift the bridge energy level so that the energies E_0 and $E_0 + U_0$ become asymmetrically arranged with respect to $E = E_F$. This asymmetry is necessary to obtain two steps in the I–V curves in the considered case of a symmetrically coupled junction.

As shown in the left panel of Fig. 3.15, the occupation numbers are sensitive to the value of the applied voltage. At low values of the bias voltage ($V < 0.4$ V), the assumed bridge energy level ($E_0 = -0.2$ eV) is situated below the chemical potentials for both leads. Therefore, the molecular bridge is able to receive an electron but unable to transfer it to another lead. Accordingly, the average

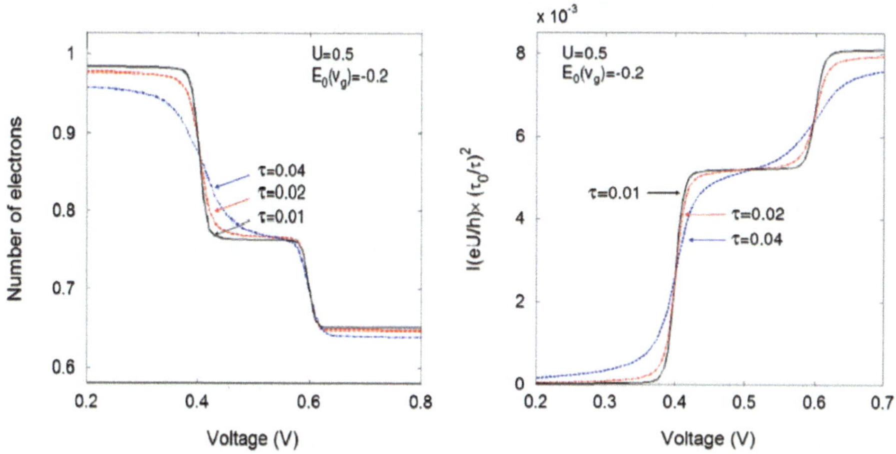

Fig. 3.15 *Left panel*: average occupation numbers in the bridge versus the voltage applied across the junction. The curves are plotted assuming $U_0 = 0.5\,\text{eV},\ k_B T = 0.00026\,\text{eV},\ E_0(V_g) = -0.2\,\text{eV},\ \tau = 0.01\,\text{eV}$ (*solid line*), 0.02 eV (*dash-dotted line*) and 0.04 eV (*dashed line*). *Right panel*: current through a junction within the Coulomb blockade regime. The curves are plotted assuming $U_0 = 0.5\,\text{eV},\ k_B T = 0.00026\,\text{eV},\ E_0(V_g) = -0.2\,\text{eV},\ \ \tau_0 = 0.04\,\text{eV},\ \tau = 0.01\,\text{eV}$ (*solid line*), 0.02 eV (*dash-dotted line*) and 0.04 eV (*dashed line*). The factor τ_0^2/τ^2 is introduced to bring all I–V curves to the same scale. The ratio of heights of the two steps revealed in the curves equals 2 : 1 [249]. Reprinted with the permission of the APS

occupation number is close to unity and the electric current through the junction takes on values close to zero (see Fig. 3.15). At $V = 0.4\,\text{V}$, the energy E_0 crosses the chemical potential μ_L, and the bridge becomes active in electronic transport. Now, the electron which arrives at the bridge from one reservoir may leave it for another reservoir. This results in a pronounced decrease in the average occupation on the bridge accompanied by an increase in the current. One more change in both average electron occupation on the molecular bridge and the current through the junction occurs at $V = 0.6\,\text{V}$ when the energy $E_0 + U_0$ ($U_0 = 0.5\,\text{eV}$) crosses μ_R. At higher voltage all curves presented in Fig. 3.15 level off. The minimum occupation number is noticeably less than one, but its value is nonzero for electrons unceasingly travel through the junction.

The current–voltage curves in the right panel of Fig. 3.15 show typical Coulomb blockade features, namely, two steps whose heights are related as 2:1. So, we see that the results concerning the electron transport through a quantum dot/molecule obtained employing the transition rates equations may be quantitatively reproduced within the NEGF formalism [35,88]. It is worthwhile to remark that particular values of the average occupation numbers on the molecular bridge are very responsive to the gate voltage value V_g (the latter determines how the bridge energy level is situated with respect to the Fermi level of the leads in the absence of the voltage applied across them) and to the Coulomb interaction energy U_0. Therefore, different values chosen for V_g and U_0 lead to different average occupation numbers. However,

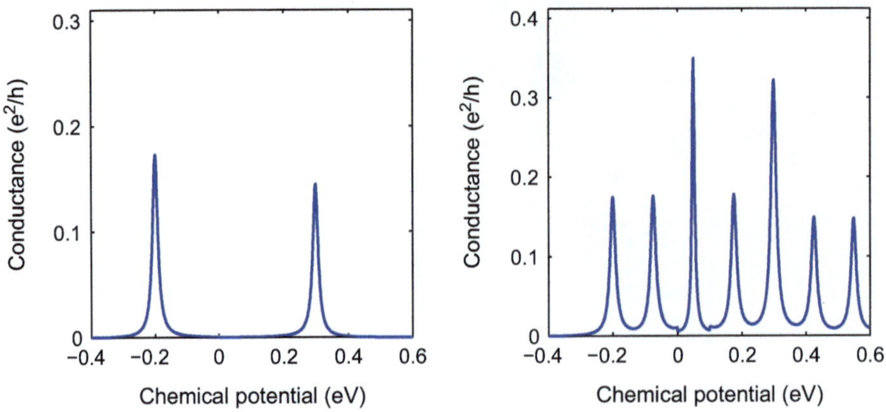

Fig. 3.16 Conductance through a single-level dot versus the chemical potential of the contacts (*left*) and conductance oscillations described by Eq. (3.37) for the dot containing five equidistant levels (*right*). The curves are plotted assuming $\tau = 0.02\,\text{eV},\ U_0 = 0.5\,\text{eV},\ k_BT = 0.00026\,\text{eV}$

under various assumptions for the V_g and U_0 values, one quantity does not vary provided the symmetric coupling of the dot to the leads. This quantity is the relative height of the subsequent steps in the average occupation numbers of the electrons on the bridge. These are revealed as the voltage across the leads sufficiently increases. As shown in Fig. 3.15, this ratio is 2:1, exactly the same as for the subsequent current steps in the I–V curves. And it is this ratio which ensures the correct shape of I–V curves. For instance, comparing Fig. 3.15 with the corresponding result reported in the work [88], one may see that the values of the occupation numbers given in [88] considerably differ from those presented here. Nevertheless, the ratio of the subsequent step heights in the occupation numbers versus voltage curves is 2:1, and this provides for the same ratio of heights of the subsequent current steps.

The linear response conductance through the junction may be presented in the form [86]:

$$g = \frac{e^2}{\hbar}\sum_\sigma \int \frac{dE}{2\pi}\frac{\partial f}{\partial E}\frac{2\Gamma_{L\sigma}\Gamma_{R\sigma}}{\Gamma_{L\sigma}+\Gamma_{R\sigma}}\text{Im}\big[G_\sigma^r(E)\big]. \tag{3.35}$$

As follows from this expression, the conductance is the sum of the coherent-transmitted probabilities associated with the conduction channels for different spin orientations. Each contribution is weighted by the derivative of the Fermi function, which is responsible for the temperature dependence of the conduction.

If the chemical potential of the unbiased junction varies, it passes E_0 and E_0+U_0, thus opening the channels for electron transport through the junction. This results in the appearance of two peaks in the g vs. μ curve, as shown in Fig. 3.16 (left panel). The width and height of peaks are determined by the coupling strengths between the leads and the molecular bridge and by the temperature. The effect of temperature on the conduction peaks shapes is determined by the relation of the thermal energy

$k_B T$ and the coupling energies Γ_β. Within the low-temperature limit $k_B T \ll \Gamma_\beta$, the temperature dependencies of conductance peaks are similar to those typical for a noninteracting single-particle resonance. The peak width is proportional to T and its height is proportional to T^{-1} [74, 86]. However, at higher temperatures when $k_B T$ exceeds Γ_β, the temperature dependencies acquire more complicated character. As indicated, under these conditions the height of the conductance peak may even increase as the temperature rises [87].

To analyze the Coulomb blockade oscillations, one must generalize consideration to multiple levels on the bridge. For this purpose, one needs to derive expressions for the relevant Green's functions and self-energy terms basing on the Hamiltonian given by Eqs. (1.34)–(1.36) which is a nontrivial and very complicated task. Research effort is applied to solve this problem (see, e.g., [277, 278]), but this research is not completed yet. Therefore, we describe these oscillations adopting a simplified model suggested by Meir and Wingreen [86] for multilevel quantum dots. Within this model, the density of states for electrons on the level "k" interacting with other electrons on the dot is approximated by the sum:

$$\sum_{m \neq k} P_k(m)\delta(E_k^0 + mU_0 - E) \tag{3.36}$$

where the weight factor $P_k(m)$ corresponds to the probability that "m" energy levels other than the level "k" are occupied. Using this expression, the conductance through the junction containing the dot/molecule coupled to the leads accepts the form

$$g = -\frac{\pi e^2}{\hbar} \sum_k \sum_{m \neq k} \Gamma_k P_k(m) \frac{\partial f}{\partial E}\bigg|_{E=E_k^0+mU_0} \tag{3.37}$$

Here, the parameters Γ_k describe the strengths of the coupling of the dot levels to the leads. As for the probabilities $P_k(m)$, they could be identified as the corresponding Boltzmann weights [74]. The conductance (3.37) as a function of the chemical potential of the leads at zero-bias voltage is plotted in the right panel of Fig. 3.16. The Coulomb blockade oscillations are well pronounced in the presented plot. One may remark that equal separations between the adjacent peaks in the conductance originate from the equipartition of the energy levels typical for quantum dots. In practical MMM junctions, the distances between the peaks in the conductance vary destroying the periodicity of the oscillations.

In the present chapter we restrict our consideration with the coherent transport through the junctions, so we do not consider the changes in the Coulomb blockade originating from the electron–phonon interactions. Nevertheless, we remark that these changes may be significant. For instance, it was demonstrated that molecular vibrational modes appear as distinct features in the current–voltage characteristics of MMM junctions in the Coulomb blockade regime [8, 34, 279, 280]. Presently, the issue is intensely studied, and we discuss it in more detail in the next chapter.

3.8 Coulomb Blockade: Multielectron Master Equations Approach

As it was mentioned before, it is hard to generalize NEGF formalism to describe the Coulomb blockade in a realistic MMM junction with a multistate molecular bridge. In particular, it is extremely difficult to implement the exclusion principle while dealing with multiple molecular orbitals. An alternative approach is based on the employment of master equations. Building up the corresponding formalism, one starts from an isolated molecule/quantum dot which is represented by its many-body spectrum. Then one adds leads weakly coupled to the molecule ($U_0 \gg \Gamma_{L,R}$) and treats electron transport as a manifold of transitions of the system between its many-electron configurations. The electron current through the junction could be computed using Eq. (2.133) where $P_{N,i}$ are occupation probabilities for N-electron excited stats and $W_{N,i;N\pm1,j}$ are transition rates between many-electron states differing by a single electron. The occupation probabilities obey the normalization condition (2.130) and the master equations of the form (2.131), whereas the transition rates are determined by Eq. (2.134). Within this approach, one may automatically take into account electron–electron correlations. However, the rate equations (2.131) were derived starting from the Liouville equation for the density matrix, and its off-diagonal elements were omitted in the process of derivation. The omission of these terms leads to the lack of coherency. Also, it is difficult to correctly introduce the levels broadening within the considered computational scheme, and the computational cost becomes quite high as we turn from a Hilbert space with the dimensions $N \times N$ (assuming that N electrons are included into the system) to a $2^N \times 2^N$ Fock space. The computational cost becomes significantly reduced assuming that identities of individual excitations are not important. Under this assumption, one can omit the induces i, j in the expression (2.133) for the electron current and reduce it the form [5]:

$$I = \pm\frac{e}{\hbar}\sum_N \left(W^L_{N\to N\pm1} - W^L_{N\to N\mp1}\right) P_N. \tag{3.38}$$

Golden rule-based calculations give the following result for the transition rates appearing in this expression for the current:

$$\begin{aligned} W^L_{N\to N+1} =& 2\pi \int dE \Gamma_L(N) D_L(E-\mu_L) f_L(E) \\ & \times D_b(E-\mu)[1 - f(E-\mu)], \\ W^L_{N\to N-1} =& 2\pi \int dE \Gamma_L(N) D_L(E-\mu_L) \\ & \times D_b(E-\mu) f(E-\mu)[1 - f_L(E)]. \end{aligned} \tag{3.39}$$

Here, $D_L(E - \mu_L)$ and $D_b(E - \mu)$ are electron densities of states on the left electrode and on the molecular bridge, respectively, and μ_L, μ are the electrochemical potential for the left lead and the local electrochemical potential of the bridge. Assuming that the electron density on the leads and the coupling parameters $\Gamma_\beta(N)$ are independent of energy, we get

$$W^{\beta}_{N\to N\pm 1} = \frac{\hbar}{R_\beta e^2}\left\{\frac{-\Delta E^{\pm}_{\beta}}{1-\exp[\Delta E^{\pm}_{\beta}/k_B T]}\right\} \tag{3.40}$$

where

$$\begin{aligned}\Delta E^{\pm}_{L} &= \Delta U^{\pm} \mp \frac{eC_R}{C_L + C_R}V \mp \frac{eC_G}{C_L + C_R}V_G,\\ \Delta E^{\pm}_{R} &= \Delta U^{\pm} \pm \frac{eC_L}{C_L + C_R}V \mp \frac{eC_G}{C_L + C_R}V_G.\end{aligned} \tag{3.41}$$

In these expressions, R_β are the resistances at the contacts, $C_{L,R}$ and C_G are capacitances of the capacitors representing the leads and the gate, respectively. Assuming that the current flows from the left to the right, one may readily identify C_L and C_R with the source and drain capacitances (see Sect. 1.2). The energy $\Delta E^{\pm}_{\beta}$ is the transition energy corresponding to adding/removing an electron at the lead β, and $\Delta U^{\pm}$ is the Coulomb offset for the electron addition/removal.

The described approach based on Eqs. (3.38)–(3.40) provides a simplest way to treat the effect of electron–electron correlations in weakly coupled MMM junctions. Following this way, one reduces the issue to solving an RC network. The relevant capacitances are determined by the geometry of the considered system and dielectric constants, whereas the resistance could be found by computing the occupations of the various levels without paying attention to individual excitations. Along with its comparative simplicity, this approach has some other significant advantages. For example, it provides means to describe ramp-like features which were experimentally observed in current–voltage characteristics of some molecules as well as plateaus and the Coulomb blockade of the current at the low bias voltage [281]. These features, shown in Fig. 3.17, are not easily captured by a NEGF-based formalism.

However, the simplicity of the described model causes its limitations. It does not provide means to capture some important effects which were repeatedly observed in experiments. These are the effects originating from individual excitations which cannot be given a proper treatment within the above-discussed simple approach. Among these are experimentally observed variations between plateaus and ramps including transitions from plateau to ramp behavior. Also, NDR caused by specific excitations could appear as well as some other effects. To capture these effects one must turn back to the original rate equations (2.131) and more carefully compute the relevant transition rates and occupation probabilities, so that the effect of individual excitations is not missed.

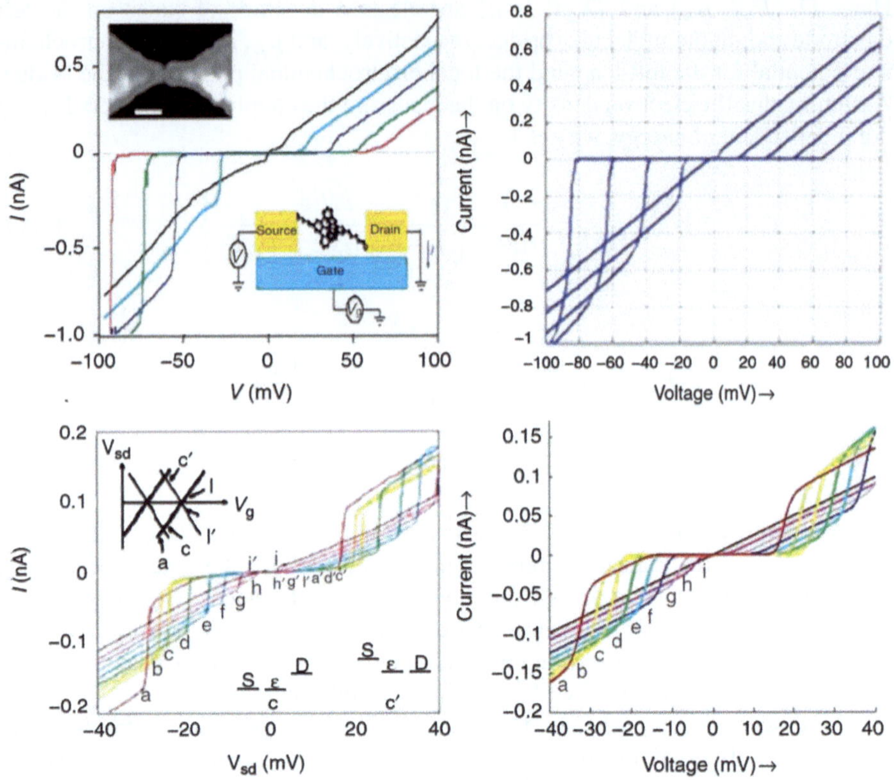

Fig. 3.17 Experimental (*left panels*) and theoretical (*right panels*) current–voltage curves within the Coulomb blockade regime. The theoretical curves are computed using the master equations (3.38)–(3.40). Reprinted from [281]. Reprinted with the permission of the APS

One can start by introducing many-body states $|i, N\rangle$ and computing the rate constants:

$$\Gamma^{ij}_{\beta r}(N) = \gamma_\beta |\langle i, N| c^\dagger_\beta | j, N-1\rangle|^2,$$
$$\Gamma^{ij}_{\beta a}(N) = \gamma_\beta |\langle i, N| c_\beta | j, N+1\rangle|^2. \tag{3.42}$$

Here, γ_β are the electron tunneling rates for the left and right electrode, respectively, $c^\dagger_\beta$, c_β are the creation and annihilation operators for an electron on the molecular end atom coupled to the corresponding electrode, and induces r, a are standing for the removal and addition of an electron on the molecular bridge/quantum dot. This leads to the expressions (2.134) for the transmission rates corresponding to electron removals ($|i, N\rangle \rightarrow |j, N-1\rangle$). Now, the energies of individual excitations $\epsilon^r_{ij}(N) = E_i(N) - E_j(N-1)$ are explicitly included into these expressions. One may get similar expressions for the transition rates corresponding to electron additions ($|i, N\rangle \rightarrow |j, N+1\rangle$). The latter depend on the individual excitations energies $\epsilon_{ij}^a(N) = E_j(N+1) - E_i(N)$. Substituting these expressions into the rate equations (2.131) and solving these equations, one may obtain occupational probabilities $P_{N,i}$ and use them to compute the current through the molecule. This approach provides means to analyze electron transport through molecules within the Coulomb blockade regime in more details. In addition to earlier discussed features, namely, the suppressed zero-bias conductance and ramp/flat current plateaus, extra features appearing due to individual excitations may be captured. Among them, there is a pronounced fine structure exhibited by the current–voltage curves and a crossover between plateaus and ramps depending on the location of the excited states [282, 283].

Also, under certain conditions, these excitations could cause the NDR. This happens as a result of the dynamics of the so-called dark states, that is, excitations whose characteristic lifetimes are significantly shorter than those typical for the direct ground-state-to-ground-state transitions. For illustration, one may adopt a simple model including only three many-body states: a ground state with N electrons (labeled "0"), an $N+1$ electron ground state (labeled "1"), and an excited state with $N+1$ electrons (labeled "2"). As follows from (2.131), the rate equations for these states have the form

$$\begin{aligned}\frac{dP_0}{dt} &= -(W_{01}P_0 - W_{10}P_1) - (W_{02}P_0 - W_{20}P_2),\\ \frac{dP_1}{dt} &= -(W_{10}P_1 - W_{01}P_0),\\ \frac{dP_2}{dt} &= -(W_{20}P_2 - W_{02}P_0).\end{aligned} \tag{3.43}$$

These equations must be supplemented with the normalization condition for the probabilities:

$$P_0 + P_1 + P_2 = 1. \tag{3.44}$$

Each transition rate included in Eq. (3.43) is the sum of contributions from the right and left contacts. Now, the current flowing through the left lead equals

$$I_L = \pm e(W_{01}^L P_0 + W_{02}^L P_0 - W_{10}^L P_1 - W_{20}^L P_1). \tag{3.45}$$

Following the standard way, one must solve the system (3.43) at steady state together with the normalization condition for probabilities and substitute the obtained results into the expression for the electron current (3.45). The resulting current–voltage curve should exhibit two plateaus where the first one is associated with the transition between the ground states ("0" and "1") and the second one corresponds to accessing the excited state "2." The NDR occurs when the height of the second step is smaller than that of the first one. Theoretical analysis carried out in [284] showed that the relative plateau heights are determined by the relations between the coupling parameters γ_β inherent to ground-state-to-ground-state transitions and γ'_β corresponding to the ground-state-to-excited-state transition ($\beta \in L, R$). These coupling parameters are closely related to the lifetimes of the relevant states. Assuming for certainty that we start from the empty state, the transition between "0" and "1" states may be referred to as LUMO transition, and the transition between the ground state and the excited state as LUMO+1 transition. In this case, it was shown that the right-sided NDR occurs when [284]

$$\left(\frac{1}{\gamma'_L}, \frac{1}{\gamma'_R}\right) > \frac{1}{\gamma_L} + \frac{1}{\gamma_R} \tag{3.46}$$

and the left-sided NDR appears if

$$\left(\frac{1}{\gamma'_L}, \frac{1}{\gamma'_R}\right) < \frac{1}{\gamma_L} + \frac{1}{\gamma_R}. \tag{3.47}$$

In the case when

$$\frac{1}{\gamma'_L} < \frac{1}{\gamma_L} + \frac{1}{\gamma_R} < \frac{1}{\gamma'_R} \tag{3.48}$$

the NDR is not manifested in the current–voltage curves behavior. Assuming that a positive bias is put on the right contact, the NDR appears when it takes more time to extract an electron from LUMO+1 state and to move it to the right contact than to fill the LUMO from the left and/or to empty it from the right. For the reversed bias-voltage polarity the NDR occurs provided that the right contact fills the LUMO+1 state faster than the LUMO could be filled/emptied. For HOMO levels, the parts of the leads are reversed. In this case, the left contact must keep HOMO-1 state empty longer than it takes to fill/empty HOMO state to make the NDR manifested. So, within the considered scenario, the NDR originates from slowly evolving many-body states (called dark states) whose existence makes it possible for the corresponding one-electron transitions to act as traps. When these traps are charged, this moves the main transport channel (operating between the ground states) out of the conducting window. As a result, the current reduces and the NDR appears, as shown in Fig. 3.18.

Finally, we remark that rate (master) equation-based approach to studies of the electron transport through molecules/quantum dots in the Coulomb blockade regime repeatedly proved its usefulness. Nevertheless, one should keep in mind that reducing of the Liouville equation for the density matrix to the set of rate equations for occupation probabilities is accompanied by the loss of coherence. To avoid this, off-diagonal matrix elements should be included in consideration. Braig and Brouwer suggested to modify the rate equations including these terms as a density matrix for the interfering subset of levels [85], but this dramatically lengthens

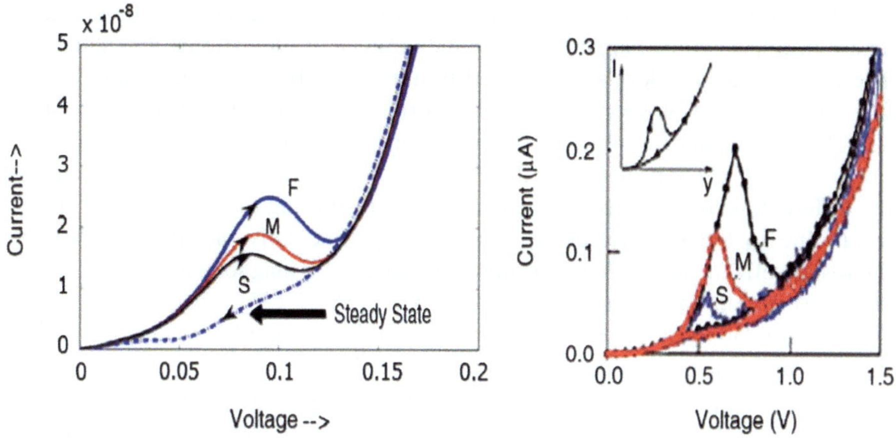

Fig. 3.18 *Left panel*: current–voltage characteristics for the electron transport through a quantum dot demonstrating the NDR due to trapping of charge by "dark states" [284]. *Right panel*: the charge storage caused NDR observed a metal–molecule–metal junction [285]. Reprinted with the permission of the APS

computations and makes them more cumbersome. Also, the range of applications of master equations is limited, as they could not be used to study transport through sufficiently strongly coupled junctions.

3.9 Transport Through Magnetic Molecules

Currently, electron transport through magnetic molecules attracts a significant interest of the research community [36,37,147–150]. Magnetic molecules are interesting for applications that combine spintronics with molecular electronics. Magnetic molecules (also called molecular magnets) are the molecules characterized by a nonzero net spin **S**. The molecular net spin magnitude may take on a significant value. For instance, a single Mn_{12} molecule, which is a prototype of molecular magnets, in the neutral state can possess S as large as 10. Usually, the net spin appears on the molecule due to the presence of several transition metal ions included in the molecule and magnetically coupled to each other [144, 145]. The molecular Hamiltonian for magnetic molecules contains a special term depending on S. The simplest expression for this term is [119, 148]:

$$H_s = DS_z^2 + g_s \mu_B B_z S_z \tag{3.49}$$

where D ($D < 0$) is the splitting constant, g_s is the spin-splitting factor, μ_B is the Bohr magneton, and B_Z is "z" component of the external magnetic field(z axis being the intrinsic anisotropy axis of the molecule). The first term in the Hamiltonian (3.49) is responsible for lifting of the degeneracy of molecular spin-multiplets, which happens even in the absence of the external magnetic field. The resulting energy levels are labeled by the magnetic quantum number M_s. A few years ago, it was shown in the transport experiments on molecular junctions [27, 28] that Zeeman splitting of the molecular energy levels could be manifested in the splitting of the peaks in the differential conductance dI/dV. Fine structure of the differential conductance originating from the splitting of the molecular energy levels due to the net molecular spin was experimentally observed [36]. Other interesting effects observed in transport experiments on magnetic molecules include negative differential resistance and spin blockade [36,37,125]. Also, one may expect spin accumulation at the biased junctions [148] and significant current-induced magnetization changes in magnetic monolayers [286].

Special transport properties of MMM junctions including magnetic molecular bridges are mostly determined by the cooperation of two factors. The first factor is the above-discussed splitting of the molecular spin-multiplets. Assuming an axial symmetry of the molecule magnetization with respect to "z" axis, one may expect the molecule to remain in a certain M_s state for a long time. However, deviations from this perfect axial symmetry give rise to perturbations in the magnetization, which lead to the mixing of various M_s states creating opportunities for transitions [144]. These transitions are controlled by selection rules depending on the particular symmetry of the disturbance, which makes them to occur. For instance, a small nonuniaxial perturbation of the form

$$\delta H = E\big(S_x^2 - S_y^2\big) \tag{3.50}$$

prohibits transitions between the magnetic states other than $M_s \rightarrow M_s \pm 2n$ where n is an integer. The second important factor is the sensitivity of the molecular spin to the electric charge on the molecule. Under certain conditions, a molecular spin could be switched from zero to maximum possible value by a relatively small extra charge put on the molecule [37].

The negative differential resistance (also called negative differential conductance in the current literature) in the magnetic molecules may appear because of the spin-selection rules prohibiting certain transitions between the molecular states. Consequently, the sequence of electron tunnelings between the sites of the magnetic molecular bridge may result in a nonequilibrium population of certain excited states. Once populated, states could be depopulated only very slowly by violation of the selection rules. This leads to hindering of the electron transport and is manifested as NDR features in the current–voltage curves, such as those observed on Mn_{12} molecules. The decrease in the current near the corresponding values of the bias voltage may be very pronounced, and in some cases the total current suppression may happen [36]. Another mechanism controlling the occurrence of negative differential resistance was suggested by Timm and Elste [148]. They showed that NDR features may appear in the current–voltage curves even though there is no blocking state where the molecule could be trapped due to the spin-selection rules. The NDR may occur simply as a consequence of the difference in

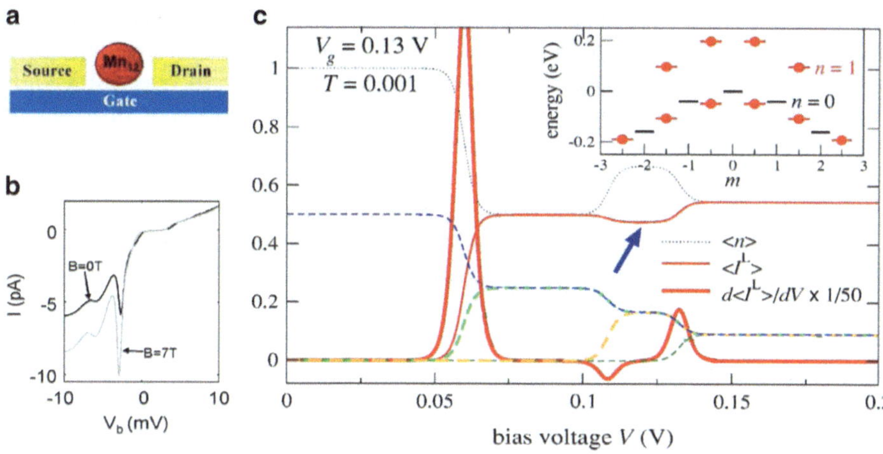

Fig. 3.19 *Left panel*: (**a**) schematic drawing of a MMM junction including Mn_{12} as the linker. A gate electrode controls the electrostatic potential on the molecule. (**b**) The current versus bias-voltage curves measured for this junction and revealing the NDR features which become better pronounced in the presence of the external magnetic field [36]. *Right panel*: (**c**) occupation, steady-state current in units of e/τ_0, dI/dV in units of e/τ_0 per volt and probabilities of various states (*dashed*) for a MMM junction including a weakly coupled magnetic molecule. The curves are plotted assuming $V_g = 0.013\,\text{V}$, $T = 0.001\,\text{K}$, and$\tau_0 = 1.6\,\text{ns}$ where τ_0 is the transition time. The *arrow* indicates the plateau with reduced current. *Inset*: molecular energy levels of states with the occupation number $n = 0$ (*black bars*) and $n = 1$ (*red bars with circles*) and magnetic quantum numbers m [148]. Reprinted with the permission of the APS

the transition rates for the allowed transitions between various molecular states. The same mechanisms could cause the spin blockade, that is, the suppression of electron current through the junction at low bias voltage. These NDR features as well as the spin blockade are shown in Fig. 3.19.

The charge sensitivity of the molecular spin is a strong correlation effect due to Nagaoka mechanism [287]. In the context of band magnetism, the efficiency of the Nagaoka mechanism is limited because of its dependence on the lattice geometry. However, in single-molecule devices, the efficiency of this mechanism increases, for advanced supramolecular structures reproducing the desirable "lattice types" could be designed and used in experiments. Therefore, the molecular spin state may be highly sensitive to the electric charge on the molecule. It was shown that the charge sensitivity of the net molecular spin could be observable in the electron tunneling current through magnetic molecules. Moreover, the spin properties of the molecule can be electrically controlled [288–290]. As was already mentioned, under certain conditions, the molecular spin could be switched from zero to maximum possible magnitude by a few extra electrons put on the molecule. A number of electrons on the molecule may be electrostatically controlled by means of the gate voltage applied to the junction, which opens up an opportunity to design a spin-switching MMM junction. Such junctions may be used as elements in building magnetic memory nanodevices.

The potentialities of magnetic molecules as memory elements have been discussed since the early 1990s [291–295]. To actualize these potentialities, it is crucial to provide means for the stored information to persist over sufficiently long time. This could be accomplished using easy axis anisotropies of magnetic molecules, which introduce energy barriers for the change of the spin direction. When such anisotropic magnetic molecule is prepared in a spin-polarized state, its relaxation to the steady state with a zero spin may be rather slow. The reason is that the relaxing molecule must pass through a sequence of intermediate states allowed by the spin-selection rules, and some of these states could be characterized by higher energies than those preceding them in this sequence. In these cases and at low values of the bias voltage, the transitions between the magnetic states need thermal activation to happen. Correspondingly, the magnetic relaxation time becomes inversely proportional to the temperature. Therefore, the relaxation is forbidden in the limit $T \to 0$, and it develops very slowly at low temperatures. So, the slow decay of magnetic states of the molecule could be provided for. Due to the same reason, the transmitted spin may be accumulated at the leads of weakly coupled MMM junctions with magnetic molecular linkers when the applied bias-voltage magnitude exceeds the Coulomb blockade threshold but remains rather close to the latter. Then the spin-polarized electron tunneling current may transmit spin through the molecular bridge for an exponentially long (in $\Delta E/k_{\mathrm{B}}T$, where ΔE is the barrier between the adjacent states) time, and this results in a giant spin amplification at the leads at low temperatures [148]. Some other interesting effects could occur when the junction has ferromagnetic leads (see, e.g., [149, 150]).

Theoretical studies of electron transport through MMM junctions with magnetic linkers based on model spin Hamiltonians such as given by Eqs. (3.49) and (3.50) certainly give an insight into physical mechanisms controlling the variety of observable effects. However, the results of these studies crucially depend on the specific choice of the parameters included in the adopted model. Therefore, the first-principles computations of the electron structure of magnetic molecules are very important. Among others, questions concerning the charged states of a biased molecule, the parameters characterizing electron–electron interactions, and the interactions between conduction electrons and ions of transition metals providing a nonzero spin of a magnetic molecule and the electrical response of a whole MMM junction could be answered by means of *ab initio* calculations. In the case of magnetic molecules, these computations are very complicated due to the interplay of various factors including the exchange interactions, which control the net spin on the molecule, the spin–orbit interactions, and Coulomb repulsion of electrons. Even a simple evaluation of the ground state of a magnetic molecule is a nontrivial problem [151, 152]. Nevertheless, first-principles calculations of the electron structure of magnetic molecules and the related characteristics are being carried out [151–155]. They are further discussed in the following chapters.

3.10 Kondo Effect

The Kondo effect has been a focus of condensed matter research for many years. It was first manifested in transport experiments on nonmagnetic metals such as Cu, Ag, and/or Au where magnetic impurities (e.g., Mn, Fe, or Co) were inserted. The effect revealed itself as an anomalous temperature dependence of resistivity at low temperatures leading to a resistivity minimum [296]. In 1964, the seminal work of J. Kondo [297] suggesting a theoretical explanation of the observed anomalies was published, and the effect derives its name from this work ever since. Kondo explained that the above-mentioned low-temperature anomalies in resistivity originated from an exchange coupling between the spins of localized magnetic impurities and delocalized conduction electrons. In the following years, the Kondo effect was considered as a paradigm for the interplay between the localized and delocalized degrees of freedom, and presently it is one of the best studied problems in the condensed matter physics.

It was first argued by Glazman and Raikh [298] and Ng and Lee [299] that since the Anderson Hamiltonian describing a Kondo impurity as a set of discrete, interacting levels coupled to a Fermi sea of conduction electrons could be applied to a system consisting of a quantum dot coupled to the pair of conducting leads, the Kondo effect must be manifested in such systems, as well. Afterwards, the Kondo peak in the conductance was repeatedly observed in experiments on low-temperature electron transport through semiconductor quantum dots [300–303] and MMM junctions [27–33]. An example is presented in Fig. 3.20. In these systems, the competition between the Coulomb repulsion of electrons on the dot/molecular bridge and band gibridization produces the Kondo effect: a crossover from weak to strong coupling between the localized and band electrons. The Kondo peak in the conductance may appear within a low-temperature regime when the ambient temperature is lower than the characteristic Kondo temperature T_K.

As the Coulomb interactions between the electrons on the quantum dot/molecular bridge are crucial for both the Kondo peak in the conductance and the Coulomb blockade to be manifested, one may theoretically analyze these effects using the same computational technique based on the NEGF [34, 35]. Employment of such complicated methods may seem excessive to study the Coulomb blockade alone. As discussed in Sect. 3.8, theoretical studies of the electron transport through a quantum dot/molecule within the Coulomb blockade regime mostly employ "master" equations for the transition rates between the states differing by a single electron [75, 83, 281–284]. While the approach based on the transition rates equations brings sound results within the Coulomb blockade regime, its generalization to the case of stronger coupled junctions is not justified. The Kondo peak appears beyond the Coulomb blockade regime, when the coupling of the leads to the bridge becomes comparable to the characteristic charging energy U_0, thus indicating that the coupling of the electrons on the bridge with the electrons on the leads strengthens. The theory of Kondo effect in quantum dots (which is equally suitable for MMM junctions) was first suggested by Meir, Wingreen, and Lee [86, 87] who

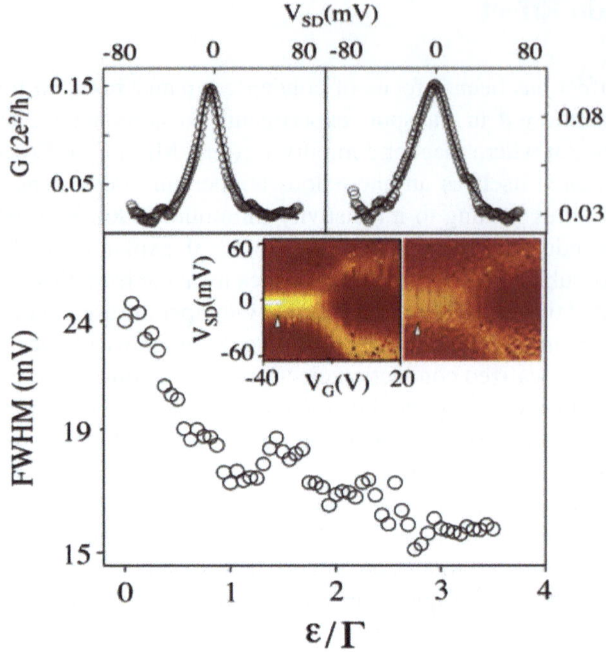

Fig. 3.20 Kondo physics observed in a metal–molecule–metal junction with a fullerene C_{60} as the linker. Reprinted from [31]. Reprinted with the permission of the ACS

used the equations of motion method to compute the relevant Green's functions. Several theoretical works followed (see, e.g., [88–92]). However, the theoretical analysis of the Kondo effect is not completed so far.

An example illustrating the equations of motion (EOM) method is given in Sect. 2.5. This technique enables us to express the relevant Green's functions in terms of higher-order Green's functions. Writing out EOM for the latter, one arrives at the infinite sequence of the equations successively involving Green's functions of higher orders. To get expressions for the necessary Green's functions, this system of EOM must be truncated. Also, higher-order Green's functions still included in the remaining EOM must be approximated to express them by means of the lower-order Green's functions. In outline, this procedure is well known and commonly used. However, in numerous existing papers important details of the above procedure vary, which brings different approximations for the relevant Green's functions. Accordingly, the results of theoretical studies of the quantum dot response vary.

It is a common knowledge that within the Hartree–Fock approximation, one cannot catch the Kondo peak. To quantitatively describe this effect one may need to include hybridization up to very high orders in calculation of the lead–bridge coupling parameters. Obviously, the Green's functions suitable to reveal the Kondo peak in the electron DOS on the dot/molecular bridge must provide the proper description of the transport through the bridge within the Coulomb blockade regime when lead–bridge couplings are weak. However, usually NEGF-based results beyond the Hartree–Fock approximation successfully employed to describe the Kondo effect and related phenomena are not applied to the limiting case of the Coulomb blockade. In those works where such application was carried out, the correct ratio of heights of the subsequent current steps, namely, **2:1**, was not obtained. For instance, Muralidharan et al. [83] reported the results of NEGF calculations which give equal heights of the two steps on the current–voltage curve within the Coulomb blockade regime, and studies of Galperin et al. [34] resulted in the height ratio of about 1.6:1. So, up to present, there exists a discrepancy between NEGF- and rate equation-based results. This discrepancy may not be easily disregarded, for its existence makes questionable general and useful results obtained in studies of the Kondo effect and based on the advanced techniques within the NEGF formalism. The above disagreement was discussed by Zimbovskaya [35]. In this work the EOM-based approach was employed to get a description of both Coulomb blockade and Kondo regimes in the electron transport through a single-site molecular bridge using identical approximations for the relevant Green's functions. These approximations were proven to give the correct ratio of the heights of the steps in the current–voltage characteristics within the Coulomb blockade limit (2:1 for a symmetrically coupled junction). The Kondo peak in the conductance was described, as well.

It seems reasonable to start the description of the Kondo effect manifestations in the transport properties of molecular junctions from the simplest case. Therefore, one may tentatively omit from consideration electron–phonon interactions which may bring significant changes to the conductance through the junction, and one may assume that the leads are made out of nonmagnetic materials. Also, one represents the bridge linking the leads by a single spin-degenerated energy level. Then the electron density of states on the bridge is given by Eq. (2.24). Using $G_\sigma^r(E)$ in the form (2.118), one may compute the electron DOS. The results are presented in Fig. 3.21 where the equilibrium DOS is shown for three values of the coupling strength τ. For a sufficiently strong coupling of the bridge to the source and drain reservoirs ($\tau = 0.2\,\text{eV}$) the sharp Kondo peak appears at $E = 0$, and the peaks at $E = E_0$ and $E = E_0 + U$ are damped. At weaker coupling strength ($\tau = 0.1,\ 0.07\,\text{eV}$), the Kondo peak is reduced to a tiny feature, but the maxima at $E = E_0$ and $E = E_0 + U_0$ which determine the conductance within the Coulomb blockade regime emerge. The weaker is the coupling, the higher become these peaks. The heights of the peaks differ. Technically, this distinction originates from the fact that the value of $\langle n_{-\sigma} \rangle$ in the expression for the retarded Green's function differs from 0.5. The physical reason of this difference in the peak heights is the same as for the discussed above difference in heights of the steps on the current–voltage characteristics within the Coulomb blockade regime. When the voltage is applied across the junction, the Kondo peak splits in two maxima whose heights are significantly smaller than the height of the original equilibrium Kondo peak, and the greater is the voltage, the lower these maxima become. This is shown in the right panel of Fig. 3.21.

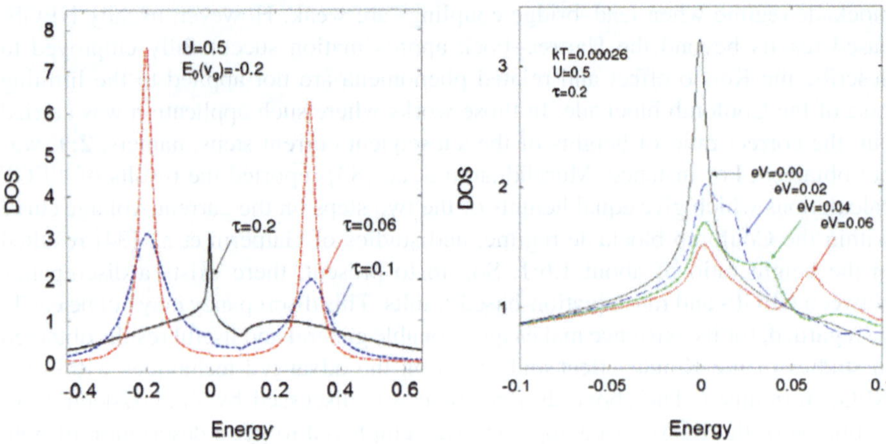

Fig. 3.21 *Left panel*: the equilibrium ($V = 0$) electron density of states on the single-state bridge. The curves are plotted for $\tau = 0.2\,\text{eV}$ (*solid line*) $0.1\,\text{eV}$ (*dashed line*) and $0.06\,\text{eV}$ (*dash-dotted line*), $k_B T = 0.00026\,\text{V}$. *Right panel*: splitting of the Kondo peak in the electron density of states with increasing bias. The curves are plotted for $\text{eV} = 0$ (*solid line*); $0.02\,\text{eV}$ (*dashed line*), $0.04\,\text{eV}$ (*dash-dotted line*) and $0.06\,\text{eV}$ (*dotted line*). The coupling strength $\tau = 0.2\,\text{eV}$ (after [35]). Reprinted with the permission of APS

If the normal (nonmagnetic) leads in the junction are replaced by ferromagnetic ones, spin degrees of freedom start to play a significant part in modifying transport properties [88, 92, 303–308]. It has been shown that the Kondo anomaly in the conductance through the symmetrically coupled junction with the ferromagnetic leads may be suppressed if the magnetic moments of the leads are parallel to each other [304–308], whereas for antiparallel configuration of the magnetic moments on the leads, the suppression of the Kondo peak does not occur. This effect was confirmed in experiments on molecular junctions (see, e.g., [303]).

Suppression of the Kondo peak may occur not only for parallel configuration of the magnetic moments but also in the case when they make a nonzero angle. Such noncollinear configuration is especially interesting, for it gives a detailed information on the variation of the effect as the angle between the magnetic moments of the leads varies. The analysis carried out in [308] shows that the Kondo peak becomes suppressed to a considerable extent at small deviations from the antiparallel configuration. For greater deviations the Kondo anomaly varies significantly slower as the angle between the magnetic moments changes.

To theoretically analyze the effect of the ferromagnetic leads on the Kondo peak in the conductance, one may start from the Hamiltonian for the molecule in the form (1.30) which describes a single-site bridge. One must keep in mind that the magnetic moments on the leads lift the spin degeneration of the bridge energy level; therefore, in Eq. (1.31) the energy E_0 is to be replaced by $E_{0\sigma}$. Different ways to describe the spin splitting of the bridge level were suggested (see, e.g., [304, 305, 308]), but they do not bring contradicting results. The most important effect of the ferromagnetic leads to the transport properties of a MMM junction manifests itself in dramatic changes happening to the coupling of the leads to the bridge. Omitting for simplicity the spin–orbit interactions, one may present the transfer Hamiltonian describing the coupling of the single-site bridge to the leads in the form [308]:

$$H_T = \sum_{r\beta} \sum_{s\sigma} \tau^*_{r\beta s\sigma} c^\dagger_{r\beta s} c_\sigma + H.C. \tag{3.51}$$

Here, operator $c^\dagger_{r\beta s}$ creates an electron with the spin s on the lead β ($s = \pm 1$ for the majority and minority electrons, respectively) and the operator c_σ annihilates an electron with the spin σ on the bridge ($\sigma = \uparrow, \downarrow$ along the quantization axis of the bridge). The coupling parameters $\tau^*_{r\beta s\beta}$ may be presented as matrix element of the matrix $\tau_{r\beta *}$:

$$\tau^*_{r\beta} = \begin{pmatrix} \tau^*_{r\beta +} \cos\left(\frac{\theta_\beta}{2}\right) & -\tau^*_{r\beta +} \sin\left(\frac{\theta_\beta}{2}\right) \\ \tau^*_{r\beta -} \sin\left(\frac{\theta_\beta}{2}\right) & \tau^*_{r\beta -} \cos\left(\frac{\theta_\beta}{2}\right) \end{pmatrix} \tag{3.52}$$

where θ_β is the angle between the spin quantization axis in the lead β and the spin quantization axis on the bridge. The matrix elements $\tau_{r\beta o}$ describe coupling strengths for an electron tunneling to the lead β (where it joins the majority/minority electrons already residing there) computed assuming $\theta_\beta = 0$. The directions of the spin quantization axes in the ferromagnetic leads are determined by the orientations of their magnetic moments. As for the local quantization axis for the bridge, its direction is governed by the exchange magnetic field which is created by the leads. For a symmetrically coupled and unbiased junction $\theta_R = -\theta_L = \theta/2$, θ being the angle between the magnetic moments of the leads.

As follows from this expression, the orientations of the leads magnetic moments strongly affect the electron transport through the junction. Correspondingly, the self-energy terms describing the coupling of the molecule to the leads accept different forms for different spin orientations, and they depend on the leads magnetization directions, namely,

$$\begin{aligned} \Gamma^\beta_{\uparrow\uparrow} &= \Gamma_{\beta +} \cos^2\left(\frac{\theta_\beta}{2}\right) + \Gamma_{\beta -} \sin^2\left(\frac{\theta_\beta}{2}\right), \\ \Gamma^\beta_{\downarrow\downarrow} &= \Gamma_{\beta +} \sin^2\left(\frac{\theta_\beta}{2}\right) + \Gamma_{\beta -} \cos^2\left(\frac{\theta_\beta}{2}\right), \\ \Gamma^\beta_{s,-s} &= \frac{1}{2}(\Gamma_{\beta +} - \Gamma_{\beta -}) \sin(\theta_\beta) \end{aligned} \tag{3.53}$$

where the terms $\Gamma_{\beta s}$ ($s = \pm 1$) are given by the expressions similar to Eq. (2.100):

$$\Gamma_{\beta s}(E) = \pi \sum_r |\tau_{r\beta s}|^2 \delta(E - \epsilon_{r\beta s}). \tag{3.54}$$

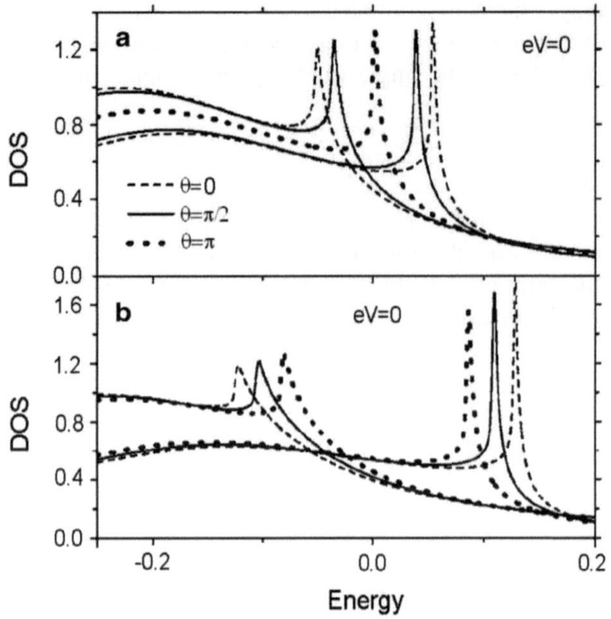

Fig. 3.22 Electron density of states on the single-level quantum dot in the Kondo regime for a symmetrically (**a**) and asymmetrically (**b**) coupled junction. A single Kondo peak typical for the antiparallel orientation of the magnetic moments on the leads ($\theta = \pi$) becomes split in two when the angle θ deviated from π (after [308]). Reprinted with the permission of APS

Using these expressions to compute the electron density of states on the molecule, one may prove that in the case of antiparallel configuration of the magnetic moments, the Kondo peak appears at the Fermi energy for both spin orientations of the electrons on the bridge. The resulting anomaly is similar to that occurring in the conduction of a junction with nonmagnetic leads. This is a consequence of the mutual compensation of the effective magnetic fields induced by the leads on the electrons on the bridge [92, 308]. As this particular magnetic configuration results in the disappearance of the spin splitting of the bridge energy level, the transport properties of the junction lose their specific features originating from the magnetization of the leads.

When the configuration departs from the antiparallel one, the spin splitting drifts apart Kondo anomalies in the electron density of states for the different spin orientations. The single Kondo maximum splits in two peaks corresponding to spin-up and spin-down electrons. As the angle between the magnetic moments varies from π (antiparallel configuration) to zero (parallel configuration), the peaks drift farther apart from each other and from the Fermi energy, thus bringing the suppression of the Kondo anomaly, as shown in Fig. 3.22. The described situation resembles that created by an external magnetic field applied to a nonmagnetic system, when a finite Zeeman splitting of the bridge level causes the splitting of

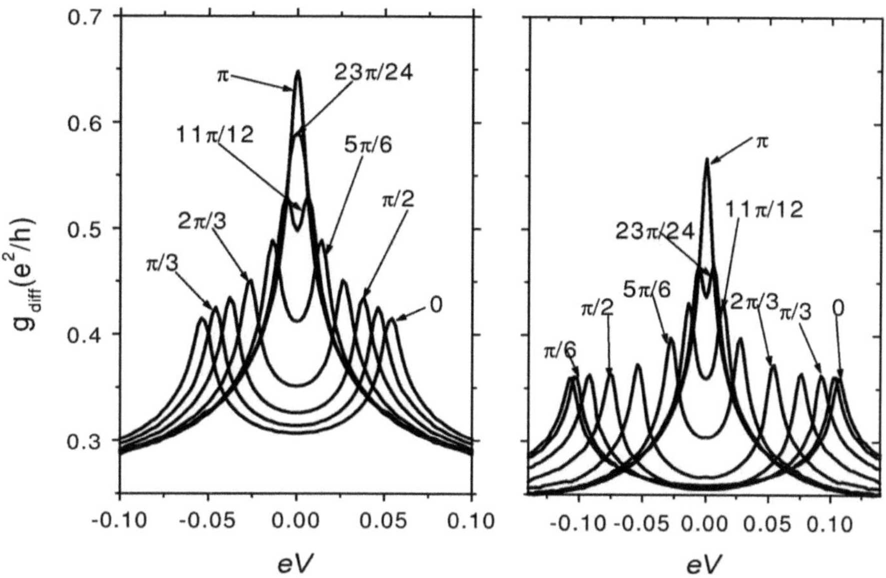

Fig. 3.23 Bias dependence of the Kondo anomaly in differential conductance dI/dV computed for several indicated values of the angle θ between the magnetic moments in the electrodes and for two different values of the spin polarization parameter P (after [308]). Reprinted with the permission of the APS

the Kondo peak. The described changes in the Kondo anomaly arising due to the magnetic moments on the leads and the corresponding magnetic field affecting the electron states on the bridge are manifested in the behavior of the differential conductance through the system, as well. The differential conductance dI/dV versus the bias voltage is shown in Fig. 3.23. The curves presented in this figure are plotted for several values of the angle θ between the magnetic moments of the leads and for two different values of the spin polarization parameter P. Again, one observes that antiparallel configuration of the magnetic moments ($\theta = \pi$) corresponds to a well-defined zero-bias anomaly in the conductance. As the angle θ decreases, the single Kondo peak in the conductance becomes split in two peaks of reduced intensities which appear at nonzero-bias voltage. The smaller the angle θ is, the farther apart these peaks are situated and the lower they become.

An interesting effect could appear if one applies an alternating voltage to the gate electrode, thus moving back and forth the bridge energy level [309]. As known, the Kondo peak in the conductance through a MMM junction or quantum dot may appear when the temperature does not exceed the Kondo temperature. The latter is given by the expression

$$k_B T_k \sim \sqrt{\frac{D\Gamma}{4}} \exp\left[-\frac{\pi|E_0|}{\Gamma}\right]. \tag{3.55}$$

Here, D equals to half bandwidth of the conduction electrons on the electrodes, and Γ characterizes the coupling of the latter to molecular bridge or the dot. As follows from this approximation, one can change T_k by changing the bridge level energy E_0. If E_0 is moved sufficiently far away from E_F, the Kondo anomaly in the conductance should disappear. Therefore, the application of an ac voltage to the gates may give rise to oscillations of the conduction, originating from the periodic changes in the Kondo temperature. The period of these oscillations is equal to the driving frequency. Also, the conductance shows an interesting behavior when the bias voltage is applied across the junction in addition to the oscillating voltage on the gate electrode [91]. As the bias increases, the time-average conductance decreases, which is consistent with the quenching of the Kondo peak by the bias voltage. However, when the bias voltage approaches the value $\hbar\Omega/|e|$ (Ω being the driving frequency), the conduction enhances, thus deviating from the monotonous decreasing behavior.

Experimental data on the electron transport through quantum dots and MMM junctions as well as theoretical research (see, e.g., [8, 22, 27–34, 91, 279]) clearly indicate that vibrational degrees of freedom could significantly affect the Kondo effect. However, these issues are beyond the scope of the present section. Some of them will be described in the following chapter along with other effects originating from electron–phonon interactions.

3.11 Kondo Effect in Nonequilibrium Quantum Dots

Almost all properties of the Kondo effect in equilibrium systems are well understood. However, the situation is very different when one is exploring how Kondo physics is affected by external currents flowing through a Kondo system. As discussed before, MMM junctions and quantum dots sandwiched in between conducting leads belong to this kind of systems along with metal–insulator–metal tunnel junctions involving electron tunneling via magnetic impurities and semiconductor–metal (Schottky) junctions. When the quantum dot is in equilibrium, a resonance state may be formed. This state is manifested as the Kondo maximum in the conductance of an unbiased system which was discussed in the previous section. The width of this peak is determined by the Kondo temperature T_K. When the finite bias voltage V is applied across the dot, it breaks down the Kondo peak provided that $|e|V > k_B T_K$. It had been argued that this happened due to the effect of inelastic processes in the leads associated with a current flowing through the system [36]. These processes exist even at low temperatures, and they destroy the coherence of electron transport through the system, thus preventing the full formation of the Kondo maximum. In sufficiently highly biased systems, the signature of Kondo effect is a logarithmic variation of diverse observables with the bias voltage. This logarithmic behavior was first explained by Kondo himself [297] within the framework of perturbation theory. The effect of perturbations was taken into account to lowest order in the exchange coupling between the electron on the dot and those on the leads. Such behavior was repeatedly observed in experiments on various Kondo systems including quantum dots.

Theoretically, Kondo effect out of equilibrium was studied using several approaches such as different forms of perturbation theory, equations of motion, and self-consistent diagrammatic techniques. Some of these methods were and still are successfully used to explore properties of Kondo systems in equilibrium. However, generalization of these methods in such a way that could make them appropriate to describe the Kondo effect out of equilibrium often meets with serious difficulties. The root of these difficulties is the nature of current-carrying states of Kondo systems. These are highly excited many-body states. Therefore, theoretical methods which are focused by construction on ground-state properties of these systems cannot be easily adapted to describe their nonequilibrium behavior [37].

Theoretical studies of Kondo effect had brought a remarkable result revealing its universal feature. It was shown that a single characteristic energy $k_B T_K$ determines the Kondo physics. As early as in 1970, Anderson suggested an efficient way known as perturbative normalization group which allows to sum up the leading logarithmic terms appearing as a result of application of perturbation theory and to establish this universal scaling behavior [310]. Further development of this approach [36, 37, 311–314] provides a complete and controlled description of transport properties of Kondo systems out of equilibrium.

It had been suggested that the immediate reason of washing out the Kondo peak in conductance through biased quantum dots is cutting off the exchange coupling between the dot and the leads. Even at sufficiently high values of the bias voltage, the cutoff may still remain within the perturbative regime. This makes the perturbation theory an appropriate method for studies of nonequilibrium Kondo physics. A "bare" perturbative approach may be developed starting from the Anderson Hamiltonian for the considered system. Modeling a quantum dot or molecular bridge in a MMM junction by a single spin-degenerated level, one can present the Hamiltonian as the sum of terms given by Eqs. (1.30), (1.31), and (1.34). Following [36], we assume that the dot/molecule is symmetrically coupled to the leads ($\tau_{kr\beta\sigma} = t$) and the dot level E_0 is situated below the chemical potentials of the leads in the absence of the bias. For certainty, we put $\mu_L = \mu_R = 0$, so E_0 takes on a negative value. Also, we assume that the difference $\mu_\beta - E_0$ significantly exceeds a characteristic energy Γ whose value is determined by the decoherence rate in the leads. Within the adopted model, this means that inequality $|E_0| \gg \Gamma$ ($\Gamma = 2\pi N_0 t^2$) is satisfied where N_0 is the electron density of states on the leads. Under these conditions and at strong Coulomb repulsion ($U_0 \to \infty$), the number of electrons on the dot takes on values close to 1. In the considered regime, the original Hamiltonian for the junction may be reduced to the form [36]

$$H_k = \sum_{k,\beta,\sigma} (\epsilon_k - \mu_\beta) c_{k\beta\sigma}^\dagger c_{k\beta\sigma} + V_0 \sum_\sigma (c_{L0\sigma}^\dagger + c_{R0\sigma}^\dagger)(c_{L0\sigma} + c_{R0\sigma}) + J \sum_{\beta,\beta'} \mathbf{S} \cdot s_{\beta\beta'}. \tag{3.56}$$

Here,

$$\mathbf{S}_{\beta\beta'} = \frac{1}{2} \sum_{k,k',\sigma,\sigma'} c^{\dagger}_{k\beta\sigma} \mathbf{S}_{\sigma\sigma'} c_{k'\beta'\sigma'}, \tag{3.57}$$

$$\mathbf{S} = \frac{1}{2} \sum_{\sigma,\sigma'} d^{\dagger}_{\sigma} \mathbf{S}_{\sigma\sigma'} s_{\sigma'}, \tag{3.58}$$

$$c_{\beta 0\sigma} = \sum_{k} c_{k\beta\sigma}, \tag{3.59}$$

$\mathbf{S}_{\sigma\sigma'}$ is the vector of Pauli matrices, vector $\mathbf{S}$ represents the spin on the quantum dot, $\beta, \beta' = L, R$, $J = 4V_0 = 2t^2/E_0$ and $\mu_{L,R} = \pm|e|V/2$. Now, the second term in the Hamiltonian H_K describes scattering processes whose intensity is characterized by the potential V_0. The last term in Eq. (3.56) is associated with the coupling between the dot and the leads. Within the accepted approximation, it is described as the coupling of the dot spin $\mathbf{S}$ to the spin-related vectors $\mathbf{S}_{\sigma\sigma'}$ which are associated with the leads. The coupling constant J characterizes the strength of this exchange interaction. For sufficiently small J the scattering term in Eq. (3.56) may be neglected.

One may use the Hamiltonian H_K to compute the relevant Green's functions employing non-crossing approximation (NCA) which corresponds to the first order of self-consistent perturbation theory with respect to the coupling parameter t. In spite of certain shortcomings concerning the analysis of low-energy states of Kondo systems within the limit of zero temperatures, the NCA brings satisfactory results in equilibrium for finite temperatures down to a fraction of T_K. Employing this perturbation theory-based approach one may obtain approximations for the exchange coupling parameter J and for the nonequilibrium decoherence rate γ. Both approximations include the energy dependence of the effective coupling constants, namely [36],

$$\begin{aligned} N_0 J_{\text{NCA}}(E) = & \left\{ \ln\left[\frac{E - |e|V/2}{k_{\text{B}}T_K} + \frac{\hbar\gamma}{k_{\text{B}}T_K}\right] \right. \\ & \left. + \ln\left[\frac{E + |e|V/2}{k_{\text{B}}T_K} + \frac{\hbar\gamma}{k_{\text{B}}T_K}\right] \right\}^{-1}, \end{aligned} \tag{3.60}$$

$$\gamma_{\text{NCA}}(E) = \frac{\pi}{8} \frac{|e|V}{\ln^2\left[\dfrac{|e|V}{2k_{\text{B}}T_K}\right]} \left(1 + \frac{2}{\ln\left[\dfrac{|e|V}{2k_{\text{B}}T_K}\right]}\right). \tag{3.61}$$

These approximations are appropriate at low temperatures ($T \ll T_K$) assuming that the bias voltage is sufficiently high $\left(\ln[|e|V] \gg k_{\text{B}}T_K\right)$, but it remains low enough to satisfy the relation $|e|V \ll D$. In this relation, D stands for the band width for the leads which takes on the part of high-energy cutoff. The perturbative expression for the effective exchange coupling parameter (3.60) reveals poles near $E = \pm|e|V/2$ provided that the decoherence rate γ is smaller than $(k_B T_K)^2/\hbar|e|V$. The breakdown scale $k_B T^*$ $(T^*/T_K \approx k_B T_K/\hbar|e|V)$ indicates that approximation (3.60) is justified only provided that $k_B T^* < \hbar\gamma < |e|V$. Within the NCA, this criterion is satisfied, as follows from Eq. (3.61). Using Eqs. (3.60) and (3.61), one must keep in mind that these analytical expressions are justified only within the limit $\ln(|e|V/k_B T_K) \gg 1$. For lower values of the bias voltage, the NCA equations must be solved numerically.

Remaining within the high bias limit, one obtains the following approximation for the conductance through the junction [36]:

$$\frac{g_{\mathrm{NCA}}}{G_0} = \frac{1}{\ln^2\left[\frac{|e|V}{2k_B T_K}\right]}\left(1 + \frac{2}{\ln\left[\frac{|e|V}{2k_B T_K}\right]}\right). \tag{3.62}$$

The results (3.60)–(3.62) derived within a simple self-consistent perturbation theory already bring logarithmic terms characterizing nonequilibrium Kondo physics. However, this approach leads to inaccurate results for the prefactors of both γ_{NCA} and g_{NCA}. In part, these inaccuracies originate from the properties of the original Anderson Hamiltonian which treats the scattering potential V_0 and the exchange coupling strength J on equal footing. To avoid these errors, one needs to emphasize the difference between these parameters. For this purpose, the Hamiltonian (3.56) should be replaced by another Hamiltonian given by [37, 312, 313]:

$$H_K = \sum_{k,\beta,\sigma} (\epsilon_k - \mu_\beta) c^\dagger_{k\beta\sigma} c_{k\beta\sigma} + \sum_{\beta,\beta'} J_{\beta'\beta} \mathbf{S} \cdot s_{\beta\beta'} \tag{3.63}$$

This Hamiltonian cannot be derived from the simple Anderson model leading to Eq. (3.56). It can be reduced to the latter by assuming that $J_{LR} = J_{RL} = \sqrt{J_{LL} J_{RR}}$ [37]. In a general case $J_{\beta'\beta}$ can be treated as an arbitrary symmetric matrix of 2×2 dimensions. In the presence of an external magnetic field $\mathbf{B} = (0, 0, B)$, an extra term proportional to BS_z appears in the Hamiltonian. In principle, a further cotunneling term which is an analog of the second term on the right-hand side of Eq. (3.56) should appear in Eq. (3.63), as well. However, such a term could be safely omitted for it does not strongly contribute to the results obtained using the perturbative renormalization group approach [37]. Now, instead of a single coupling parameter $N_0 J$, one may introduce two dimensionless coupling constants (for a symmetrically coupled dot) namely, $g_d = N_0 J_{LL} = N_0 J_{RR}$ and $g_{LR} = N_0 J_{LR} = N_0 J_{RL}$.. These parameters could take on different values depending on specific properties of the considered system.

Another and even more important root of the discussed inaccuracies lies in the low efficiency of the "bare" perturbative approach when applied to Kondo systems. Even for small coupling parameters g_d and g_{LR} and sufficient values of the bias voltage (or external magnetic field magnitude) when the inequalities $(g_d, g_{LR}) \ln D/|e|V < 1$, $|e|V \ll D$ are satisfied, the perturbation theory converges very slowly. To get accurate results one needs to sum the leading logarithmic contributions up to high-order terms. Only after such resummation is carried out, the universality of Kondo model which lies in the fact that its whole physics depends on a single energy scale indicated by Kondo temperature becomes clearly manifested [37]. The perturbative renormalization group approach gives a way to efficiently sum up the leading logarithmical corrections. This method is based on a fundamental idea that small changes in the cutoff energy D may be accounted for by renormalization of coupling parameters g_d and g_{LR}. While a coupling parameter remains small, its change indicated by an infinitesimal change of D, $\partial g_{d,LR}/\partial(\ln D)$ may be computed using the perturbation theory.

Within a very simplified version of the renormalization group perturbative approach, the behavior of coupling parameters is determined by the system of equations [36]:

$$\frac{dg_D}{d(\ln D)} = -\left(g_D^2 + g_{LR}^2\right);$$
$$\frac{dg_{LR}}{d(\ln D)} = -2g_D g_{LR} \tag{3.64}$$

which are justified at $|e|V \ll D$. For a strongly biased quantum dot ($|e|V \gg k_B T_K$), these equations remain appropriate down to $D \sim |e|V$, and one obtains

$$g_d = \frac{1}{2}\left\{\left(\ln\left[\frac{|e|V}{k_B T_K}\right]\right)^{-1} + \left(\frac{1}{\alpha_0} + \ln\left[\frac{|e|V}{k_B T_K}\right]\right)^{-1}\right\}. \tag{3.65}$$

$$g_{LR} = \frac{1}{2}\left\{\left(\ln\left[\frac{|e|V}{k_B T_K}\right]\right)^{-1} - \left(\frac{1}{\alpha_0} + \ln\left[\frac{|e|V}{k_B T_K}\right]\right)^{-1}\right\}. \tag{3.66}$$

In these expressions, α_0 is a dimensionless parameter which characterizes the difference between the coupling constants g_d and g_{LR} ($\alpha_0 = 0$ if $g_d = g_{LR}$). When α_0 is sufficiently small ($\alpha_0 \ll 1/\ln[|e|V/k_B T_K]$),the effective exchange parameter $\hbar\gamma \approx |e|V g_{LR}$ takes on the form

$$\hbar\gamma \approx |e|V \Big/ \left(2\ln\left[\frac{|e|V}{k_B T_K}\right]\right)^2 \tag{3.67}$$

whereas within the opposite limit ($\alpha_0 \gg 1/\ln[|e|V/k_B T_K]$), one gets

$$\hbar\gamma \approx |e|V \Big/ \left(4\alpha_0^2 \ln^4\left[\frac{|e|V}{k_B T_K}\right]\right). \tag{3.68}$$

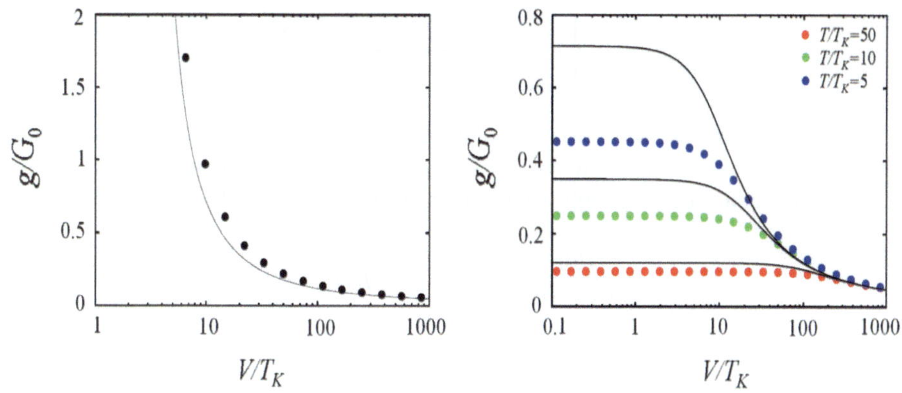

Fig. 3.24 Conductance through a biased quantum dot computed using approximation (3.69) (*solid lines*) and renormalization group perturbative approach (*dots*). The curves are plotted within the limit $T \to 0$ (*left panel*) and for several finite temperatures (*right panel*) (after [314]). Reprinted with the permission of the APS

Comparison of these approximations with the previous result (3.61) demonstrates the renormalization of the coupling parameters.

Summarizing the presented above results one may conclude that for highly biased quantum dots ($k_B T \ll |e|V \ll D$), nonequilibrium relaxation processes allow to destroy the Kondo resonance in conductance even at $T = 0$ as their rate increases. The rise of the decoherence rate accompanies the increase of the bias voltage and leads to the drop in the conductance as follows from Eqs. (3.61) and (3.62). Nevertheless, the "bare" perturbative theory gives inaccurate results for the relevant quantities which need renormalization appearing as a result of summation of leading logarithmical corrections up to high orders of perturbation theory. Even the simplified equations for the coupling parameters (3.64) give a significant renormalization. A thorough mathematical formalism which allows to completely avoid errors and inaccuracies in deriving expressions for the decoherence rate and other important characteristics of Kondo systems out of equilibrium is described in [37, 311–314]. We do not reproduce it here, for it is beyond the scope of the present book. However, in Fig. 3.24 we present some results illustrating the dependence of the conductance through the dot of the bias voltage computed using a "bare" perturbation theory compared to the results obtained applying the renormalization group approach in its fully developed form. We remark that the leading term in the expression for the conductance computed within a perturbation theory differs from the corresponding term in the NCA approximation (3.62) by a prefactor 3/4. This prefactor appears as a result of improving of the NCA first performed in [36]. Taking into account this improvement along with the influence of a nonzero temperature, one may approximate the leading term of the conductance as follows [313]:

$$\frac{g}{G_0} = \frac{3\pi^2}{16} \left\{ \ln \left[\frac{\sqrt{(|e|V/2)^2 + (k_B T)^2}}{k_B T} \right] \right\}^{-2}. \tag{3.69}$$

It is important to show how decoherence processes may be captured within the fully developed renormalization group perturbative formalism. Within this approach, the equations for coupling parameters (3.64) are generalized to the form [37]

$$\begin{aligned} \frac{\partial g_D}{\partial(\ln D)} &= -(g_{LR}^2 + g_D^2), \\ \frac{\partial g_{LR}}{\partial(\ln D)} &= -g_{LR} g_D (\theta_v + \theta_\gamma) \end{aligned} \tag{3.70}$$

where θ_v and θ_γ are step functions, namely, $\theta_v = \theta(D - |e|V)$ and $\theta_\gamma = \theta(D - \hbar\gamma)$. As follows from these equations, the flow of coupling constants is not completely cut off when $|e|V$ reaches the value of D. The flow may continue until the condition $D > \hbar\gamma$ remains satisfied. This emphasizes the importance of decoherence processes by showing that the coupling between the dot and the leads is not cut off by the bias voltage itself. As mentioned before, it is stopped by decoherence processes induced by the current flowing through the biased quantum dot/molecular bridge in a MMM junction.

Chapter 4
Inelastic Transport

4.1 Vibration-Induced Features in the Electron Conductance and Current Through MMM Junctions

Interaction of electrons with molecular vibrations is known to be an important source of inelastic contribution to the electron transport through molecules. Theoretical studies of vibrationally inelastic electron transport through molecules and other similar nanosystems (e.g., carbon nanotubes) were carried out over the past decade by a large number of authors [24–26, 98–106]. Also, manifestations of the electron–vibron interactions were experimentally observed (see [2, 29, 110–117, 315, 316]). To analyze vibration-induced effects in the electron transport through molecular bridges, one must assume that molecular orbitals are coupled to the phonons describing vibrations. While on the bridge, electrons may participate in the events generated by their interactions with vibrational phonons. These events involve virtual phonon emission and absorption. For sufficiently strong electron–phonon interactions this leads to the appearance of metastable electron levels which could participate in the electron transport through the junctions bringing an inelastic component to the current. As a result, vibration-induced features occur in the differential molecular conductance dI/dV and in the inelastic tunneling spectrum d^2I/dV^2. This was observed in the experiments (see, e.g., [29, 111]). Sometimes even current–voltage curves themselves exhibit extra steps originating from the electron–vibron interactions [110]. An illustrating example is presented in Fig. 4.1.

As was discussed above, particular manifestations of electron–vibron effects in the transport characteristics are determined by the relation of three relevant energies. These are the coupling strengths of the molecule to the electrodes $\Gamma_{L,R}$, the electron–phonon coupling strength Λ, and the thermal energy $k_B T$. When the molecule is weakly coupled to the electrodes ($\Gamma_{L,R} \ll \Lambda, \hbar\Omega$) and the temperature is low ($k_B T \ll \Gamma_{L,R}$), the electron transfer through the junction may give rise to a strong vibrational excitation, and one may expect a pronounced vibrational resonance structure in the electron transmission to appear. Correspondingly, extra steps should occur in the I–V curves as shown in Fig. 4.1. Proper theoretical

N.A. Zimbovskaya, *Transport Properties of Molecular Junctions*, Springer Tracts in Modern Physics 254, DOI 10.1007/978-1-4614-8011-2_4,

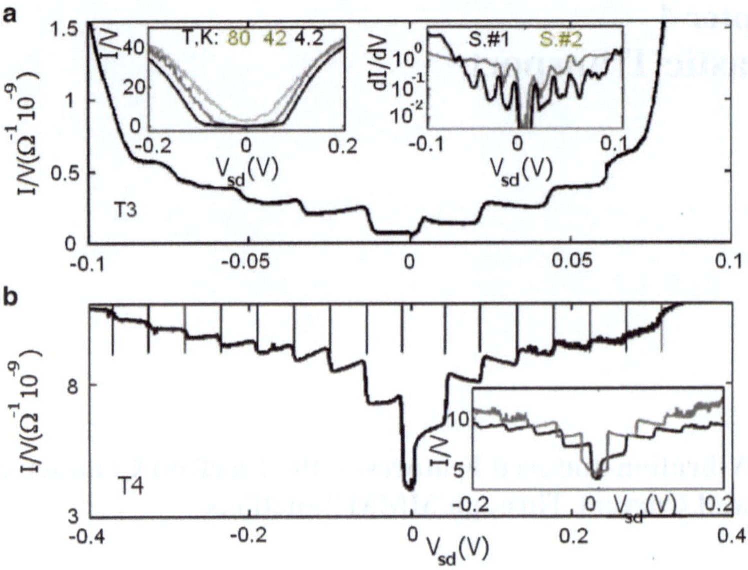

Fig. 4.1 Steplike features appearing in the conductance and current flowing through two MMM junctions at low temperatures ($T = 4.2\,\text{K}$). The features appear due to electron–vibron interactions. Spacings between the adjacent steps give information concerning the frequencies of vibrational modes coupled to electrons (after [29]). Reprinted with the permission of APS

consideration of the electron transport in this regime could be carried out within the approach proposed by Wingreen et al. [205]. Here, we employ a very simple semiquantitative approximation which, nevertheless, allows to qualitatively describe this structure including the effect of higher phonon harmonics on the transport characteristics.

We consider a junction including a single-site bridge which is coupled to a single vibrational mode with the frequency Ω. An electron on the bridge may virtually absorb several phonons which results in the creation of a set of metastable states with the energies $E_n = \tilde{E}_0 + n\hbar\Omega$ $(n = 0, 1, 2, \ldots)$. Here, the energy $\tilde{E}_0$ is shifted with respect to E_0 due to the electron–phonon interaction. The difference in these energies E_p is called a polaronic shift and could be estimated as $E_p = \Lambda^2/\hbar\Omega$ [102,205,317]. At weak coupling of the bridge state to the electrodes, the lifetime of these metastable states is long enough for them to serve as channels for the electron transmission. Therefore, one may roughly approximate the transmission as a sum of contributions from all these channels. The terms in the sum have the form similar to the well-known expression for the coherent transmission [see Eq. (1.12)]. However, every term includes the factor $P(n)$ which corresponds to the probability of the metastable state to appear. So, we obtain

$$T(E) = \Gamma_L\Gamma_R \sum_n \frac{P(n)}{(E - \tilde{E}_0 - n\hbar\Omega)^2 + \left[(\Gamma_L + \Gamma_R)/2\right]^2}. \tag{4.1}$$

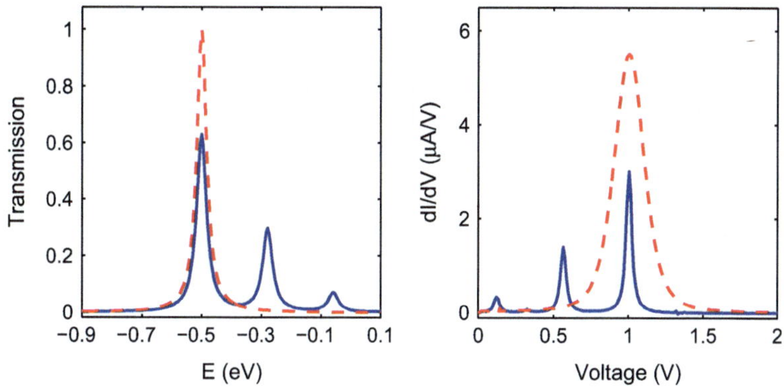

Fig. 4.2 The electron transmission versus energy (*left panel*) and the conductance versus voltage (*right panel*) for a junction with the molecular bridge simulated by a single electronic state weakly coupled to the leads: $\Gamma_L = \Gamma_R = 0.01\,\text{eV},\ E_0 = 0.5\,\text{eV}$. *Solid lines* are plotted assuming that the bridge is coupled to a single phonon mode ($\hbar\Omega = 0.22\,\text{eV},\ \Lambda = 0.3\,\text{eV}$). *Dashed lines* correspond to the coherent electron transport

Here [107]:

$$P(n) = \frac{1}{n!}\left(\frac{\Lambda^2}{2\hbar^2\Omega^2}\right)^n \exp\left(-\frac{\Lambda^2}{2\hbar^2\Omega^2}\right). \tag{4.2}$$

The phonon-induced peaks in the transmission are displayed in Fig. 4.2 along with the transmission peak for the coherent transport through a single-site bridge. As expected, the coupling of electronic degrees of freedom to the vibrational motion splits the single peak in the coherent transmission into the set of smaller peaks associated with vibrational levels. The peaks could be resolved when $\Gamma_{L,R} \ll \hbar\Omega$. This agrees with the results of the earlier theoretical work of Wingreen et al. [205] as well as with the experiments [29, 115]. Phonon-induced peaks in the transmission give rise to the steps in the I–V curves and rather sharp features (peaks and dips) in the inelastic tunneling spectrum (IETS). The latter are shown in Fig. 4.3, and they resemble those obtained using proper NEGF-based calculations [25].

The visibility of the vibron-induced peaks in the differential conductance and the corresponding steps in the current–voltage curves generally depends on the heights and spacings of these peaks/steps as well as on their broadening. In the considered case of a weakly coupled junction, the broadening is obviously controlled by the temperature and the vibrational dissipation. Also, in practical metal–molecular junctions an additional broadening mechanism occurs, which originates from anharmonicities inherent to the molecular vibrations and from the fact that the vibrational frequencies depend on the molecular charge state [241]. These factors can cause significant alterations in the transport characteristics. First, the vibron-induced conductance peaks (and steps in I–V curves) cease to be equidistant. Secondly, it appears that the strength of the direct vibrational relaxation is an

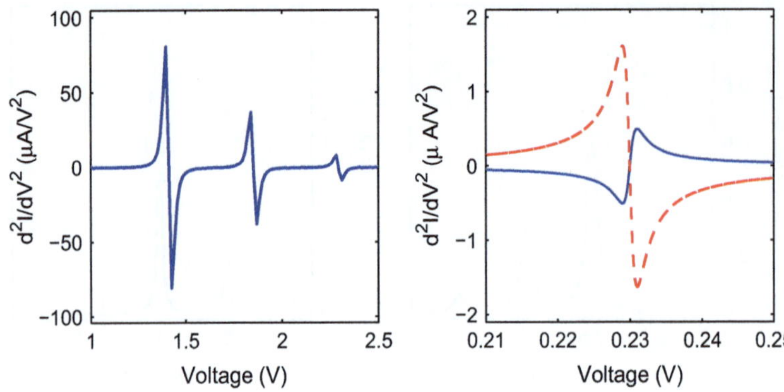

Fig. 4.3 The inelastic electron tunneling spectrum plotted against the bias voltage at $\hbar\Omega = 0.22\,\text{eV}$, $\Lambda = 0.3\,\text{eV}$. *Left panel*: (the bridge is weakly coupled to leads) $\Gamma_L = \Gamma_R = 0.01\,\text{eV}$, $E_0 = 0.5\,\text{eV}$. *Right panel*: (strongly coupled junction) $\Gamma_L = \Gamma_R = 0.5\,\text{eV}$, $E_0 = 0.5\,\text{eV}$ (*solid line*), $E_0 = -0.5\,\text{eV}$ (*dashed line*)

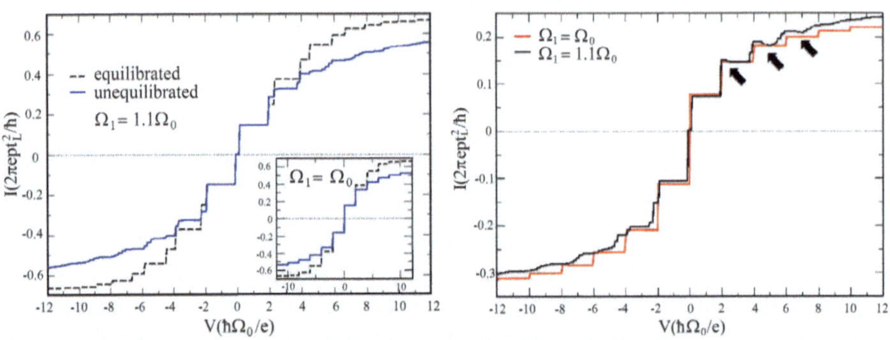

Fig. 4.4 Alterations in the current–voltage curves computed for a MMM junction with a single-site spin-degenerated bridge which occur due to the dependence of the vibrational frequency on the charge on the bridge for a symmetrically (*left*) and asymmetrically (*right*) coupled systems. The curves are plotted assuming $\Omega_1 = 1.1.\Omega_0$ (Ω_1, Ω_0 are the vibrational frequencies for the charged and neutral bridge state, respectively). $U_0 \to \infty$, $\Lambda/\hbar\Omega_0 = 1.2$, $k_B T = 0.02\hbar\Omega_0$, $E_0 = 0$, $\Gamma_L = \Gamma_R$ (*left panel*), and $\Gamma_R = 0.3\Gamma_L$ (*right panel*). For the asymmetrically coupled system, weak peak-like features are observed at the onset of phonon steps (marked with *arrows*). These features appear due to the combination of the asymmetric coupling and differing vibrational frequencies [241]. Reprinted with the permission of the APS

important factor noticeably affecting the I–V characteristics. When the relaxation is strong, it merely causes shifts in the steps positions. However, in the limit of the weak vibrational relaxation, the vibrational steps reveal a bias-dependent broadening arising due to the steps splitting into several closely spaced substeps (see Fig. 4.4).

The difference originates from the distinction between the vibrational phonon distribution functions corresponding to the above-described regimes. Within the strong vibrational relaxation regime, the phonon distribution is distinctly peaked for the phonon ground state where the number of excited phonons is $m = 0$ and the phonons are nearly equilibrated. The shifts in the step positions and variations in the step spacings are caused by the energy spectrum of an anharmonic oscillator (which is not equidistant) and by the vibron frequency variations occurring as the charge on the molecule varies. On the contrary, the weak vibrational relaxation limit corresponds to a nonequilibrium state of the phonon system described by the nonequilibrium distribution function. A multitude of excitation and de-excitation processes changing the number of unequilibrated phonons from m to m' is permitted. The values of contributing m, m' are determined by the form of the nonequilibrium distribution function. One may expect that as the bias voltage magnitude increases, the nonequilibrium distribution widens, thus widening the range where the relevant m, m' could belong. This corresponds to splitting of a vibrational step into several substeps. The number of the substeps increases as the bias voltage rises, and the spacings between the adjacent substeps become minute, which results in the vibrational steps broadening.

It was first suggested by the results of numerical simulations reported by Mitra et al. [22] that the vibrational nonequilibrium state becomes more pronounced as the electron–phonon coupling weakens. Considering the electron transport through a symmetrically coupled junction where the molecular bridge is represented by a single orbital coupled to a sole vibrational mode, these authors observed that the width of the phonon excitations distribution grows as the electron–phonon coupling parameter $\tilde{\Lambda} = \Lambda/\hbar\Omega$ decreases. Later, these conclusions were confirmed by the analysis based on the same model for the junction and using the rate equations (2.141) for the occupation probabilities of the molecular states with N electrons and m phonons [239]. Neglecting the effects of the direct vibrational relaxation, it was shown that the width of the phonon distribution is proportional to $\tilde{\Lambda}^{-\alpha}$ where the parameter α takes on positive values depending on the applied bias magnitude:

$$\alpha = \frac{2\big[|e|V/2\hbar\Omega\big]}{\big[|e|V/2\hbar\Omega\big] + 1}. \tag{4.3}$$

Here, a denotation $[|e|V/2\hbar\Omega]$ is introduced to indicate the largest integer which does not exceed the magnitude of $|e|V/2\hbar\Omega$. Therefore, the width of the phonon distribution for a biased junction diverges as $\tilde{\Lambda}$ approaches zero as shown in Fig. 4.5. This striking conclusion is to be modified if the direct vibrational relaxation processes described by the last term in Eq. (2.141) are taken into account. It appears that when $\tilde{\Lambda}$ becomes smaller than a certain crossover value $\tilde{\Lambda}_0$, the width of the phonon distribution becomes controlled by the direct phonon relaxation, and this keeps its width finite.

The transition rates in Eq. (2.141) are proportional to Franck–Condon matrix elements

$$M_{m \to m'} = \int_{-\infty}^{\infty} dx \Phi_{m'}(x) \Phi_m\left(x - \sqrt{2}\tilde{\Lambda} l_{\text{osc}}\right) \tag{4.4}$$

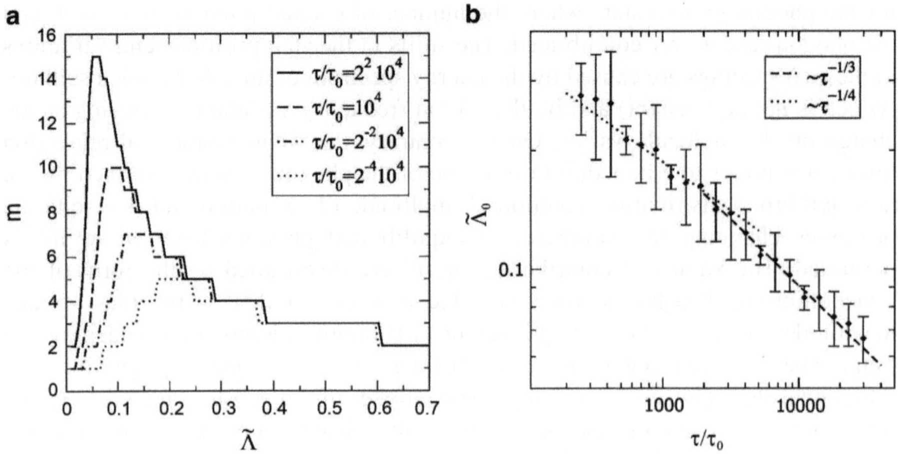

Fig. 4.5 *Left panel*: width of the phonon distribution as a function of electron–phonon coupling strength for several vibrational relaxation times τ for the symmetrically coupled ($\Gamma_L = \Gamma_R = \Gamma$) metal–molecule junction with a single-site bridge. The curves are plotted assuming $|e|V = 3\hbar\Omega$, $E_0 = 0$, $\tau_0\Gamma/\hbar$, $U_0 \to \infty$, $k_B T = 0.05\hbar\Omega$. For $\tilde{\Lambda} > 0.25$ all curves show the approximate $m \sim \tilde{\Lambda}^{-1}$ scaling. Below a relaxation-rate-dependent crossover point $\tilde{\Lambda}_0$, the processes changing the number of unequilibrated phonons are strongly suppressed due to the direct relaxation. *Right panel*: crossover point $\tilde{\Lambda}_0$ versus relaxation time τ [239]. Reprinted with the permission of the APS

which are the overlaps of the harmonic oscillator functions shifted with respect to each other by a distance $\sqrt{2}\tilde{\Lambda} l_{osc}$. Here, $l_{osc} = (\hbar/M\Omega)^{1/2}$ is the vibrational oscillator length, and M denotes the mass associated with the nuclear motion. If an electron–vibron interaction is strong, the Franck–Condon physics may lead to a suppression of a current through a junction at low bias voltage, which is called the Franck–Condon blockade [240]. This interesting phenomenon could occur in weakly coupled junctions ($\Gamma_{L,R} \ll \Lambda, \hbar\Omega, U_0$) and at low temperature ($k_B T \ll U_0, \hbar\Omega$). The current suppression originates from the Franck–Condon matrix elements behavior. As follows from Eq. (4.4), at weak electron–phonon coupling, the phonon transmissions mostly occur in such a way that keeps the number of excited phonons the same ($m = m'$). As the electron–phonon coupling parameter $\tilde{\Lambda}$ increases, and the distribution of transition rates widens, the transitions with $m \neq m'$ become permitted. Finally, at $\tilde{\Lambda} \gg 1$, a gap of exponentially suppressed transitions between low-lying phonon states opens. Assuming that the junction is symmetrically coupled, a tunneling event can increase the number of the excited phonons by $\Delta m = [|e|V/2\hbar\Omega]$ due to the energy conservation. Therefore, when the bias voltage is low and the electron–phonon coupling is strong, the system remains trapped in a region of exponentially small transition rates, which is the immediate reason for the current suppression. The direct vibrational relaxation may noticeably affect the current–voltage curves in the Franck–Condon blockade regime. This is discussed in detail in [240].

4.2 Vibration-Induced Features in the Inelastic Electron Tunneling Spectra (IETS)

When the molecule is strongly coupled to the electrodes ($\Gamma_{L,R} \gg \Lambda$) and the temperature is still low ($k_B T \ll \Lambda, \hbar\Omega$), electron–vibron interaction effects are less pronounced. Both current and conductance are weakly affected by the electron–phonon coupling [24–26, 318]. However, the IETS features remain distinguishable. These features appear at the threshold $V = \hbar\Omega/|e|$ which corresponds to the opening of a channel for inelastic transport. To analyze IETS in a simplest way one may use Eq. (2.87) for electron transmission derived within the Buttiker model where the parameter ϵ is expressed in terms of the relevant energies. The explicit expression for ϵ could be found by comparison of Eq. (2.87) with the NEGF-based expression for the inelastic contribution to the electron current given by Eq. (2.125), as was suggested by Zimbovskaya [52]. It has the form

$$\epsilon = \frac{\Gamma_{\text{ph}}}{2(\Gamma_L + \Gamma_R) + \Gamma_{\text{ph}}} \tag{4.5}$$

where $\Gamma_{\text{ph}} = 2\text{Im}(\Sigma_{\text{ph}})$ and Σ_{ph} is the self-energy term originating from the electron–phonon interaction. Basing on the nonequilibrium Green's function formalism, the expression for Γ_{ph} was derived in the form given by Eq. (2.126).

To better analyze electron–vibron interactions we turn to the model where the molecule is simulated by a chain of identical hydrogen-like atoms. This model was already discussed in Sect. 2.2 of present work. We assume that the chain includes an odd number of elements ($N = 2m - 1$), and we set ionization energies $\alpha_k = E_0$ and the nearest neighbor coupling strengths $\beta_k = \beta$ for all elements of the chain. Now, we suggest that the chain vibrates at a fundamental frequency Ω, and we study the electron interactions with the corresponding transverse mode. In this case the maximum displacement of the sites occurs in the middle of the chain. Therefore we assume as a first approximation that only the middle site in the chain is coupled to the vibrational mode as shown in Fig. 4.6. To describe this coupling within the Buttiker approach, we maintain the contact of the middle site with an electron reservoir detaching the reservoirs from all remaining sites ($\epsilon_j = 0$ for $j \neq m$.) In this particular case the number of channels connecting the junction to the surroundings is reduced to four regardless of the number of sites on the bridge. Accordingly, the most of the scattering matrix elements turn zero, and the $2N \times 2N$ scattering matrix $S^{(2N-1)}$ given by Eq. (2.76)–(2.78) may be replaced by 4×4 matrix S whose elements are computed below. In this case the transmission $T(E)$ is described by Eqs. (2.70) and (2.71) as well as in the case of a single-level bridge, but the expressions for the matrix elements T_{ik} have a different form.

To reach the middle site in the molecular chain an injected electron must first tunnel from the left electrode to the chain, and then it tunnels through $m - 1$ barriers separating the sites. The tunneling through this set of barriers may be described by the 2×2 transfer matrix W_L whose elements are expressed in terms of transmission and reflection amplitudes for the barriers:

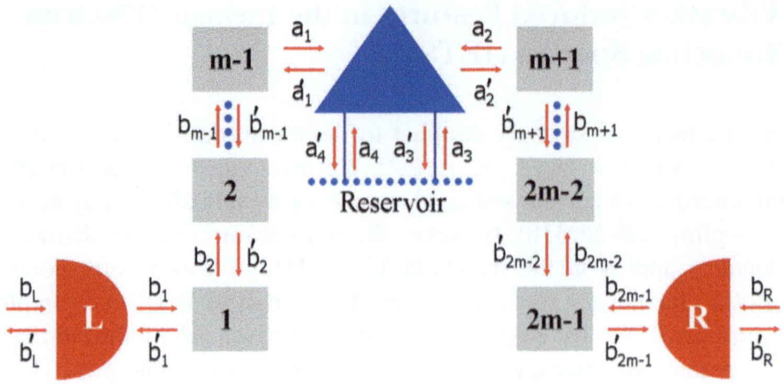

Fig. 4.6 Schematic drawing illustrating electron transport through the junction with the bridge represented by a chain of $2m - 1$ identical sites. The dissipative electron reservoir is attached to the middle site

$$W_L = \begin{pmatrix} u_m^L & -v_m^L \\ v_m^L & u_m^L \end{pmatrix}; \qquad (u_m^L)^2 + (v_m^L)^2 = 1. \tag{4.6}$$

Assuming all barriers separating the sites to be identical, we arrive at the following expression for this matrix:

$$W_L = \begin{pmatrix} \frac{1}{t^L} & -\frac{r^L}{t^L} \\ \frac{r^L}{t^L} & \frac{1}{t^L} \end{pmatrix} \begin{pmatrix} \frac{1}{t} & -\frac{r}{t} \\ \frac{r}{t} & \frac{1}{t} \end{pmatrix}^{m-1} \tag{4.7}$$

where t and r are the transmission and reflection coefficients for a single barrier separating the sites on the left half of the chain, respectively. The electron tunneling through the right half of the chain could be described in a similar way using the transfer matrix W_R :

$$W_R = \begin{pmatrix} u_m^R & -v_m^R \\ v_m^R & u_m^R \end{pmatrix} \equiv \begin{pmatrix} \frac{1}{t} & -\frac{r}{t} \\ \frac{r}{t} & \frac{1}{t} \end{pmatrix}^{m-1} \begin{pmatrix} \frac{1}{t^R} & -\frac{r^R}{t^R} \\ \frac{r^R}{t^R} & \frac{1}{t^R} \end{pmatrix} \tag{4.8}$$

Combining Eqs. (2.71), (1.33), and (4.8) we obtain the expression for the scattering matrix S. For a symmetrically coupled junction (the left and right barriers being identical), we get

$$S = \frac{1}{Z} \begin{pmatrix} u_m^L v_m^R + \alpha_m^2 u_m^R v_m^L & \alpha_m & u_m^R \beta_m & \alpha_m \beta_m v_m^R \\ \alpha_m & u_m^R v_m^L + \alpha_m^2 u_m^L v_m^R & \alpha_m \beta_m v_m^L & u_m^L \beta_m \\ \beta_m u_m^R & \alpha_m \beta_m v_m^R & \beta_m^2 u_m^R v_m^L & \alpha_m (v_m^L v_m^R - u_m^L u_m^R) \\ \alpha_m \beta_m v_m^R & \beta_m u_m^L & \alpha_m (v_m^L v_m^R - u_m^L u_m^R) & \beta_m^2 u_m^L v_m^R \end{pmatrix}. \tag{4.9}$$

Here, $Z = u_m^L u_m^R - \alpha_m^2 v_m^L v_m^R$; $\alpha_m = \sqrt{1-\epsilon_m}$; $\beta_m = \sqrt{\epsilon_m}$. For a single-site bridge symmetrically coupled to the electrodes ($m = 1$, $t^L = t^R \equiv t'$; $r^L = r^R \equiv r'$) $u_m = 1/t'$, $v_m = -r'/t'$, and Eq. (4.9) agrees with Eq. (2.74). Inserting the expressions for matrix elements T_{jk} determined by Eq. (4.9) into Eqs. (2.70) and (2.71), we may present the electron transmission in the form

$$T(E) = \frac{(1+\alpha_m^2)\left[w_m^2 - \alpha_m^2(w_m^2-1)\right]}{2\left[w_m^2 + \alpha_m^2(w_m^2-1)\right]^2}. \tag{4.10}$$

where w_m could be expressed in terms of matrix elements of the matrices $W_{L,R}$. The adopted model is appropriate when the bridge is not too long ($m \lesssim 5$). For long bridges, we must put in contact with the reservoirs not only the middle site but a few adjacent sites as well. Nevertheless, following a way described in Sect. 2.4, we may derive the expression for the electron transmission function suitable for long molecular bridges.

Our next step is to find the relation of the quantities w_m to the relevant Green's functions. Disregarding for a while the effects of nuclear motions we may write the relevant matrix element of the retarded Green's function $G_{1,2m-1}^r(E)$ in the form given by Eq. (2.53). The electron transmission for the elastic coherent transport through the considered junction may be presented as

$$T = \Gamma_L \Gamma_R |G_{1,2m-1}|^2 \tag{4.11}$$

In the absence of phonons ($\alpha_j = 1$), the expression for the electron transmission takes on the form

$$T(E) = \frac{1}{(2w_m^2-1)^2} \equiv \frac{\theta_m^4}{(2-\theta_m^2)^2} \tag{4.12}$$

where $\theta_m^2 = 1/w_m^2$. Comparing this result with Eq. (4.11), we obtain

$$\theta_m^2 = \frac{2g_m}{1+g_m} \tag{4.13}$$

where the function $g_m(E)$ is closely related to the Green's function on the chain, namely,

$$g_m(E) = \sqrt{\Gamma_L \Gamma_R}\left|G_{1,2m-1}(E)\right|. \tag{4.14}$$

Returning back to the chosen model ($\epsilon_j = 0$ for $j \neq m$) and putting $\epsilon_m = \epsilon$, we may rewrite the general expression for the electron transmission (4.10) as follows:

$$T(E) = \frac{g_m(E)(1+\alpha^2)\left[1+g_m(E) - \alpha^2(1-g_m(E))\right]}{\left[1+g_m(E)+\alpha^2(1-g_m(E))\right]^2}. \tag{4.15}$$

This expression is the starting point in further analysis.

In the case when the considered junction includes a single-site bridge the parameter ϵ given by Eq. (4.5) must depend on the self-energy correction arising due to the electron–phonon interaction ($\Gamma_{\text{ph}} = -2\text{Im}(\Sigma_{\text{ph}})$) and of the self-energy terms describing the coupling of the bridge to the electrodes ($\Gamma_{L,R}$). Presently, we are considering a multisite bridge, but we assume that only one site in the middle of the latter is coupled to the vibrational mode. Therefore we approximate ϵ by the expression similar to Eq. (4.5), namely,

$$\epsilon = \frac{\Gamma_{\text{ph}}}{2\Delta + \Gamma_{\text{ph}}} \tag{4.16}$$

where the energy Δ depends on both the bridge ends coupling to the leads ($\Gamma_{L,R}$) and site to site couplings on the bridge itself (β). Also, considering the situation when several bridge sites are linked to dephasing reservoirs, one may express the relevant dephasing parameters ϵ_j in terms of energies characterizing the electron transport through the junction. However, this complicated and nontrivial task is beyond the scope of the present work.

The form of Γ_{ph} varies depending on the characteristics of the electron–phonon interaction in the system and on the kind of phonons contributing to these interactions. For instance, the electron coupling to the thermal phonon bath results in the expression for the Γ_{ph} which strongly differs from that obtained while considering the electron coupling to a few vibrational phonon modes. Here, we assume that a single vibrational mode with the frequency Ω exists in the system and that the temperature is low ($k_{\text{B}}T \ll \hbar\Omega, \Gamma_{L,R}, \beta$). Then we can write the following expressions for the phonon density of states ρ_{ph}:

$$\rho_{\text{ph}}(\omega) = \frac{1}{\pi\hbar} \frac{\gamma}{(\omega - \Omega)^2 + \gamma^2}. \tag{4.17}$$

The parameter γ characterizes the broadening of the maximum in ρ_{ph} at $\omega = \Omega$ due to the interaction of the vibronic mode with the environment. The corresponding linewidth $\hbar\gamma$ has the same order as the thermal energy $k_{\text{B}}T$. At low temperatures we may significantly simplify the Eq. (2.126) for Γ_{ph}. Within the conduction window $\mu_R < E < \mu_L$, we get

$$\begin{aligned} \Gamma_{\text{ph}}(E) \approx & \pi\Lambda^2 \Bigg\{ \int_0^{(\mu_L - E)/\hbar} d\omega \rho_{\text{ph}}(\omega) D(E + \hbar\omega) \\ & + \int_0^{(E-\mu_R)/\hbar} d\omega \rho_{\text{ph}}(\omega) D(E - \hbar\omega) \Bigg\}. \end{aligned} \tag{4.18}$$

Omitting from consideration the coupling of the phonon mode to the environment ($\gamma \to 0$), we may easily carry out integration over ω in Eq. (4.18), and we arrive at the result:

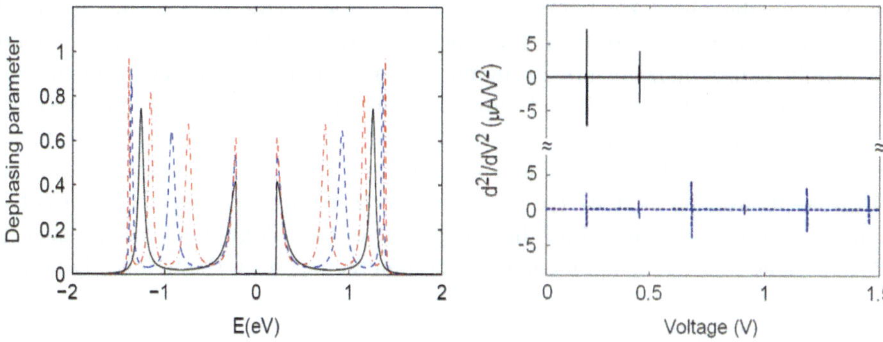

Fig. 4.7 *Left panel*: energy dependencies of the parameter ϵ. The curves are plotted at $N = 11$ (*dash-dotted line*), $N = 9$ (*dashed line*), $N = 7$ (*solid line*). *Right panel*: Inelastic electron spectrum d^2I/dV against V for a molecular junction with a single-site (*top*) and multisite (*bottom*) bridge including seven sites. The curves are plotted assuming $\Gamma_L = \Gamma_R = 0.2\,\text{eV}$, $\beta = 0.4\,\text{eV}$, $E_o = 0$, $\hbar\Omega = 0.22\,\text{eV}$, $\Lambda = 0.3\,\text{eV}$ [109]. Reprinted with the permission of the AIP

$$\Gamma_{\text{ph}}(E) \approx \pi\Lambda^2\Big\{D(E+\hbar\Omega)\theta(\mu_L - \hbar\Omega - E) + D(E-\hbar\omega)\theta(E-\mu_R-\hbar\Omega)\Big\} \tag{4.19}$$

where $\theta(x)$ is the step function.

The electron density of states (DOS) on the middle site of the chain is described by the expression $D(E) = -\frac{1}{\pi}\text{Im}[G_{mm}(E)]$. Within the adopted model we obtain

$$G_{mm}(E) = \frac{1}{\zeta}\frac{(\lambda+\zeta)^{m-1}(\lambda+\zeta+i\Gamma) - (\lambda-\zeta)^{m-1}(\lambda-\zeta+i\Gamma)}{(\lambda+\zeta)^{m-1}(\lambda+\zeta+i\Gamma) + (\lambda-\zeta)^{m-1}(\lambda-\zeta+i\Gamma)}. \tag{4.20}$$

Here, as before, $\lambda = E - E_0$, $\zeta = \sqrt{\lambda^2 - 4\beta^2}$, $\gamma = \Gamma_L + \Gamma_R$. Now, we may apply the obtained approximation for the parameter ϵ to compute the vibration-induced electron transmission, and then we employ the latter to calculate the electron current through the junction as well as its derivatives with respect to the bias voltage. The current includes the vibration-induced inelastic contribution. The latter is brought in there by the expression (4.15) for the electron transmission $T(E)$. As shown above, the parameter $\alpha^2 = 1 - \epsilon$ in Eq. (4.15) depends on the characteristics of the electron–vibron interaction, and this leads to the occurrence of the inelastic term in the current along with the elastic one. The proposed formalism enables us to qualitatively analyze the effects of the electron–vibron interactions in the IETS for multisite molecular bridges.

The electron–vibron interaction caused self-energy Γ_{ph} strongly depends on the tunnel energy E, and this results in a well-pronounced energy dependence of the parameter ϵ shown in the left panel in Fig. 4.7. We remark that ϵ remains zero within the range $E_0 - \hbar\Omega < E < E_0 + \hbar\Omega$. This fits into the accepted model where electron–vibron interactions with a single phonon mode are assumed to mainly cause inelastic effects in the electron transport. An electron on the bridge may virtually absorb and emit phonons of the energy $\hbar\Omega$, thus creating metastable states around the energy states of the molecular chain. In general, these bridge energy states themselves are shifted from their original positions due to the electron–phonon interactions, but we tentatively disregard these polaronic shifts for simplicity. Within our model for the bridge, one of the bridge states has the energy E_0, the same as the ionization energy of noninteracting sites. Being initially at this energy level, an electron could make transitions to the states $E = E_0 \pm n\hbar\Omega$ as a result of its interaction with the vibrational mode. So, change in the energy of the electron being initially at the state with the energy E_0 and participating in the events occurring due to the their interactions with vibrational phonons cannot be less than $\hbar\Omega$. Otherwise, the electron transport must be coherent and elastic ($\epsilon = 0$). At $E = E_0 \pm \hbar\Omega$ the parameter ϵ sharply increases, revealing peaks which correspond to the appearance of the metastable vibration-induced states around $E = E_0$. The same consideration could be applied to another electronic states on the bridge. As expected, ϵ versus E plot displays a sequence of peaks whose number is related to the number of elements included in the chain. All these features are located in the energy range $E_0 - 2\beta < E < E_0 + 2\beta$, and ϵ again becomes zero outside of this range, as shown in Fig. 4.7.

Due to the interaction with the phonon mode the electron transmission gets extra features which could be revealed in the observable characteristics of the electron transport through molecular junctions. The specifics of these manifestations are determined by the relative values of the relevant energies. Here, we concentrate on the case of a strongly coupled junction, assuming that $\Gamma_{L,R} \sim \Lambda$. In this case inelastic effects do not cause significant changes in the I–V characteristics. However, they cause the appearance of a distinct signal in the inelastic tunneling spectrum at $|e|V \approx \hbar\Omega_0$ similar to that occurring on a single-site bridge interacting with the vibrational mode which is shown in the right panel of Fig. 4.3. The size of the IETS signal at $|e|V \approx \hbar\Omega$ depends on the number of sites in the bridge chain, as shown in Fig. 4.8 (top right panel). For $N = 1$ (a single-site bridge), the signal size takes on the value of the same order as that reported by Galperin et al. [25] for a symmetrically coupled junction with relatively close values of the relevant parameters. For multisite bridges the IETS reveals more signals which indicate inelastic electron transport through the channels associated with the bridge energy levels situated apart from $E = E_0$. An example of the vibron-induced IETS for a multisite bridge ($N = 7$) is presented in the right panel of Fig. 4.7. Some IETS signals are separately displayed in Fig. 4.8 to show their shape and size for different values of N.

The analysis carried out in [109] shows that the signal indicating the phonon excitation threshold is accompanied with another feature appearing at $|e|V \approx 2\hbar\Omega$, and this happens for multisite and single-site bridges in the same way. The multiplication of IETS signals is known for molecular junctions with weak coupling of the bridge to the leads, and it indicates the contribution of higher

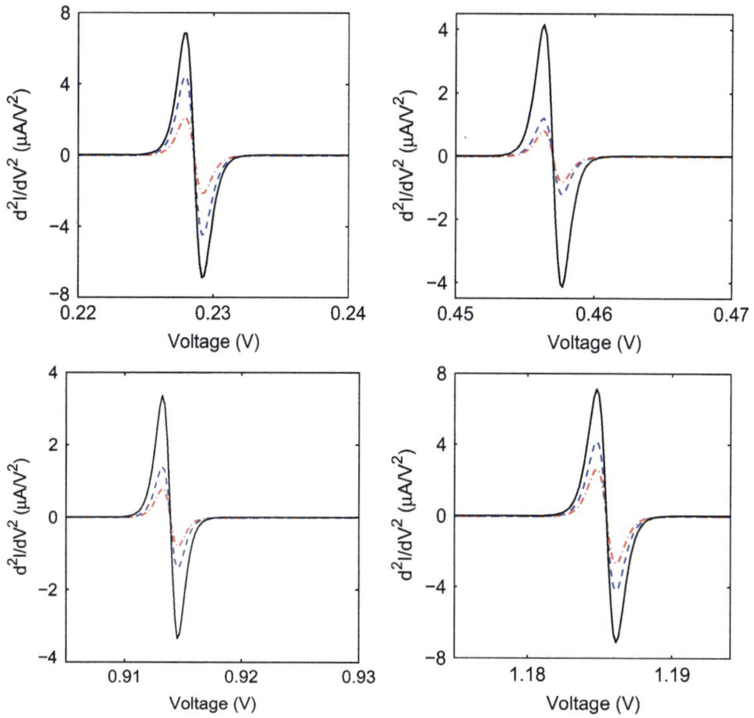

Fig. 4.8 The IETS features in d^2I/dV^2 for the several resonant levels model (multisite bridge) plotted, assuming $\Gamma_L = \Gamma_R = 0.2\,\text{eV}$, $\beta = 0.4\,\text{eV}$, $\hbar\Omega = 0.22\,\text{eV}$, $\Lambda = 0.3\,\text{eV}$, $E_0 = 0$. *Top panels*: $N = 11$ (*dash-dotted lines*), $N = 3$ (*dashed lines*), $N = 1$ (*solid lines*). *Bottom panels*: $N = 11$ (*dash-dotted lines*), $N = 7$ (*dashed lines*), $N = 3$ (*solid lines*) [109]. Reprinted with the permission of the AIP

phonon harmonics into the electron transport. However, this explanation is hardly justified for the presently considered junction with rather strong coupling of the bridge to the leads. To clarify the nature of the obtained IETS signal duplication, one may compute d^2I/dV^2 for a junction with a slightly asymmetric coupling. It appears that the extra signal position varies depending on the ratio Γ_L/Γ_R, (see Fig. 4.9), and the stronger is the asymmetry, the closer it moves to the phonon excitation threshold. In general, the extra signal appears at $|e|V \approx \hbar\Omega(1+\Gamma_L/\Gamma_R)$. One may expect that within the scanning tunneling microscopy (STM) junction configuration ($\Gamma_L \ll \Gamma_R$), this signal would be superimposed on that indicating the excitation threshold for the vibrational phonons. This gives grounds to conjecture that the effect of the electron–vibron interactions on the IETS is twofold. The mere opening of the channel for inelastic transport gives rise to the signal in the inelastic tunneling spectrum of electrons. This signal always occurs at the phonon excitation threshold regardless of the characteristics of the bridge coupling to the leads. Another signal indicates that the corresponding metastable electronic state appears in the conduction window between μ_L and μ_R and its position is determined by the bias voltage distribution in the junction. The two coincide within the STM

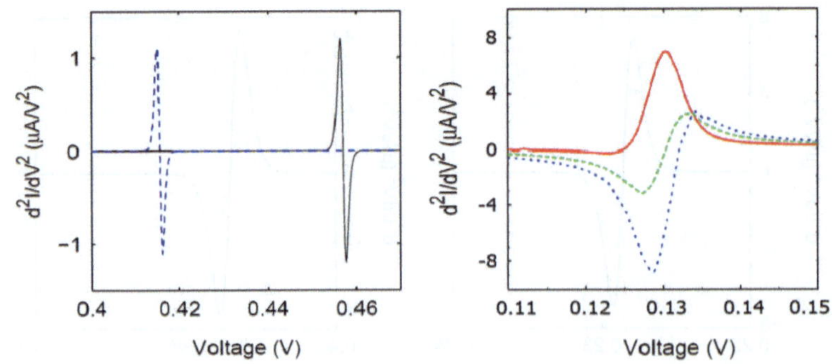

Fig. 4.9 *Left panel*: the effect of asymmetric coupling of the bridge to the leads on the IETS feature in d^2I/dV^2. The curves are plotted for $N = 7$, $\Gamma_L = \Gamma_R = 0.2\,\text{eV}$ (*solid line*) and $\Gamma_L = 0.18\,\text{eV}$, $\Gamma_R = 0.22\,\text{eV}$ (*dashed line*). *Right panel*: IETS threshold feature for the junction with a single-site molecular bridge computed using the NEGF formalism assuming $\hbar\Omega = 0.13\,\text{eV}$, $\Gamma_L = \Gamma_R = 0.5\,\text{eV}$, $\Lambda = 0.3\,\text{eV}, k_B T = 0.0009\,\text{eV}$, $E_0 = 0.7\,\text{eV}$ (*solid line*), $0.6\,\text{eV}$ (*dashed line*), and $0.55\,\text{eV}$ (*dotted line*) [26]. Reprinted of the permission of the Institute of Physics (IOP)

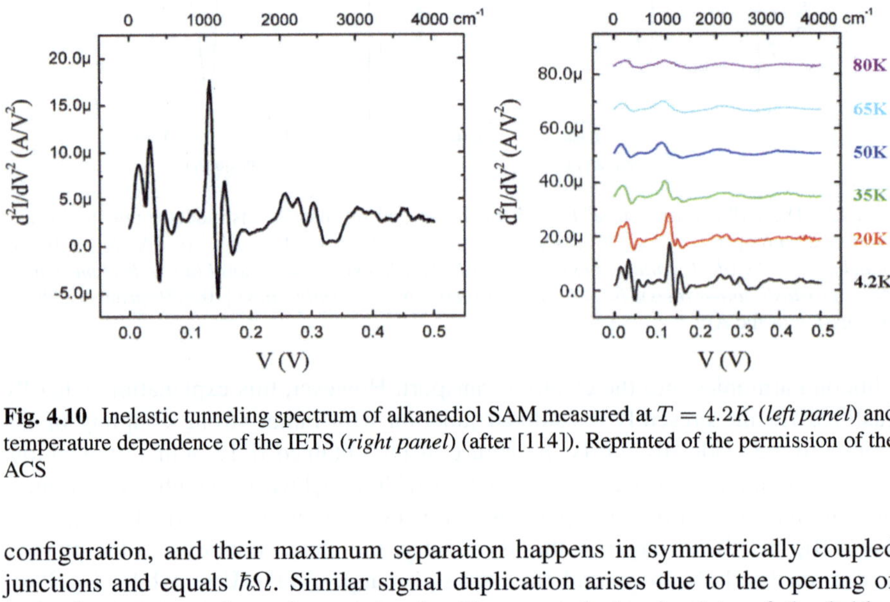

Fig. 4.10 Inelastic tunneling spectrum of alkanediol SAM measured at $T = 4.2K$ (*left panel*) and temperature dependence of the IETS (*right panel*) (after [114]). Reprinted of the permission of the ACS

configuration, and their maximum separation happens in symmetrically coupled junctions and equals $\hbar\Omega$. Similar signal duplication arises due to the opening of inelastic transport channels associated with every electronic state of the bridge chain.

Simplifications introduced into the employed approach do not prevent it from producing reasonably good results for sizes and positions of the IETS features. Nevertheless, the simplified semiquantitative method described above fails to bring correct results for the shapes of these features. More thorough analysis based on the NEGF formalism shows that the IETS signals may accept diverse shapes appearing as peaks, dips, or combinations of the two as presented in right panel of Fig. 4.9. The shape variability of these signals is reminiscent of Fano resonances lineshaps. Experiments [29, 110, 112, 114] corroborate this variability, as shown in Fig. 4.10.

At first glance, the emergence of a dip calls for an explanation because it is natural to expect the net current through the junction to increase at the threshold. Indeed, the inelastic contribution to the current increases from zero to a certain nonzero value at this threshold for the channel for inelastic transport opens up. However, more thorough studies show that both elastic and nonelastic contributions to the net current undergo changes at the inelastic tunneling threshold, and the elastic current could decrease there, as was first shown by Persson and Baratoff [319]. Moreover, this decrease in the elastic current may overweigh the contribution coming from the inelastic channel. Depending on the relative value of the elastic and inelastic contributions to the net current near threshold, the IETS reveals a peak, a dip, or a derivative-like feature at the corresponding voltage.

The shape of the signal is very sensitive to the characteristics of the junction such as the position of the electronic state, electron–phonon and molecule-to-leads coupling strengths, and the vibrational frequency. For instance, at a very strong coupling of the molecular bridge to the leads, it could so happen that the backscattered by the negatively biased electrode electrons whose energies belong to the conduction window between μ_L and μ_R are locally depleted near the junction. In this situation, the opening of the inelastic channel may cause the increased reflection (otherwise forbidden by the Fermi exclusion), leading to the decrease in the conduction. Consequently, the described scenario should result in the dip in IETS signal. The results reproduced in the right panel of Fig. 4.9 give evidence concerning the effect of the position of the bridge level (for a single-site bridge) with respect to the Fermi energy on the IETS signal shape. Three curves presented in this figure are plotted for three different values of E_0. One sees that the signal is transformed from a peak to a dip through an intermediate derivative-like shape as the bridge level is being shifted with respect to the Fermi energy. Also, as concluded in the work of Ryndyk and Cuniberti [102], the above-discussed sideband phonon-induced features in the electron spectral density could give rise to the corresponding features in the differential conductance dI/dV and IETS, assuming that the molecule coupling to the leads is not too strong. Contributions from these sideband features may be responsible for the shape of the IETS signal at the threshold of the inelastic tunneling channel. These contributions could produce an extra inelastic signal as well. The latter appears as an additional peak or dip in the differential conductance.

The question of the current decrease/increase at the phonon excitation threshold, which corresponds to the peak/dip in the IETS, is not completely answered so far, and the appropriate theory is being developed [101,320]. Nevertheless, it is presently understood that four relevant energies, namely, molecule-electrode couplings $\Gamma_{L,R}$, electron–phonon coupling strength Λ, the bridge state position, and the phonon energy $\hbar\Omega$ play crucial parts in determining the shape of the IETS signal. Varying these parameters one may convert the IETS signal from a peak to a dip and vice versa.

4.3 The Effect of Molecular Vibrations on the Coulomb Blockade and Kondo Anomaly

The effects of electron–electron and electron–phonon interactions on the electron transport through MMM junctions are often considered separately. In part, this separation is interpreted basing on the assumption that the small size of molecular bridges makes their charging by more than one electron too energetically costly. Consequently, the Pauli exclusion principle alone could suffice to take care of restrictions relevant to single electron transport through molecules. Basing on this philosophy, electron–electron interactions were often disregarded in earlier works concerning electron transport through molecules. However, in the last decade both Coulomb blockade and Kondo peak in conductance were observed in experiments on single-molecule junctions. These are often accompanied by vibrational features indicating the effects of electron–phonon couplings as shown in Fig. 4.11. These observations have made obvious the necessity of treatment of the effects originating from electron–electron and electron–phonon interactions in combination. To qualitatively analyze these effects, one may again employ a simplified model where a single orbital represents the molecular bridge in the MMM junction and a sole mode with the frequency Ω represents vibrational phonons. The Hamiltonian for the junction may be rearranged using a polaron (Lang–Firsov) transformation [93]:

$$H = e^{-s}\tilde{H}e^{s};$$

$$S = \left(\frac{\Lambda}{\hbar\Omega}\right)\sum_{\sigma} c_\sigma^\dagger c_\sigma (a^\dagger - a). \tag{4.21}$$

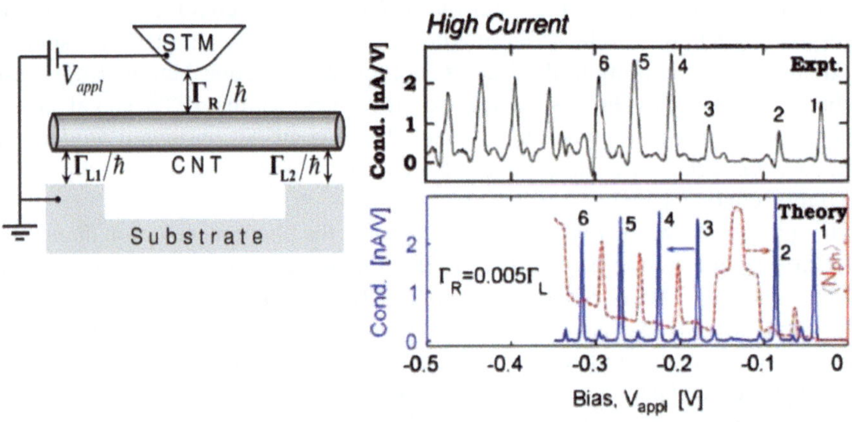

Fig. 4.11 Measurement geometry (*left panel*) and measured and calculated conductance through the carbon nanotube (*right panel*) showing Coulomb blockade peaks of fourfold degeneracy (two spins and two isospins) accompanied by equiseparated phonon-assisted tunneling sidebands (after [103]). Reprinted with the permission of the APS

Here, as before, $c_\sigma^\dagger, c_\sigma, a^\dagger, a$ are the operators of creation and annihilation of the electrons on the bridge and the vibrational phonons, respectively. This transformation gives means to decouple electron and phonon operators in the transformed Hamiltonian, which accepts the form

$$\tilde{H} = \tilde{H}_M + \tilde{H}_L + \tilde{H}_R + \tilde{H}_T + \tilde{H}_{\text{ph}} \tag{4.22}$$

where the term $\tilde{H}_M$ describing the molecular bridge is given as follows [26]:

$$\tilde{H}_M = \sum_\sigma \tilde{E}_\sigma^0 c_\sigma^\dagger c_\sigma + \tilde{U}_0 c_\sigma^\dagger c_\sigma c_{-\sigma}^\dagger c_{-\sigma}. \tag{4.23}$$

In this expression, $\tilde{E}_\sigma^0 = E_\sigma^0 - \Lambda^2/\hbar\Omega$, and $\tilde{U}_0 = U_0 - 2\Lambda^2/\hbar\Omega$. So, besides other effects, the Hamiltonian transformation brings the polaronic shift of the bridge energy level and the renormalization of the charging energy U_0. The expression for the renormalized energy $\tilde{U}_0$ gives the evidence of the effective attractive interaction between the electrons on the bridge induced by the electron–phonon coupling. This bipolaronic attraction may bring significant changes in the transport properties of metal–molecular junctions and other similar systems [241, 321].

The transfer term in the Hamiltonian (4.22) takes on the form

$$\tilde{H}_T = \sum_{k,r,\beta,\sigma} \tilde{\tau}_{kr\beta\sigma}^* c_{r\beta\sigma}^\dagger c_{k\sigma} + H.C. \tag{4.24}$$

where

$$\tilde{\tau}_{kr\beta\sigma}^* = \tau_{kr\beta\sigma}^* X \tag{4.25}$$

and the phonon shift generator X could be presented as follows [27]:

$$X = \exp\left[\frac{\Lambda}{\hbar\Omega}(a - a^\dagger)\right] \tag{4.26}$$

To complete the decoupling of electron and phonon subsystems, one may replace the phonon operator by its expectation value at thermal equilibrium:

$$X \to \langle X \rangle = \exp\left\{-\frac{\Lambda}{\hbar\Omega}\left(N + \frac{1}{2}\right)\right\} \tag{4.27}$$

where N denotes the equilibrium phonons population. So, the contact to molecule couplings now depends on additional variables associated with the vibrational mode.

The model described by Eqs. (4.21)–(4.26) or its variations were repeatedly employed to analyze vibrational effects on the electron transport through molecules/quantum dots in the Coulomb blockade and Kondo regimes. However, considerable computational difficulties which one meets applying the Hamiltonian (4.22) to transport calculations make further simplifications necessary and unavoidable. The analysis could be carried out assuming that the considered system remains in a nearly equilibrium state [98] or master equation-based approaches could be applied [103, 239, 242, 321]. Also more advanced approach using equations of motion method within the NEGF formalism was developed and successfully employed [34, 322, 323].

As known, the Kondo peak may appear in the conduction of a molecule and/or quantum dot strongly coupled to the electrodes. Electron–vibron interactions can significantly affect the Kondo anomaly in the molecular conductance [34, 321–326].To analyze the effect of electron–vibron interactions, one may start from the expressions (2.96), (2.101), (2.102) for the electron current. Basing on these expressions, the current may be presented as follows:

$$I = \frac{ie}{2h} \int dE \mathrm{Trace}\big[(\Gamma_L - \Gamma_R) G^< + (f_L \Gamma_L - f_R \Gamma_R)(G^> - G^<)\big] \tag{4.28}$$

where $G^>(E)$ and $G^<(E)$ are the Fourier transforms of the greater and less Green's functions for electrons. It may be shown that $G^>(E)$ and $G^<(E)$ are related to the less and greater Green's functions associated with the transformed Hamiltonian $\tilde{H}$ in the following way [322]:

$$\begin{aligned} G^<(E) &= \sum_{n=-\infty}^{\infty} L_n \tilde{G}^<(E + n\hbar\Omega), \\ G^>(E) &= \sum_{n=-\infty}^{\infty} L_n \tilde{G}^>(E - n\hbar\Omega). \end{aligned} \tag{4.29}$$

Here,

$$L_n = \exp\left[-\left(\frac{\Lambda}{\hbar\Omega}\right)^2 (2N+1) + \frac{n\hbar\Omega}{2k_B T}\right] J_n\left(\frac{2\Lambda}{\hbar\Omega}\sqrt{N(N+1)}\right), \tag{4.30}$$

and $J_n(z)$ denotes the Bessel function. The Green's functions $\tilde{G}^<(E)$, $\tilde{G}^<(E)$ are determined by the Hamiltonian $\tilde{H}_{\mathrm{el}} = \tilde{H}_L + \tilde{H}_R + \tilde{H}_T + \tilde{H}_M$ which is similar to the Hamiltonian of a single-site molecular bridge or quantum dot coupled to the leads. In this Hamiltonian, the effect of the phonons is reduced to the renormalization of the relevant energies E^0 and U^0 and of coupling terms Γ_β. So, one may apply the NEGF formalism to find the expressions for the retarded and advanced Green's functions $\tilde{G}^{r(a)}(E)$. The resulting expressions could be used to compute $\tilde{G}^<(E)$ and $\tilde{G}^>(E)$ which are related to $\tilde{G}^{r(a)}(E)$ by the Keldysh equations. Computing the renormalized Green's functions $\tilde{G}^>(E)$ and $\tilde{G}^<(E)$, one may substitute them into Eq. (4.29) and obtain the expressions for the greater and less Green's functions. Inserting these expressions into Eq. (4.28), one may compute the electron current and use the result to analyze the effect of phonons on the electron transport within the Coulomb blockade and/or Kondo regime.

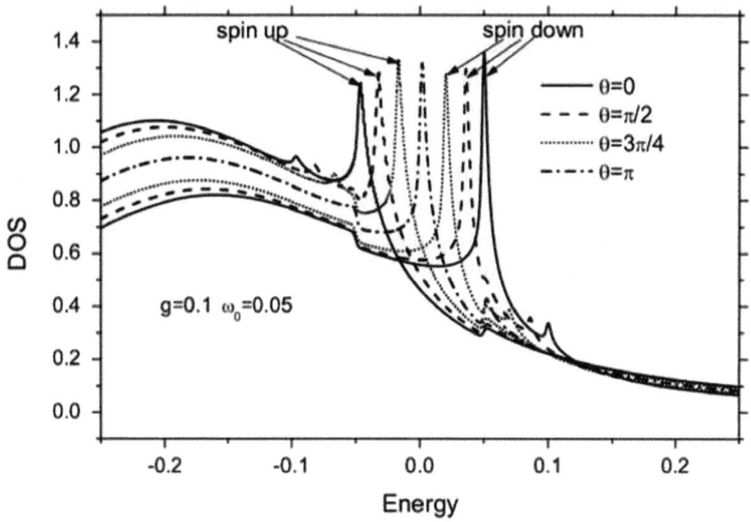

Fig. 4.12 Kondo peaks in the electron DOS on the single-level quantum dot placed in between ferromagnetic leads. The peaks are shown for both spin orientations and for the indicated values of the angle θ between the magnetic moments on the leads. Every peak is accompanied by small satellite features appearing due to the interaction of electrons on the dot with a single vibrational mode (after [322]). Reprinted with the permission of the IOP

It could be shown that vibrational sidebands may accompany the Kondo peak. This is illustrated in Fig. 4.12. These satellite peaks are arranged on both sides of the Kondo maximum, and they are situated at the distances $\pm\hbar\Omega$ from the latter. Also, the electron–phonon interaction leads to jumps in the electron density of states at the energies $\pm n\hbar\Omega$. These steps may be very sharp at low temperatures when the Fermi distribution functions have distinct stepwise profiles. When the temperature rises, these steps are being washed out. The number of these features manifested in the electron DOS versus energy profiles depends on the electron–vibron coupling strength. When the ratio $\Lambda/\hbar\Omega$ is small one may expect only the features corresponding to $n = 1$ to appear. As the electron–vibron coupling strengthens, the main Kondo peak becomes reduced due to the effect of vibrations [322]. At the same time, the satellite peaks remain well pronounced, and more steps in the electron DOS corresponding to these peaks may occur. The appearance of the satellite peaks accompanying the main Kondo maximum results in the additional peaks in the conductance through the junction. For a symmetrically coupled junction these extra maxima occur at $|e|V = \pm 2\hbar\Omega$. They could be interpreted as the onset of the phonon-assisted channel for the electron tunneling.

Another interesting effect of electrons interactions with molecular vibrations is an asymmetry in the Coulomb blockade peaks in the differential conductance dI/dV observed in experiments (see, e.g., [8, 31]). In these experiments, the measurements of the differential conductance as a function of the bias and gate voltage showed Coulomb diamonds with significantly different peak intensities at the opposite sides of the charge degeneracy point. The experiments were carried out on single-molecule junctions with specific linkers (such as di-(di-pyridil-pyrrilato) cobalt) which exhibit a tendency towards Jahn–Teller distortion. As suggested by Elste and von Oppen [327], the electron transport through easily distorted molecules may be accompanied by pronounced deformations of the latter provided that the electron–vibron coupling is sufficiently strong. Under these conditions, one should go beyond the common assumption of harmonic molecular vibrations by including anharmonicities into consideration. The authors used a simple model where the decoupling of the electron and phonon variables by means of the unitary transformation (4.21) was not applied. Instead, they supplemented the electron part of the Hamiltonian given by Eqs. (1.31) and (2.104) by the vibrational contribution:

$$H_{\rm ph} = \frac{P^2}{2M} + V_n(x) \tag{4.31}$$

describing the kinetic and potential energies for the collective vibrational mode characterized by the displacement x, the momentum P, and the mass M. The electron–phonon coupling is responsible for the dependence of the potential energy $V_N(x)$ on the molecule charge state N. Accordingly, the potential term was written out in the form

$$V_N(x) = \sum_n v(x - \sqrt{2}n\tilde{\Lambda}l_{\rm osc})|N|\rangle\langle N| \tag{4.32}$$

where $|N\rangle$ denoted the electron state of the system with N electron on the bridge. The anharmonicities may be included by choosing the corresponding profile for the potential $v(x)$. In [327], this potential was simulated by a Morse potential:

$$v(x) = D\left(e^{-2\beta x} - 2e^{-\beta x}\right). \tag{4.33}$$

Within the Coulomb blockade regime, electron current through the junction could be computed using rate equations (2.141) for the probabilities P_m^N of the molecular states with electronic occupancy N and the number of vibrational phonons m. In the considered case, the Franck–Condon matrix elements (which to a significant extent determine the values of transition rates) depend on the parameters of the Morse potential. This brings serious consequences for the Franck–Condon matrix loses its symmetry with respect to parity transformations. The Franck–Condon elements with $m' < m$ become strongly suppressed in comparison with those with $m' > m$ due to the anharmonicity of the molecular potential. Ultimately, this anharmonicity leads to the asymmetry of the Coulomb blockade peaks in the differential conductance shown in Fig. 4.13. The fine structure appearing near the peaks results from excitations of bound states of Morse oscillator, and these features occur when the bias voltage exceeds certain limits.

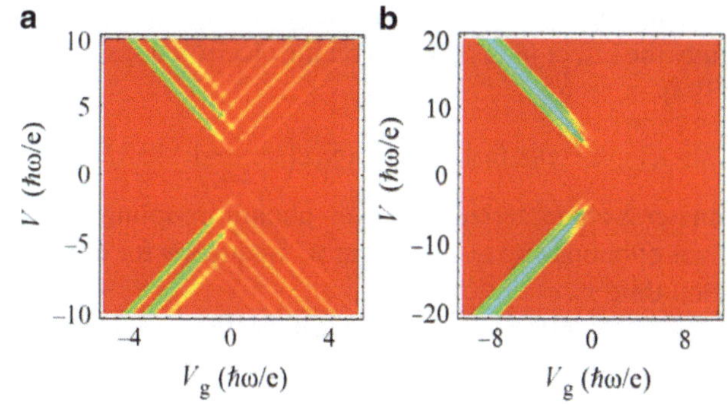

Fig. 4.13 Asymmetry in the Coulomb blockade peaks in differential conductance through a MMM junction with easily distorting linker which appear due to the effect of the vibrational mode with the frequency Ω for ($k_B T = 0.1\hbar\Omega$ (**a**) and $k_B T = 0.2\hbar\Omega$ (**b**) (after [327]). Reprinted with the permission of the IOP

4.4 Dissipative Transport: Direct Coupling of the Molecular Bridge to the Phonon Bath

At finite temperatures molecular vibrations always occur in the presence of stochastic nuclear motions. These motions could be described as a phonon thermal bath. Coupling of vibrational modes to this bath further affects the electron transport causing energy dissipation. The dissipative processes must be taken into account to properly analyze the effects of electron–phonon interactions in the electron transport. The importance of dissipative effects mostly depends on the temperature and the size and complexity of the molecular bridge. The temperature determines the intensity of the nuclear motions, and the size of the molecule determines the contact time that is the time for electron to travel through the molecular bridge. For small molecules at low temperatures the contact time is shorter than characteristic times for fluctuations in the environment. So, the effect of the latter on the electron transport is not very significant. On the contrary, in large molecules such as proteins and DNA, electron transport is accompanied by strong energy dissipation. The significance of the system-environment interactions in macromolecules was recognized long ago in studies of long-range electron transfer reactions. In these reactions electrons travel between distant sites on the molecule. It was established that when an electron initially localized on the donor site moves to the acceptor site with a lower energy, the energy difference must be dissipated to the environment to provide the irreversibility of the transfer [328, 329].

A usual way to theoretically analyze dissipative effects in the electron transport through molecules is to introduce a phonon bath representing the random motions in the environment. In general, there is no one/several dominating modes in the bath. Instead, the bath is characterized by a continuous spectral function of all relevant phonon modes $\rho_{\rm ph}(\omega)$. The electrons are supposed to be coupled to the phonon bath, and this coupling is specified by the spectral function. The particular form of

$\rho_{\text{ph}}(\omega)$ may be found basing on the molecular dynamics simulations. However, to qualitatively study the effect of dissipation on the electron transport, one can employ the expression [93]

$$\rho_{\text{ph}}(\omega) = \lambda \frac{\omega}{\omega_c} \exp\left(-\frac{\omega}{\omega_c}\right) \tag{4.34}$$

where the parameter λ characterizes electron–phonon coupling strength for interactions with thermal phonons and ω_c is the cutoff frequency for the bath related to its thermal relaxation time $\tau_c = \omega_c^{-1}$.

To illustrate possible effects of dissipation on the electron transport, we return back to our simple model where the molecular bridge is represented by a single state. Now, we assume that this state is coupled to the phonon bath. This model is hardly appropriate to properly analyze dissipative effects in the electron transport through practical molecules for the molecule length is very important for dissipative effects to be pronounced. Nevertheless, it still could serve to basically outline main features of the dissipative electron transport through molecular junctions. Also, the proposed model could be useful to analyze electron transport in doped polyacetylene/polyaniline–polyethylene oxide nanofibers [222]. As discussed above, these materials could be treated as some kind of granular metals where metallic-like regions (grains) are embedded in the poorly conducting medium of disorderly arranged polymer chains. While in metallic state, the intergrain electron transport in these nanofibers is mostly provided by electron tunneling through intermediate states on the polymer chains between the grains. In this case the contact time could be long enough for the effects of dissipation to be well manifested which justifies the adoption of the above model.

Again, one may carry out transport calculations using Eq. (2.87) for the electron transmission and expressing ϵ in terms of the relevant energies. Substituting Eq. (4.34) in the expression (2.126) we may calculate $\Gamma_{\text{ph}}(E)$. We obtain

$$\begin{aligned} \Gamma_{\text{ph}} =& 2\pi\lambda \int d\omega \frac{\omega}{\omega_c} \exp\left(-\frac{\omega}{\omega_c}\right) \\ & \times \big\{N(\omega)[D(E-\hbar\omega) + D(E+\hbar\omega)] \\ & + [1-n(E-\hbar\omega)]D(E-\hbar\omega) \\ & + n(E+\hbar\omega)D(E+\hbar\omega)\big\}. \end{aligned} \tag{4.35}$$

Here, $n(E) = \big(\Gamma_L f_L + \Gamma_R f_R\big)/(\Gamma_L + \Gamma_R)$, $N(\omega)$ is the Bose–Einstein distribution function for the phonons at the temperature T. The asymptotic expression for the self-energy term Γ_{ph} depends on the relation between two characteristic energies, namely, $\hbar\omega_c$ and $k_B T$. At moderately low or room temperatures, $k_B T \sim$ 10–30 meV. This is significantly greater than typical values of $\hbar\omega_c$ ($\hbar\omega_c \sim 1$ meV [25]). Therefore, in further calculations, we assume $\hbar\omega_c \ll k_B T$. Under this assumption, the main contribution to the integral over ω in Eq. (4.35) originates from the region where $\omega \ll \omega_c \ll k_B T/\hbar$, and we can use the following approximation:

$$\Gamma_{\text{ph}}(E) = \frac{2\Gamma M(\lambda, \omega_c, T)}{(E-E_0)^2 + \Gamma^2}. \tag{4.36}$$

Here, $\Gamma = \Gamma_L + \Gamma_R + \frac{1}{2}\Gamma_{\text{ph}}(E)$

$$M = \frac{4\lambda}{\hbar\omega_c}(k_B T)^2 \zeta\left(2; \frac{k_B T}{\hbar\omega_c} + 1\right) \tag{4.37}$$

where $\zeta(2; k_B T/\hbar\omega_c + 1)$ is the Riemann ζ function:

$$\zeta = (2; k_B T/\hbar\omega_c + 1) = \sum_{n=1}^{\infty} \frac{1}{(n + k_B T/\hbar\omega_c)^2}. \tag{4.38}$$

Under $\hbar\omega_c \ll k_B T$, we may apply the estimation $M \approx 4k_B T\lambda$.

Solving Eq. (4.36), we obtain a reasonable asymptotic expression for Γ_{ph} :

$$\Gamma_{\text{ph}} = \frac{\Gamma_L + \Gamma_R}{2} \frac{\rho^2(1 + \sqrt{1+\rho^2})}{4\left(\frac{E - E_0}{\Gamma_L + \Gamma_R}\right)^2 + (1 + \sqrt{1+\rho^2})^2} \tag{4.39}$$

where $\rho^2 = 32k_B T\lambda/(\Gamma_L + \Gamma_R)^2$. Substituting this result into the expression (4.5) for the dephasing/dissipative parameter ϵ, we obtain [222]:

$$\epsilon = \frac{1}{2} \frac{\rho^2(1 + \sqrt{1+\rho^2})}{4\left(\frac{E - E_0}{\Gamma_L + \Gamma_R}\right)^2 + \frac{1}{2}(1 + \sqrt{1+\rho^2})^3} \tag{4.40}$$

This expression shows how ϵ depends on the temperature T, the electron–phonon coupling strength λ, and the energy E. In particular, it follows from Eq. (4.40) that ϵ reaches its maximum at $E = E_0$ and the peak value of this parameter is given by

$$\epsilon_{\max} = \frac{\sqrt{1+\rho^2} - 1}{\sqrt{1+\rho^2} + 1}. \tag{4.41}$$

The maximum value of the dissipative strength is determined with two parameters, namely, T and λ. As illustrated in Fig. 4.14, $\epsilon_{\max}$ increases when the temperature rises, and it takes on greater values when the electron–phonon interaction is getting stronger. This result has a clear physical sense. Also, as follows from Eq. (4.40), the dissipation parameter exhibits a peak at $E = E_0$ whose shape is determined by the product $k_B T\lambda$. When either λ or T or both enhance, the peak becomes higher and its width increases. The manifested energy dependence of the dissipation strength allows us to resolve a difficulty occurring when the inelastic contribution to the electron transmission function is estimated using the simplified approximation of the parameter ϵ by a constant. Within this model one must assume low values of ϵ to secure the peak in the transmission from being washed out, and this gives rise to doubts. When the energy dependence of ϵ is accounted for, the peak in the electron transmission at $E = E_0$ may be still distinguishable when $\epsilon_{\max}$ takes on values as big as 0.5.

The obtained result enables us to analyze the temperature dependencies of the electric current and conductance through molecular bridges. Current–voltage

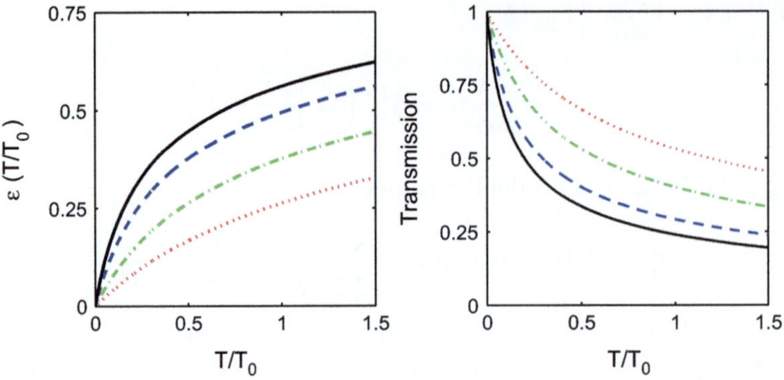

Fig. 4.14 Temperature dependencies of the maximum dissipation parameter ϵ (*left panel*) and the electron transmission (*right panel*). The curves are plotted at $T_0 = 50\,\mathrm{K}$, $\Gamma_L = \Gamma_R = 4\,\mathrm{meV}$, $E = E_0 = 0$, $\lambda = 9.0\,\mathrm{meV}$ (*solid line*), $6.0\,\mathrm{meV}$ (*dashed line*), $3\,\mathrm{meV}$ (*dash-dotted line*), $1.5\,\mathrm{meV}$ (*dotted line*) [222]. Reprinted with the permission of the AIP

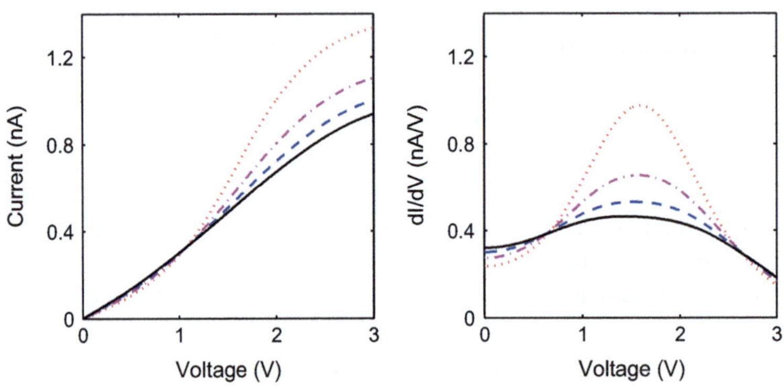

Fig. 4.15 Current (*left panel*) and conductance (*right panel*) versus bias voltage. The curves are plotted at $T = 30\,\mathrm{K}$, $E_1 = 40\,\mathrm{meV}$, $\Gamma_1 = \Gamma_2 = 4\,\mathrm{meV}$, assuming $\lambda = 9\,\mathrm{meV}$ (*solid line*), $\lambda = 6\,\mathrm{meV}$ (*dashed line*), $\lambda = 3\,\mathrm{meV}$ (*dash-dotted line*), and $\lambda = 1.5\,\mathrm{meV}$ (*dotted line*) [222]. Reprinted with the permission of the AIP

characteristics and voltage dependencies of the conductance $g = dI/dV$ computed using Eq. (4.40) for the dissipation strength ϵ are presented in Fig. 4.15. We see that at low values of the applied voltage, the electron–phonon coupling brings an enhancement in both current and conductance. The effect becomes reversed as the voltage grows above a certain value. This happens because the phonon-induced broadening of the bridge energy level assists the electron transport at small bias voltage. As the voltage rises, this effect is surpassed by the scattering effect of phonons which hinders the electron transport. When the electron–phonon coupling strengthens, the I–V curves lose their specific shape typical for the elastic tunneling through the intermediate state. They become closer to straight lines corresponding to the Ohmic law. At the same time, the maximum in the conductance originating from the coherent tunneling through the bridge gets eroded. These are the obvious

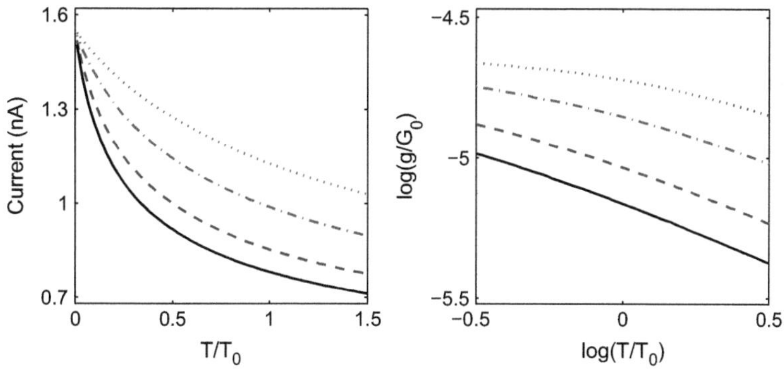

Fig. 4.16 Temperature dependence of current (*left panel*) and conductance (*middle panel*) at low voltage bias ($V = 1.5\,\mathrm{mV}$) and (*right panel*) temperature dependence of the electron current at higher bias voltage ($V = 10\,\mathrm{mV}$). The curves are plotted at $T_0 = 50\,\mathrm{K}$, $\lambda = 9\,\mathrm{meV}$ (*solid line*), $\lambda = 6\,\mathrm{meV}$ (*dashed line*), $\lambda = 3\,\mathrm{meV}$ (*dash-dotted line*), and $\lambda = 1.5\,\mathrm{meV}$ (*dotted line*). The values of the remaining parameters are the same as in Fig. 4.15. The current I_0 is computed using $\lambda = 3\,\mathrm{meV};\ T_0/T = 1.5$

results discussed in some earlier works (see, e.g., [96]). The relative strength of the electron–phonon interaction is determined by the ratio of the electron–phonon coupling constant λ and the self-energy terms describing the coupling of the intermediate state (bridge) to the leads Γ_β. The effect of phonons on the electron transport becomes significant when $\lambda > \Gamma_{L,R}$. Otherwise, the coherent tunneling between the electrodes prevails in the electron transport, and the influence of thermal phonon bath is weak. Again, we may remark that λ and T are combined as $k_B T\lambda$ in the expression (4.40). Therefore, an increase in temperature at a fixed electron–phonon coupling strength enhances the inelastic contribution to the current in the same way as the previously discussed increase in the electron–phonon coupling.

Now, we consider temperature dependencies of the electric current and conductance. These dependencies are shown in Fig. 4.16. The curves in the figure are plotted at low bias voltage so that $|e|\Delta V < k_B T$. This regime is chosen to compare the obtained temperature dependencies with those typical for the phonon-assisted hopping transport. It is clear that the temperature dependence of the electron current presented in the left panel of Fig. 4.16 differs from that described by the well-known Mott's expression for the hopping conductivity $\sigma(T)$ which has the form

$$\sigma(T) = \sigma(0)\exp\left[-(T/T_0)^p\right] \tag{4.42}$$

where T_0 is the characteristic temperature of a particular material and the parameter p takes on values 0.25, 0.33, and/or 0.5 depending on the dimensions of the hopping process. On the contrary, the current shown in the left panel of Fig. 4.16 is proportional to $(T_0/T)^\beta$, and the exponent β takes on values close to minus one.

The drop in the conductivity upon heating a sample was observed in polymers and carbon nanotubes. However, such metallic-like behavior could originate from various dc transport mechanisms. Correspondingly, the specific features of temperature dependencies of the conductivity and/or current vary depending on the responsible conduction mechanism. The particular temperature dependence of the electron tunneling current shown in Fig. 4.16 differs from those occurring due to other transport mechanisms. Such dependence was observed in the experiments on the electron transport in a single low-defect-content carbon nanotube rope, whose metallic-like conductivity was manifested within a wide temperature range ($T \sim 35$–$300\,\mathrm{K}$), as reported in [95]. The observed temperature dependence of conductivity was approximated as $\sigma(T)/\sigma(300) \sim a + bT_0/T$ where A, b are dimensionless constants. The approximation includes the temperature-independent term, which corresponds to the Drude conductivity. The second term is inversely proportional to the temperature in agreement with the results for the current shown in Fig. 4.16 (left panel). It is also likely that a similar approximation may be adopted to describe experimental data obtained for chlorate-doped polyacetylene samples at the temperatures below 100 K [95].

The conductance due to the electron tunneling current reduces when the temperature increases, as shown in the middle panel of Fig. 4.16. Regardless of the electron–phonon coupling strength, we may approximate the conductance by a power law $g \sim T^{\alpha}$ where α takes on values close to -1. This agrees with the results for the current. At higher bias voltage, the temperature dependence of the current changes, as shown in the right panel of Fig. 4.16. The curves shown in this panel could be approximated as $\log(I/I_0) \sim c + dT_0/T$, c, d being dimensionless constants. This resembles typical temperature dependencies of the tunneling current in quasi-one-dimensional metals (see, e.g., [95]).

4.5 Dissipative Transport: Indirect Coupling of the Molecular Bridge to Thermal Phonons

Dissipative electron transport through large DNA molecules was studied both theoretically and experimentally [20, 183, 317, 328]. Theoretical studies were based on a model where the molecule was simulated by a tight-binding chain of sites linking the electrodes and attached side chains. Electrons were allowed to travel along the bridge chain and to hop to the nearby side chains. In studies of dissipative transport, these chains were supposed to be coupled to the phonon bath providing the energy dissipation. Although first proposed for the specific kind of poly(dG)-poly(dC) molecules, this model seems to be quite generic and useful for a larger class of macromolecules. Here, for simplicity we apply this model to a MMM junction where the molecular bridge is represented by a single orbital with the single side chain attached to it. Within the adopted model the retarded Green's function for the bridge acquires the form [20, 330]:

$$G^{-1}(E) = E - E_0 - \Sigma_L - \Sigma_R - w^2 P(E). \tag{4.43}$$

The first four terms in this expression represent the inversed Green's function for the resonance site coupled to the leads, and the factor w is the hopping integral between the bridge and the attached side chain. The term $P(E)$ represents the effect of the phonons and has the form

$$P(E) = -i \int_0^\infty dt \exp\left[it(E - \tilde{E}_0 + \lambda + i\eta)\right] \times \left\{(1-f)\exp[-F(t)] + f\exp[-F(-t)]\right\} \tag{4.44}$$

with $\tilde{E}_0$ being the on-site energy for the side site, which is close to the bridge site energy E_0, $\exp[-F(t)]$ being a dynamic bath correlation function, and f taking on values 1 or 0 when the attached site is occupied or empty, respectively.

Characterizing the phonon bath with a continuous spectral density $\rho_{\rm ph}(\omega)$ given by Eq. (4.34), one may write out the following expressions for the functions $F(t)$ and λ (the latter describes the effect of electron–phonon coupling):

$$F(t) = \int_0^\infty \frac{d\omega}{\omega^2}\rho_{\rm ph}(\omega)\left[1 - e^{-i\omega t} + \frac{2[1-\cos(\omega t)]}{\exp(\hbar\omega/k_{\rm B}T) - 1}\right], \tag{4.45}$$

$$\lambda = \int_0^\infty \frac{d\omega}{\omega}\rho_{\rm ph}(\omega). \tag{4.46}$$

Within the short time scale ($\omega_c t \ll 1$) the function $F(t)$ could be presented in the form

$$F(t) = \frac{\lambda}{\hbar\omega_c}\left\{\frac{1}{2}\ln[1 + (\omega_c t)^2] + i\arctan(\omega_c t) + K(t)\right\} \tag{4.47}$$

where

$$K(t) = (k_{\rm B}T)^2 t^2 \zeta\left(2; \frac{k_{\rm B}T}{\hbar\omega_c} + 1\right). \tag{4.48}$$

Here, again, $\zeta\left(2; k_{\rm B}T/\hbar\omega_c + 1\right)$ is the Riemann ζ function. The asymptotic expression for $K(t)$ depends on the relation between two parameters, namely, the temperature T and the cutoff frequency ω_c of the phonon bath. Assuming $k_{\rm B}T \gg \hbar\omega_c$,

$$K(t) \approx \frac{k_{\rm B}T}{\hbar\omega_c}(\omega_c t)^2. \tag{4.49}$$

In the opposite limit $\hbar\omega_c \gg k_{\rm B}T$ we obtain

$$K(t) \approx \frac{\pi^2}{6}\left(\frac{k_{\rm B}Tt}{\hbar}\right)^2 \tag{4.50}$$

Also, we may roughly estimate $K(t)$ within the intermediate range. Taking $k_B T \approx \hbar\omega_c$, we arrive at the approximation $K(t) \approx a^2(k_B T t/\hbar)^2$ where a^2 is a dimensionless constant of the order of unity. Correspondingly, within the short time scale, we can omit the first term in the expression (4.47), and we get

$$F(t) \approx \frac{\lambda}{\hbar\omega_c}\{i\omega_c t + K(t)\} \tag{4.51}$$

where $K(t)$ is given by either Eq. (4.49) or (4.50), depending on the relation between $\hbar\omega_c$ and $k_B T$.

Within the longtime scale $\omega_c t \gg 1$ and provided that temperatures are not very low ($k_B T \gg \hbar\omega_c$), we may present the function $K(t)$ as

$$\begin{aligned} K(t) &= \frac{2k_B T t}{\hbar}\int_0^\infty \frac{dz}{z^2}(1-\cos z)e^{-z/\omega_c t} \\ &\approx \frac{\pi k_B T t}{\hbar}. \end{aligned} \tag{4.52}$$

Now, the term $K(t)$ is the greatest addend in the expression for $F(t)$, so the latter could be approximated as $F(t) \approx \pi k_B T t\lambda/\omega_c\hbar^2$. The same approximation holds within the low-temperature limit when $\omega_c \gg k_B T/\hbar \gg t^{-1}$.

Using the asymptotic expression (4.51), we may calculate the contribution to $P(E)$ coming from the short time scale ($\omega_c t \ll 1$). It has the form

$$\begin{aligned} P_1(E) &= -\frac{i}{2}\sqrt{\frac{\pi}{\lambda k_B T}}\exp\left[-\frac{(E-\tilde{E}_0)^2}{4\lambda k_B T}\right] \\ &\times \left\{1 + \Phi\left[\frac{i(E-\tilde{E}_0)}{2\sqrt{\lambda k_B T}}\right]\right\} \end{aligned} \tag{4.53}$$

where $\Phi(z)$ is the probability integral. When both $\hbar\omega_c$ and $k_B T$ have the same order of magnitude, the expression for $P(E)$ still holds the form (4.53). At $k_B T \ll \hbar\omega_c$, the temperature $k_B T$ in the expression (4.53) is to be replaced by $\hbar\omega_c$. We remark that under the assumption $k_B T \gg \hbar\omega_c$, the function $P_1(E)$ does not depend on the cutoff frequency ω_c, whereas at $\hbar\omega_c \gg k_B T$, it does not depend on temperature. The longtime ($\omega_c t \gg 1$) contribution to $P(E)$ could be similarly estimated as follows:

$$P_2(E) = \frac{1}{E - \tilde{E}_0 + \lambda + i\pi\lambda k_B T/\hbar\omega_c}. \tag{4.54}$$

Comparing these expressions (4.53) and (4.54), we see that the ratio of the peak values of $P_2(E)$ and $P_1(E)$ is of the order of $(\hbar^2\omega_c^2/\lambda k_B T)^{1/2}$. Therefore, the term $P_1(E)$ predominates over $P_2(E)$ when the temperatures are moderately high

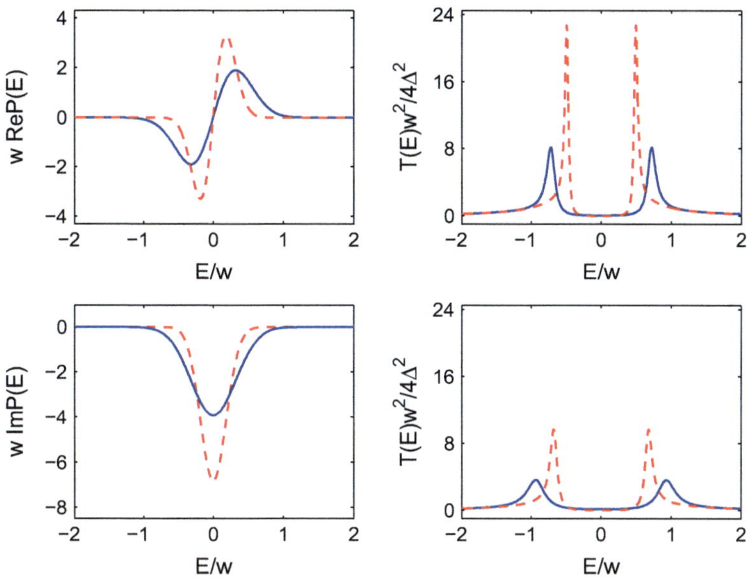

Fig. 4.17 *Left panels*: energy dependence of the real (*top panel*) and imaginary (*bottom panel*) parts of $P(E)$. The curves are plotted at $\lambda = 20\,\mathrm{meV}$, $k_B T \gg \hbar\omega_c$. *Right panels*: the renormalized electron transmission function versus energy. The constant λ equals $20\,\mathrm{meV}$ (*top panel*) and $50\,\mathrm{meV}$ (*bottom panel*). All curves are plotted assuming $w = 100\,\mathrm{meV}$, $E_0 = \tilde{E}_0 = 0$, $\Gamma = \Gamma_L + \Gamma_R$, $T = 100\,\mathrm{K}$ (*dashed lines*), and $T = 300\,\mathrm{K}$ (*solid lines*) [330]. Reprinted with the permission of the AIP

$(\hbar\omega_c < k_B T)$ and the electron–phonon interaction is not too weak $\lambda/\hbar\omega_c \sim 1$. Assuming that the temperature is not too low $(\hbar^2\omega_c^2/\lambda k_B T)^{1/2} \ll 1$, the term $P_2(E)$ could be omitted, so $P(E) \approx P_1(E)$. As shown in Fig. 4.17, the imaginary part of $P(E)$ exhibits a dip around $E = \tilde{E}_0$, and the width of the latter is determined by the product of the temperature $k_B T$ (or $\hbar\omega_c$) and the constant λ characterizing the strength of the electron–phonon interaction. When either factor increases, the dip becomes broader and its magnitude reduces.

The most referred regime for dissipative effects to appear is the strong-coupling limit defined by the condition $\lambda/\hbar\omega_c > 1$. Within this regime, the characteristic time for the electron–bath interactions is much shorter than typical electron time scales. Consequently, the bath makes a significant impact on the molecule electronic structure. New bath-induced states appear in the molecular spectrum inside the HOMO–LUMO gap. However, these states are strongly damped due to the dissipative action of the bath [20]. As a result, a small finite density of phonon-induced states appears inside the gap supporting electron transport at low bias voltage. So again, the environment induces incoherent phonon-assisted transport through molecular bridges. For illustration we show here the results of calculations carried out for a model with a single-site bridge and a side chain attached to the latter [330]. The side chain is supposed to be coupled to the phonon bath. We see that the original bridge state at $E = 0$ is completely damped, but two new phonon-induced states emerge nearby which could support electron transport.

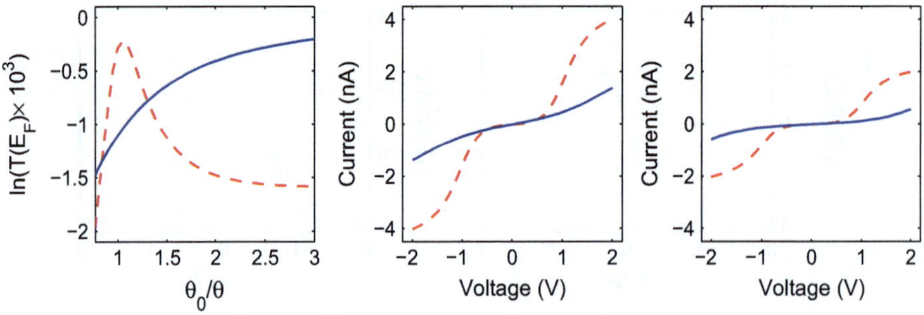

Fig. 4.18 *Left panel*: Arrhenius plot of the peak value of the electron transmission function for $\lambda = 20\,\text{meV}$, $w = 100\,\text{meV}$, $\Gamma_L = \Gamma_R = 0.5\,\text{meV}$, and $T_0 = 100\,\text{K}$. The *dashed line* is plotted assuming the indirect coupling of the bridge to the phonon bath. The *solid line* is plotted assuming that the bridge is directly coupled to the phonons. The current–voltage characteristics (nA–V) plotted at $\Gamma_L = \Gamma_R = 0.5\,\text{meV}$, $w = 100\,\text{meV}$, $\lambda = 20\,\text{meV}$ (*middle panel*), and $\lambda = 50\,\text{meV}$ (*right panel*) at $T = 100\,\text{K}$ (*dash lines*) and $T = 300\,\text{K}$ (*solid lines*) (after [330]). Reprinted with the permission of the AIP

Assuming $E_0 = \tilde{E}_0 = 0$, the new states are situated at [330]:

$$E = \pm 2\sqrt{\lambda k_B T |\ln(2\lambda k_B T / w^2)|}. \tag{4.55}$$

These new states are revealed in the structure of the electron transmission $T(E)$. The structure of $T(E)$ is shown in Fig. 4.17. Two peaks in the transmission are associated with the phonon-induced electronic states. Their positions and heights depend on the temperature and on the coupling strengths λ and w. The important feature in the electron transmission is the absence of the peak associated with the bridge site itself. This happens due to the strong suppression of the latter by the effects of the environment. Technically, this peak is damped for it is located at $E = 0$ where the imaginary part of $P(E)$ reaches its maximum in magnitude. To provide the damping of the original resonance, the contribution from the environment (including the side chain attached to the bridge) to the Green's function (4.43) must exceed the terms $\Gamma_{L,R}$, describing the effect of the leads. This occurs when the inequality

$$\Gamma_{L,R} < w^2/\sqrt{\lambda k_B T} \tag{4.56}$$

is satisfied. When the coupling of the bridge to the attached side site is weak, the influence of the environment slackens, and the original peak associated with the bridge at $E = E_0$ may emerge. At the same time, the features originating from the phonon-induced states become small compared to this peak.

The phonon-induced peaks in the electron transmission determine the shape of the current–voltage curves at reasonably high values of the bias voltage applied across the junction. Current–voltage characteristics are shown in Fig. 4.18. The I–V curves exhibit a nonlinear shape even at room temperature despite the fact that the original state for the resonance tunneling is completely suppressed. This occurs because the electron transport through the junction is supported by new phonon-induced electron states.

It is worthwhile to discuss the temperature dependence of the peak value of the electron transmission which follows from the present results. Using the expression (4.43) for the Green's function and the expression (4.53) for $P(E)$, we may show that at low temperatures the transmission accepts small values and exhibits rather weak temperature dependence. At higher temperatures ($T \sim 100K$), the transmission increases fast as the temperature rises, and then it reduces as the temperature further increases. The peak in the transmission is associated with the most favorable conditions for the environment-induced states to appear when all remaining parameters (such as λ and w) are fixed. At high temperatures the peaks associated with the environment-induced states are washed out, as usual.

We may compare this result with the temperature dependence of the electron transmission function occurring when the molecular bridge is directly coupled to the phonon bath. In this case the transmission peak value may be presented in the form determined by Eqs. (2.87) and (4.40):

$$T(E) = 1 - \frac{\rho^4}{2(1 + \sqrt{1 + \rho^2})^2}. \tag{4.57}$$

The temperature dependencies are shown in the left panel of Fig. 4.18. Both curves are plotted at the same value of the electron–phonon coupling strength λ. Comparing them we conclude that at higher temperatures the dependencies significantly differ. While the temperature rises, we observe a peak in the electron transmission assuming the indirect coupling of the bridge to the phonon bath, and we see the transmission to monotonically decrease when we consider the bridge directly coupled to the bath. Correspondingly, we may expect qualitative diversities in the temperature dependencies of the current as well. These diversities originate from the difference in the effects of environment on the electron transport through a molecular junction in the cases of direct and indirect coupling of the bridge site to the phonon bath. When the bridge is directly coupled to the bath, the stochastic motions in the environment only cause washing out of the peak in the electron transmission, and the higher is the temperature, the less distinguishable is the peak. However, when the bridge is screened from the direct coupling with the phonons due to the presence of the nearby sites, the stochastic nuclear motions in the ambience may take a very different part in the electron transport through MMM junctions at moderately low and room temperatures. Due to their influence, the original intermediate state for the electron tunneling may be completely suppressed, but new phonon-induced states may appear to support the electron transport through the system.

4.6 Current Fluctuations

Electron current fluctuations (noise) provide an important source of information concerning transport properties of MMM junctions and other similar systems. Standard analysis of current noise usually considers its spectrum $S(\omega)$ which is defined as follows:

$$S(\omega) = 2\int_{-\infty}^{\infty} dt S(t) e^{i\omega t} \tag{4.58}$$

where

$$S(t) = \frac{1}{2}\{\Delta I(t)\Delta I(0) + \Delta I(0)\Delta I(t)\} \tag{4.59}$$

and

$$\Delta I(t) = I(t) - I \tag{4.60}$$

The quantity $\Delta I(t)$ introduced by Eq. (4.60) describes the current fluctuations with respect to its average value I.

The current fluctuations (noise) occur due to several reasons including thermal excitations in the contacts and the so-called shot noise. The latter originates from the discrete nature of the electron transport through a molecule/quantum dot, and it can last even at low temperatures when thermal excitations cease to significantly contribute to the current fluctuations [331]. Also, electron–phonon interactions accompanying electron transfer through the molecular bridge may bring a substantial contribution to the noise which could be reflected in the noise spectrum $S(\omega)$. This contribution was first studied in connection with nanoelectromechanical systems [332, 333], and the studies of features appearing in $S(\omega)$ due to electron–vibron interactions continue ever since. Presently, several methods are used to theoretically analyze the noise spectrum. In some works the analysis is based on the scattering theory approaches, whereas in other works the NEGF formalism or Monte Carlo simulations are applied.

A common experimental situation suggests that the measurement time is much longer than a typical characteristic time for electron transfer through the molecule. Under these conditions, one may restrict the analysis of the noise spectrum to zero-frequency noise $S(0)$. Accordingly, noise intensity is characterized by means of the so-called Fano factor F. The Fano factor is defined as the ratio of $S(0)$ and the average current I. While studying the effects of electron–vibron interactions, the relevant characteristics ($S(0)$, I, and F) are usually normalized by their counterparts S_0, I_0, and F_0 computed in the absence of electron–phonon interactions.

Omitting for a while electron–phonon interactions, the noise intensity is determined by the temperature and by electrostatic properties of the electrodes described by their capacitances $C_{L,R}$. Also important are the division parameter η which determines the bias voltage distribution over the system [see Eq. (3.4)] and the asymmetry in the lead-molecule couplings described by the parameter α. The capacitances $C_{L,R}$ are important, for their values control the response of the corresponding lead-molecule interfaces to the charge accumulation, which determine the instantaneous current through the junction. Following [334], one may right the current in the form

$$I(t) = \frac{C_L I_L(t) + C_R I_R(t)}{C_L + C_R} \tag{4.61}$$

where the currents $I_{L,R}(t)$ are flowing through the left and right lead, respectively. The parameter α describing the lead-molecule coupling asymmetry may be defined as follows [26]:

$$\begin{aligned} \alpha &= \Gamma_L/\Gamma, \\ \Gamma &= \Gamma_L + \Gamma_R. \end{aligned} \tag{4.62}$$

Comparing this definition to that chosen for the division parameter ($\eta = \Gamma_R/\Gamma$), one may erroneously conclude that there exists a very simple relation between these two parameters ($\alpha = 1 - \eta$) which makes the parameter α unnecessary if η is already introduced. However, this seemingly obvious conclusion is valid only when one ignores the current fluctuations. When these fluctuations are taken into consideration, the relationship between α and η becomes more complicated. Basing on the equivalent circuit arguments, one can derive the following relationship between the relevant parameters [334]:

$$\frac{\alpha}{\alpha - 1} = \frac{1 - \eta}{\eta} \frac{C_R}{C_L}. \tag{4.63}$$

It was shown that the noise intensity depends on the bias voltage applied across the junction. The profile of the differential noise intensity $dS(0)/dV$ is determined by the inherent asymmetry of the considered MMM junction. For nearly symmetrically coupled junctions ($\alpha \approx 0.5$) the profile is a double-peak structure. At low temperatures $k_B T \ll \Gamma$ when the effect of thermal excitations in the leads becomes negligible, the peaks have identical height. At higher temperatures their heights differ due to the thermal noise contributions. For strongly asymmetrically coupled junctions ($\alpha \ll 1$), the differential noise profile is converted from the double-peak structure to a single peak, as shown in Fig. 4.19. Within a simplified model simulating the molecular bridge by a single orbital, the profile transformation occurs when the asymmetry parameter α satisfied the equation $\alpha^2 - \alpha + 0.125 = 0$ [334].

When the electron interacts with the vibrational phonons, this considerably affects the noise spectrum. In the presence of electron–vibron couplings, $S(0)$ can no longer be treated as a sum of thermal and shot noise contributions [25,26]. Also, in this case the noise spectrum strongly depends on the electron–vibron coupling strength in addition to the junction characteristics $C_{L,R}$ and α. Again employing the model where the bridge is represented by a single orbital and vibrational phonons by a single mode with the frequency Ω, it was shown that the noise intensity $S(0)$ and Fano factors vary depending on the relationship between the electron–vibron coupling strength Λ and the molecule–leads coupling parameters $\Gamma_{L,R}$. Another factor controlling the effect of electron–vibron interactions on the noise intensity is the separation between the bridge level E_0 and the equilibrium Fermi energy of the

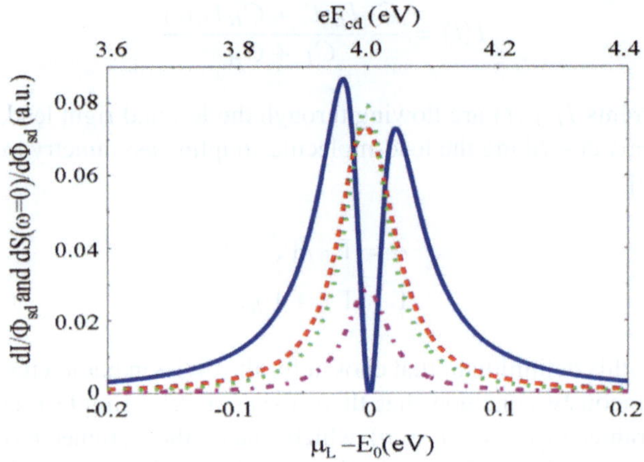

Fig. 4.19 Differential conductance dI/dV (*dash-dotted line*) and noise intensity $dS(0)/dV$ (*solid and dashed lines*) versus bias voltage for a MMM junction with a single-site bridge in the absence of electron–vibron interactions. The differential noise profiles are plotted assuming $\alpha = 0.5$ (*solid line*) and $\alpha = 0.1$ (*dashed line*) for $T = 10\,\mathrm{K}$, $\Gamma_L = \Gamma_R = 0.04\,\mathrm{eV}$ (after [334]). Reprinted with the permission of the APS

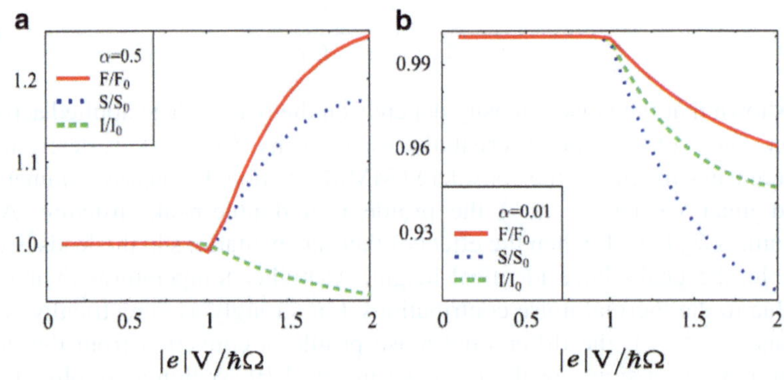

Fig. 4.20 Ratios of Fano factors (*solid lines*) and the noise intensities $S(0)$ (*dotted lines*) with and without electrons coupling with the vibrational mode with the frequency Ω plotted against bias voltage V, assuming $\alpha = 0.5$ (*left*) and $\alpha = 0.1$ (*right*) (after [334]). Reprinted with the permission of the APS

system E_{F}. Within the resonant tunneling regime ($|E_0 - E_{\mathrm{F}}| < \Gamma_{L,R}$), the current fluctuations are more responsive to the effect of vibrons coupled to the traveling electrons. Within the off-resonance tunneling regime ($|E_0 - E_{\mathrm{F}}| > \Gamma_{L,R}$), the effect of vibrons is significantly weakened.

Two illustrating examples are presented in Fig. 4.20. The displayed curves show the dependencies of renormalized noise intensity $S(0)/S_0$ and the Fano factor F/F_0 of the bias voltage. The curves are plotted assuming that the interaction of electron with the vibrational mode is strong ($\Lambda > \sqrt{(E_0 - E_{\mathrm{F}})^2 + \Gamma^2/4}$). Usually, this condition is satisfied in resonance tunneling situations provided that coupling of

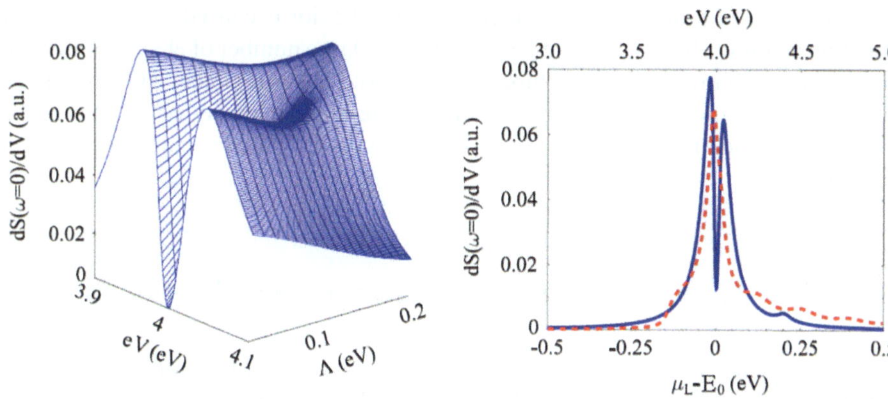

Fig. 4.21 *Left panel*: differential noise as a function of applied bias voltage V and the electron–vibron coupling strength Λ for a symmetrically coupled junction ($\alpha = 0.5$). *Right panel*: differential noise plotted against the gate potential for weak ($\Lambda/\hbar\Omega = 0.2$, *solid line*) and strong ($\Lambda/\hbar\Omega = 1.5$, *dashed line*) electron–vibron coupling. The presented results are obtained within the model where the molecular bridge is simulated by a single orbital and the molecular vibrations are represented by a single vibrational mode (after [334]). Reprinted with the permission of the APS

the bridge to the leads is moderate. The effect of the vibrational mode becomes distinguishable at the excitation threshold $|e|V/\hbar\Omega = 1$. At low bias voltage the renormalized characteristics remain independent on the bias ($S(0)/S_0 = F/F_0 = 1$). When the voltage V exceeds the threshold value, these quantities show pronounced dependencies of V, indicating the effect of the molecular vibrations. The character of these dependencies is determined by the specifics of the considered system. In particular, the value of the asymmetry parameter α is important, as illustrated in Fig. 4.20.

In the presence of electron–vibron interactions, the differential noise profile displays the central elastic feature at the energy of the bridging orbital which could be accompanied by phonon sidebands. As well as in the previously considered case when there is no vibrons in the system, the central feature could be shaped either as a double-peak structure or as a single peak. As discussed before, the transformation of the double-peak to the single-peak structure may be generated by varying the asymmetry parameter α. Electron–vibron coupling strength may take on a similar part. Assuming that the junction is symmetrically coupled, it was proven that the increase in the electron–vibron coupling strength Λ may stipulate the conversing of the double-peak structure into the single peak. This is shown in Fig. 4.21.

Very strong electron–phonon interactions may give rise to the variety of other noise-related phenomena. For example, sufficiently strong electron–phonon coupling may lead to multistable behavior of the system that can be manifested as intermittent noise in the electron current [335]. Moreover, strong electron–vibron interactions may lead to pronounced conformational fluctuations in the molecular bridge, and these can noticeably contribute to the observed noise. Also, it was shown that strong electron–phonon coupling associated with the Franck–Condon blockade leads to electron transport by avalanches. This behavior is caused by sequences of junction heating which is followed by transfer of a large number of electrons through the junction occurring once the blockade threshold is surmounted. This avalanche behavior is manifested as a sharp increase of the noise intensity $S(0)$, so that the Fano factor could reach values of the order of 10^2–10^3, which is much greater than unity [238].

4.7 Heat Transfer, Thermal Conductance, and Thermopower

It is known that in insulating materials heat is transferred by atomic vibrations, whereas in metals, conduction electrons are the dominant carriers. In metal–molecular junctions carriers of both kinds contribute to the heat conduction. Therefore, a unified description of the electrons and phonons dynamics is needed to thoroughly analyze thermoelectric effects in transport through these systems. The means for the general analysis of these effects are provided by NEGF. The starting point for analysis is given by the Hamiltonian (1.34)–(1.36), (1.40)–(1.42). The electron and phonon Green's functions defined by Eqs. (2.86)–(2.90) are to be used to compute the relevant particle and energy fluxes. At steady state, the net electronic energy flux into the junction may be estimated as the sum of the energy fluxes $Q_{\rm el}$ at the left and right terminals (leads):

$$P = Q_L + Q_R. \tag{4.64}$$

One may express the energy fluxes in terms of renormalized currents at the electrodes $\tilde{I}_{L,R}(E)$ which are defined as follows [4]:

$$I_{L,R}^{\rm el} = \frac{e}{\pi\hbar}\int \tilde{I}_{L,R}(E)dE. \tag{4.65}$$

Then $Q_{L,R}$ may be presented in the form

$$Q_{L,R}^{\rm el} = \frac{1}{\pi\hbar}\int E\tilde{I}_{L,R}(E)dE. \tag{4.66}$$

We remark that the currents $I_{L,R}$ in Eq. (4.65) are related to the corresponding leads and their signs are accordingly defined. An outgoing current is supposed to be positive, whereas an incoming one is negative for each lead. To provide electric charge conservation in the junction one must require that $I = I_L = -I_R$ for the chosen direction of the current flow. Therefore the energy fluxes also have different signs. As for the $Q_{L,R}^{\rm el}$ magnitudes, they may differ only if the renormalized currents $\tilde{I}_L(E)$ and $\tilde{I}_R(E)$ are distributed over energies in different ways. This cannot happen in the case of elastic transport, for in this case $\tilde{I}_L(E) = -\tilde{I}_R(E)$. However, if the transport process is accompanied by the energy dissipation, the energy distributions for $\tilde{I}_L$ and $\tilde{I}_R$ may differ. In this case, electrons lose some energy while moving through the junction, and this gives rise to the differences in the renormalized currents energy distributions. For instance, in the case when the electrodes are linked with a single-state bridge, the energy distribution of $\tilde{I}_L(E)$ has a single maximum whose position is determined by the site energy E_0 (the polaron shift is omitted for simplicity) and the applied bias voltage V. Assuming that the average energy loss due to dissipation could be estimated as ΔE, the maximum in the $\tilde{I}_R(E)$ distribution is shifted by this quantity, so the current $\tilde{I}_R(E)$ flows at lower energies compared to $\tilde{I}_L(E)$. This results in the power loss and Joule heating of the junction [21]. When the system is unbiased but the leads are kept at different temperatures, Eqs. (4.64)–(4.66) might be used to compute the heat carried through the junction by the thermoelectric (Seebeck) current.

Now, we write down a general expression for the phonon thermal flux into the junction. Considering the electrodes as phonon thermal reservoirs, one may derive the following expression for the energy fluxes from the leads to the junction [26]:

$$Q_{L,R}^{\text{ph}} = \frac{\hbar^2}{2\pi} \int_0^\infty d\omega\omega \text{Trace} \left\{ \Pi_{L,R}^{\text{ph},<}(\omega) D^>(\omega) - \Pi_{L,R}^{\text{ph},>}(\omega) D^<(\omega) \right\} \tag{4.67}$$

where $D^>(\omega)$ and $D^<(\omega)$ are the Fourier transforms of the relevant Green's functions describing the phonons on the bridge and the quantities $\Pi_{L,R}^{\text{ph},>}(\omega)$ and $\Pi_{L,R}^{\text{ph},<}(\omega)$ are the less and greater self-energy matrices describing the coupling of these phonons to the corresponding thermal bath. Equations (4.64)–(4.67) present a general nonequilibrium Green's function-based formalism for analysis of thermoelectric effects in the transport through metal–molecular junctions. However, the application of this formalism to realistic molecular junction models is extremely difficult. Several simplified approaches were developed and used to analyze the heat transfer and other related phenomena in molecular junctions. Some of these approaches are described in the review [26].

Omitting electron–phonon interactions and restricting ourselves within the linear response theory, we may write out the following expressions for the charge current and heat flux through the system [93]:

$$I = e^2 M_0 \Delta V + e M_1 \frac{\Delta T}{T}, \tag{4.68}$$

$$Q = -e M_1 \Delta V - M_2 \frac{\Delta T}{T} \tag{4.69}$$

where ΔV and ΔT are the voltage drop and the temperature difference between the junction leads, respectively. The coefficients M_n included in Eqs. (4.68) and (4.69) are given by

$$M_n = -\frac{1}{h} \int dE \frac{df}{dE} (E - \mu)^n T(E). \tag{4.70}$$

Here, the electron transmission function is defined by Eq. (2.66). The commonly used characteristics of the junction thermal properties such as the thermopower S and the thermal conductance κ may be expressed in terms of these coefficients.

The thermopower is defined as the ratio $\Delta V/\Delta T$ calculated on the condition that the charge current does not flow through the system. Putting $I = 0$ in Eq. (4.68), we find

$$S = \frac{\Delta V}{\Delta T} = -\frac{1}{eT}\frac{M_1}{M_0}. \tag{4.71}$$

Also, using the definition for the thermal conductance $Q = \kappa\Delta T$ and Eqs. (4.68) and (4.69), we obtain

$$\kappa = \frac{1}{T}\left(M_2 - \frac{M_1^2}{M_0}\right). \tag{4.72}$$

The thermoelectric efficiency of the system is described by a dimensionless quantity ZT known as the figure of merit. The figure of merit is defined in terms of measurable coefficients S, κ, and g (g being the electric conductance), namely,

$$ZT = \frac{gS^2T}{\kappa} = \frac{M_1^2}{M_0M_2 - M_1^2}. \tag{4.73}$$

Functionality and reliability of molecular devices crucially depend on their thermal properties. Typically small molecular heat capacities combined with inefficient heat transfer away from the device might cause a significant temperature increase, threatening to destroy the stability and integrity of molecular nanodevices. At present, there exists a very significant interest to studies of energy (heat) transfer through nanodevices of various types including MMM junctions [15, 16, 18, 21, 336–348]. In particular, thermopower had been experimentally studied for junctions including quantum dots and molecules [348–359]. Besides its importance for providing reliability of nanoelectronic devices, the research of thermal properties of metal–molecular junctions is one of a notable fundamental interest. Quantization of molecular bridge levels and the Coulomb interactions between the electrons on the bridge give rise to novel effects which are specific for molecular and other similar nanodevices. Some effects of this kind were observed in heat transport through quantum dots. These include the thermopower oscillations [348, 349] and the oscillations of the thermal conductance κ [353, 354] occurring when the gate voltage varies.

The gate voltage variations lead to the shifts of the Fermi energy, thus providing favorable conditions for quantum oscillations of the thermal conductance. These oscillations occur at low temperature when the thermal energy k_BT is much smaller than energy-level separation on the molecule/quantum dot. The physical reason for these oscillations is the same as for the Coulomb blockade oscillations of the electric conductance discussed in the previous chapter. It is necessary to remark that in the case of practical molecules where the energy levels are not equidistant, the variations of the thermal (as well as electric) conductance appearing when the gate voltage varies do not qualify as oscillations for they do not show periodicity. Strictly speaking, periodical oscillations of the thermoelectric characteristics could be observed only in quantum dot with equidistant energy levels.

To properly compute the quantities M_n given by Eq. (4.70), one needs to calculate the retarded and advanced Green's functions for a multilevel dot, taking into account electron–electron interactions. This is an extremely complicated task. To circumvent these difficulties, one may adopt a simplified model approximating the density of states for electrons on the level "p" of the dot by the expression similar to (3.36). Using this approximation, one may calculate the coefficients M_n. One obtains

$$M_0 = -\frac{\pi}{\hbar}\sum_p \sum_{N \neq p} \Gamma_p P_p(N) \frac{\partial f}{\partial E}\bigg|_{E=E_p^0+NU_0} \tag{4.74}$$

where U_0 is the charging energy, the factors

$$\Gamma_p = \frac{\Gamma_L^p \Gamma_R^p}{\Gamma_L^p + \Gamma_R^p} \tag{4.75}$$

describe the coupling strengths of the dot levels to the leads, and $P_p(N)$ represent the probabilities that N energy levels in the dot other than the level "p" are occupied. Deriving similar expressions for $M_{1,2}$ and substituting them into Eqs. (4.71) and (4.72), one arrives at the approximate results for the thermopower and thermal conductance. For instance,

$$\kappa = \frac{1}{T}\sum_p \sum_{N \neq p} \Gamma_p P_p(N)(E_p^0 + NU_0 - E_{\mathrm{F}})^2 \frac{\partial f}{\partial E}\bigg|_{E=E_p^0+NU_0}. \tag{4.76}$$

More thorough analysis carried out in [354] brings qualitatively similar results shown in Fig. 4.22.

Also, the energy-level discreteness leads to a fine structure manifested in the thermal transport characteristics for small quantum dots [350]. Besides, the thermopower appears to be sensitive to the Kondo correlations, which results in the logarithmic temperature dependencies of the thermoelectric coefficients [355]. Some experiments reveal significant effects of spin correlations on the thermal characteristics [351].

The combined effect of Kondo correlations and Fano interference on the thermal conductance is shown in Fig. 4.23. The curves displayed there are plotted using the model of a single-level dot with the level energy E_0 coupled to the leads. As presented in the top panel, in the absence of the direct tunneling between the electrodes ($t = 0$), the Kondo screening makes the dot transparent to electrons when temperature is well below the Kondo temperature ($k_{\mathrm{B}}T = 10^{-9}D$; D being the halfwidth of the electrodes conduction band). This pushes the thermal conductance up so that the renormalized thermal conductance κ/T approaches the

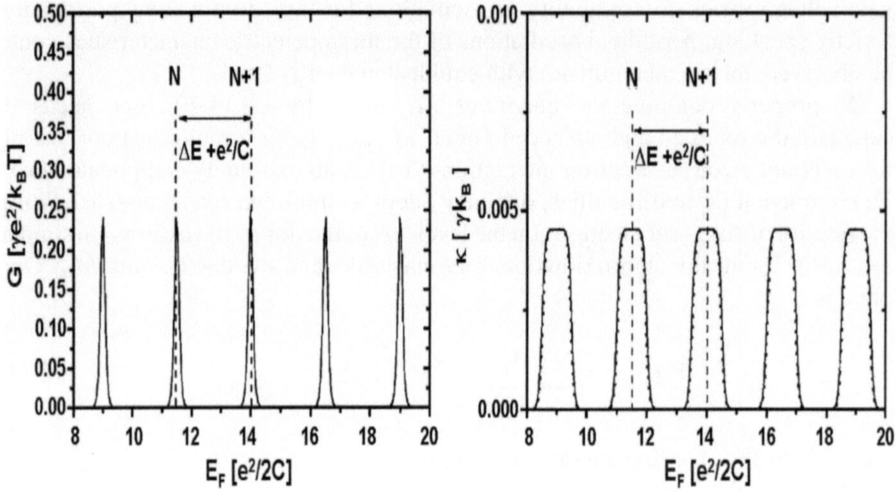

Fig. 4.22 Electron electrical (*left*) and thermal (*right*) conductance oscillations for a series of equidistant nondegenerate levels with separation $\Delta E = 0.5e^2/C$ (C being the effective capacitance of the dot) at low temperature ($k_B T = 0.1\Delta E$). The curves are plotted assuming $\Gamma_p = \Gamma = 0.2\Delta E$ (after [354]). Reprinted with the permission of the APS

maximum value allowed by the Wiedemann–Franz law, namely, $\kappa/T = 2\pi^2 e^2/3h$. It remains close to this maximum value while the energy varies within the range $E_0 < E < E_0 + U_0$. At higher temperatures the Kondo cloud evaporates, and the thermal conductance steeply drops in the middle of this energy interval, displaying two maxima at $E = E_0$ and $E = E_0 + U_0$. When the direct tunneling between the leads becomes possible ($t \neq 0$), the current through the dot tends to interfere with the current bypassing the dot, and this creates opportunities for Fano resonances to appear (see bottom panel of Fig. 4.23).

So, Coulomb interactions between the electrons on the molecular bridge/quantum dot may significantly affect thermoelectric transport coefficients. Under certain conditions, these interactions may lead to violation of the Wiedemann–Franz law [353]. In such systems, the Lorentz ratio $L = \kappa/gT$ strongly depends on the gate voltage, and it may accept values which significantly differ from the standard Lorentz number $\pi k_B^2/3e^2$ typical for bulk materials. Murphy and coworkers [360] suggested that the departure from the Wiedemann–Franz law occurring in the metal–molecular junctions is the main reason for their enhanced thermoelectric efficiency which gives grounds to conjecture that MMM junctions could be used as elements in energy-conversing devices. Also, possible potentialities of molecular devices are indicated by the theoretical prediction of a giant thermopower in single-molecular systems [361].

As suggested in [361], large values of both thermopower and the figure of merit could be manifested in MMM junctions when Fano resonances are moved close to the Fermi energy. This happens regardless of the specific physical mechanism

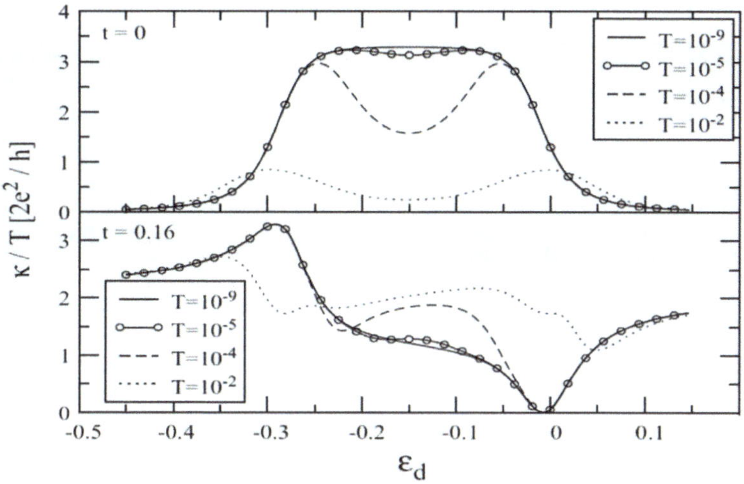

Fig. 4.23 Electron thermal conductance as a function of energy. The displayed curves are plotted at $E_0 = -0.3D$, $U_0 = 0.3D$ and the dot level width $\Gamma = 0.01D$ for two different values of the transfer amplitude t corresponding to the direct tunneling of electrons between the leads passing the dot (after [359]). Reprinted with the permission of the ELSEVIER

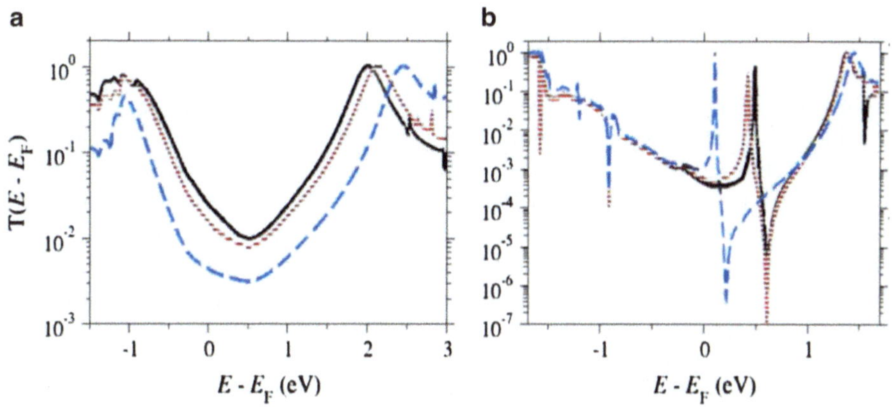

Fig. 4.24 Transition coefficient for biphenyldithiol (*left*) and CSW-470 (*right*) molecules for the twist angle of 0° (*solid lines*), 30° (*dotted lines*), and 60° (*dashed lines*) (after [361]). Reprinted with the permission of the APS

producing the shift. For example, rotation of aromatic side groups in CSW-470-bipyridine molecules (CSW-479) serving as linkers in MMM junctions may bring the desired closeness of Fano resonances to the Fermi energy as shown in Fig. 4.24. One may see that Fano resonances in transmission appear to be near the Fermi energy at certain values of the twist angle between the side group and the backbone of the molecule, and these same angles are corresponding to the giant thermopower (see Fig. 4.25). The effect becomes more pronounced as the temperature increases.

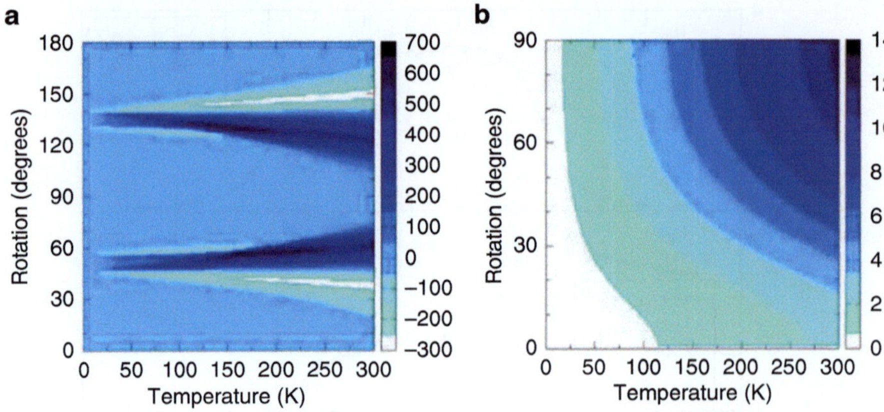

Fig. 4.25 Thermopower as a function of temperature and the twist angle computed for a MMM junction with the CSW-470 linker (*left*) and the biphenyldithiol linker (*right*; after [361]). Reprinted with the permission of the APS

Giant thermopower and figure of merit indicate potentialities of high efficiency of the corresponding single-molecule thermoelectric device. However, in practice, the values of S and ZT strongly depend on the phonon contribution to the thermal conductance which is not included in Eqs. (4.71)–(4.73). The effect of phonons on thermoelectric properties of weakly coupled MMM junctions within an interesting regime where $\Gamma_{L,R}$ are much smaller than the phonon exchange rate γ was considered in [360]. Recently, nonlinear thermoelectric properties of a junction with a single-state bridge coupled to a single vibrational mode were theoretically analyzed using the rate equations [362]. It was shown that the optimal efficiency and power output of a single-molecule device converting heat to electric power could be qualitatively changed by the presence of the vibrational mode. Nevertheless, at present the studies of phonon contribution to thermoelectric properties of MMM junctions and similar systems are not completed, and further research is necessary.

Currently, the research community becomes increasingly interested in studies of interplay between the heat transport and spin effects. The spin Seebeck effect was experimentally observed, and the spin voltage generated by a temperature gradient in metallic magnetic contacts was measured [363]. This gives grounds to expect that the observed effect may be used to obtain spin current in magnetoelectronic nanodevices. Carrying on studies of Peltier and Seebeck effects in magnetic monolayers, Hatami and coworkers [363] came to the conclusion that thermoelectric effects showed a significant dependence on the relative orientations of the magnetization in the adjacent magnetic layers. Recently, thermoelectric effects in quantum dots coupled to ferromagnetic leads were discussed and analyzed [342, 364].

Experimental manifestations of the heat transfer in molecular junctions include current-induced conformational and chemical changes in molecular bridges [365, 366]. These changes originate from the combined effect of the energy transfer from electronic to vibrational degrees of freedom in the junction and of the heat conduction out of it. These processes could lead to the energy accumulation in the phonon subsystem, triggering chemical reactions and/or changes in the molecular bridge geometry. Moreover, when the energy accumulates at a sufficient rate, it may result in the molecular decomposition and the consequent junction disintegration. It is natural to expect the increasing energy contents as temperature rises; therefore, one needs to properly define the local temperature in the junction which is a nontrivial task [34]. However, while dealing with steady-state processes in the junction, one may characterize the latter with three values of the temperature, namely, temperatures of the electrodes $T_{L,R}$ and the bridge temperature $T_{Br} = \frac{1}{2}(T_L + T_R)$, as was suggested by DiVentra and coworkers [367].

4.8 Polaron Effects: Hysteresis, Switching and Negative Differential Resistance

While studying electron transport through molecular junctions, hysteresis in the current–voltage characteristics was observed in some systems [368]. Multistability and stochastic switching were reported in single-molecule junctions [10] and in single metal atoms coupled to a metallic substrate through a thin ionic insulating film [369–371]. Coupling of an electron belonging to a certain atomic energy level to the displacements of ions in the film brings a possibility of polaron formation in there. This leads to the above-discussed polaron shift in the electron energy. It was noticed that multistability and hysteresis in molecular junctions mostly occurred when the molecular bridges included centers of long-living charged electronic states (redox centers). On these grounds, it was suggested that hysteresis in the I–V curves observed in molecular junctions appeared due to formation of polarons on the molecules [18].

The presence of the polaron shift in the energy of a charged (occupied) electron state creates a difference between the latter and the energy of the same state while it remains unoccupied. Assuming for simplicity a single-state model for the molecular bridge coupled to a single optical phonon mode, we may write the following expression for the renormalized energy:

$$\tilde{E}_0(n_0) = E_0 - \frac{\Lambda^2 n_0}{\hbar\Omega} \tag{4.77}$$

where the electronic population on the bridge n_0 is given by

$$n_0 = \frac{1}{\pi} \int dE \frac{f_L(E)\Gamma_L + f_R(E)\Gamma_R}{\left[E - \tilde{E}_0(n_0)\right]^2 + (\Gamma_L + \Gamma_R)^2}. \tag{4.78}$$

So, as follows from Eq. (4.77), the polaron shift depends on the bridge occupation n_0, and the latter is related to $\tilde{E}_0$ by Eq. (4.78). Therefore, the derivation of an explicit expression for $\tilde{E}_0(n_0)$ is a nontrivial task even within the chosen simple model. Nevertheless, it could be shown that two local minima emerge in the

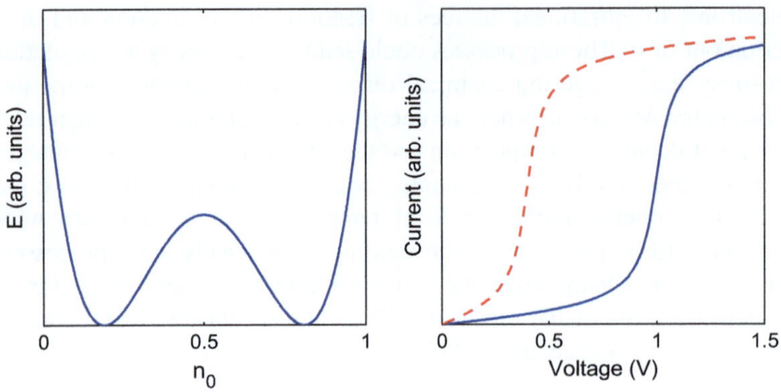

Fig. 4.26 *Left panel*: schematics of the potential energy of the molecular junction versus the occupation number n_0. *Right panel*: hysteresis in the current–voltage characteristics. The *solid line* corresponds to the transport via the unoccupied bridge electronic state, and the *dashed line* corresponds to the transport via the occupied state shifted due to the polaron formation

potential energy profile of a molecular junction including two electrodes linked by the molecular bridge with the occupation number n_0. These minima are located near $n_0 = 0$ and $n_0 = 1$, and they correspond to the neutral (unoccupied) and charged (occupied) states, respectively. This is illustrated in Fig. 4.26 (left panel). These states are metastable, and their lifetime could be limited by the quantum switching [335, 372]. When the switching time between the two states is longer than the characteristic time for the external voltage sweeping, one may expect the hysteresis to appear for the states of interest live long enough to maintain it. Within the opposite limit, the average washes out the hysteresis. When the states are especially short-lived, this could even result in a telegraph noise at finite bias voltage which replaces the controlled switching.

Further we consider long-lived metastable states, and we concentrate on the I–V behavior. Let us for certainty assume that the bridge state at zero bias voltage is situated above the Fermi energy of the system and remains empty. As the voltage increases, one of the electrodes chemical potentials crosses the bridge level position, and the current starts to flow through the system. The I–V curve reveals a step at the bias voltage value corresponding to the crossing of the unoccupied bridge level with the energy E_0 by the chemical potential. However, while the current flows through the bridge, the level becomes occupied and, consequently, shifted due to the polaron formation. When the bias voltage is reversed the current continues to flow through the shifted bridge state of the energy $\tilde{E}_0$, until the recrossing happens. Due to the difference in the energies of neutral and charged states, the step in the I–V curves appears at different values of the voltage, and this is the reason for the hysteresis loop to appear, as shown in the right panel of Fig. 4.26. One could also trace the hysteresis in the I–V characteristics starting from the filled (and shifted) bridge state.

Again, the hysteresis loop in the I–V curves may occur when both occupied and unoccupied states are rather stable which means that the potential barrier separating the corresponding minima in the potential energy profile is high enough, so that quantum switching between the states is unlikely. This happens when the bridge is weakly coupled to the electrodes ($\Gamma_{L,R} \ll \Lambda^2/\hbar\Omega$) which is an obvious requirement for the involved states (neutral or charged) to be distinguishable. In other words, the broadening of the relevant levels due to the coupling to the electrodes must be much smaller that the polaron shift. As was recently shown, the stronger is the electron–phonon coupling, the less the probability of the switching becomes. At large values of Λ, the tunneling between the charged and neutral states is exponentially suppressed. Also, it was shown that the symmetry/asymmetry in the coupling to the electrodes could significantly affect the hysteresis behavior [243, 373]. For asymmetric junctions ($\Gamma_L \neq \Gamma_R$), two nearly stable states exist at zero bias voltage. When the asymmetry is very strong, both states could appear stable at one bias voltage polarity and unstable when the polarity is reversed. It was suggested that under appropriate choice of parameters, one may create a situation when the instability regions for the involved states do not overlap. These properties give grounds to conjecture that such strongly asymmetric junctions could reveal memory functionalities, which make them potentially useful in fabrication of nanodevices.

Among various potentially important properties of the electron transport through metal–molecule junctions, one may separate out the negative differential resistance (NDR), that is, the decrease in the current I while the bias voltage across the molecule increases. As was discussed in the previous chapter, several possible scenarios were proposed to explain the NDR occurrence in the electron transport through molecules, and various mechanisms could play a major part in the latter. Here, we discuss the NDR features in the current–voltage characteristics which originate from the electron coupling with the molecular vibrational modes. As was shown [18,374], the polaron formation could give rise to the NDR. This may happen if the polaron shift in the energy of the occupied state moves the energy level away from the conduction window. As the shift depends on the electronic population, and the latter changes as the bias voltage increases, the occupied state falls out from the window between the chemical potentials of the electrodes at certain value of the voltage. If the transport is being conducted via this occupied state, then it stops when this voltage value is reached. Correspondingly, the current value drops to zero, revealing a distinctive NDR feature. This is a realistic scenario based on the main features of the electron transport through molecules under polaron formation.

In conclusion, electron–vibron coupling could lead to the formation of a polaron which results in the energy differences between occupied and unoccupied states on the molecular bridge. At weak interaction between the bridge and the electrodes and strong electron–phonon coupling, these charged and neutral states are metastable and could serve for electron conduction. This results in such interesting and potentially useful effects as the hysteresis and NDR features in the current–voltage characteristics.

4.9 Molecular Junction Conductance and Long Range Electron Transfer Reactions

Theoretical and experimental studies of long-range electron transfer reactions last more than four decades, but they still remain within a very active research area. In the intramolecular electron transfer reactions the electric charge moves from one section of a molecule to another section of the same molecule. Long-range transfer typically occurs in large molecules such as proteins and/or DNA, so that these two sections are situated far apart from each other. A common setup for the transfer reactions includes a donor, a bridge, and an acceptor. Due to the large distances between the donor (where an electron leaves) and the acceptor (where it arrives) typical for the long-range transfer reactions, a direct coupling between the two is negligible. Therefore, the electron participating in the transfer needs a molecular bridge providing a set of intermediate states for the electron transport.

Essentially, the intramolecular electron transfer is a combination of nuclear environment fluctuations and electron tunneling. Electron transfer reactions result from the response of a molecule to environmental polarization fluctuations which accompany nuclear fluctuations. The molecule responds by redistribution of the electronic density, thus establishing opportunities for the charge transfer to occur. The main characteristic of the electron transfer processes is the transfer rate K_{et} which is the inversed time of the reaction. The transfer rate for the electron tunneling in the molecule interior from the donor to the acceptor depends both on the electronic transmission amplitudes and the electron–vibron coupling. The latter provides the energy exchange between the electronic and nuclear systems.

Viewing fast electronic motions in the background of the slowly moving nuclei (nonadiabatic electron transfer) and applying the Fermi golden rule of perturbation theory, it was shown that K_{et} could be written in the form first suggested by Marcus [375]:

$$K_{et} = \frac{2\pi}{\hbar} |H_{DA}|^2 (FC). \tag{4.79}$$

Here, the first cofactor is the electron transmission coefficient H_{DA}, being the effective matrix element between the donor and the acceptor. The second term is the density of states weighted Franck–Condon factor which describes the effect of nuclear motions in the environment.

There exists a noticeable resemblance between the long-range electron transfer and the molecular conductance, and this resemblance was analyzed in several theoretical works [45–52]. The fundamental part in both molecular conduction and electron transfer reactions is taken by the electron tunneling inside the molecule via the set of intermediate states. In large molecules the electron involved in the transfer reaction or contributing to the conductance, with a high probability follows a few pathways which work as molecular bridges. The rest of the molecule do not significantly participate in the transfer/conduction and may be omitted from consideration in calculations of the electron transmission [73, 246]. On these grounds, it was suggested that the original molecule could be reduced to a simpler chain-like structure, and the latter may replace the former in calculations of molecular conductance and/or transfer rates [376]. Such simplification significantly eases computations of the molecular conductance and the electron transfer rates in macromolecules.

The resemblance between the molecular conductance and long-range electron transfer does not mean that these processes are nearly identical. Along with the similarities there are substantial differences between the former and the latter. For instance, the continuum of states causing the charge transport arises from the multitude of electronic states on the electrodes in the case of molecular conduction and from the vibrations and fluctuations in the environment in the case of electron transfer reactions. Also, the driving force which puts electrons in motion originates from the external bias voltage applied across the junction in the case of molecular conduction. In the electron transfer situation, this force appears due to the electron–phonon interactions in the system, or the electron transfer reaction starts as a result of photoexcitation of the donor part of the molecule. The observables, namely, the molecular conductance g and/or current I, and the transfer rate K_{et} differ as well.

Notwithstanding these differences, the electron transport through molecular junctions and the electron transfer could be theoretically analyzed using the common formalism, as was proposed by Yeganeh, Ratner, and Mujica [49]. Further we follow their approach. The starting point is that one may simulate the donor-bridge-acceptor system as some kind of molecular junction. In this system, one represents the donor as a single state $|i\rangle$, and this state mimics the left lead (assuming the transport from the left to the right). The donor is coupled to the bridge, and the latter is coupled to the continuum of the final states $|f\rangle$, simulating the right lead. These couplings are described by self-energy terms $\Gamma_{i,f}$. For simplicity, one may assume that there exists a single state on the bridge, and this state is coupled to a vibrionic mode of the frequency Ω. The latter mimics environment fluctuations. The "leads" are assumed to be weakly coupled to the bridge ($\Gamma_{L,R} \ll \Lambda$) which is typical for electron transfer situations. Also, the representation of the environment motions by the single vibrionic mode is justified only at low temperatures when the thermal energy $k_{\mathrm{B}}T$ is much smaller than the electron–vibron coupling parameter Λ.

Now, we can write the Landauer expression for the current flowing through the "junction" using Eqs. (2.54) and (2.66). Assuming that the bridge includes only one state, Eq. (2.66) accepts the form

$$\begin{aligned} T(E) &= 2\mathrm{Trace}\{\Gamma_i G \Gamma_f G^\dagger\} \\ &= -2\frac{\Gamma_i \Gamma_f}{\Gamma_i + \Gamma_f}\mathrm{Im}(G). \end{aligned} \tag{4.80}$$

Here, the subscripts i/f label initial and final states. Within the Fermi golden rule regime which allows to introduce the transfer rate, the bridge must be coupled to the final states much stronger than to the initial state ($\Gamma_f \gg \Gamma_i$). Therefore,

$$\frac{\Gamma_i \Gamma_f}{\Gamma_i + \Gamma_f} \approx \Gamma_i, \tag{4.81}$$

and the coupling to the final states falls out of the expressions for the electron transmission and current. Within the chosen model, this is a physically reasonable result, for the final state reservoir (the right "electrode" in the system) was merely introduced to impose a continuum of states, maintaining the transfer process at a steady-state rate. Also, considering the current flow one may suppose that the initial state is always filled ($f_i(E) = 1$) and the final states are empty ($f_f(E) = 0$). Therefore, the current flow through the "junction" is given by

$$I = -\frac{2e}{\pi\hbar}\int dE\Gamma_i \mathrm{Im}(G). \tag{4.82}$$

Both the current and the transfer rate are fluxes closely related to each other, namely, $K_{et} = I/e$. So, one may write

$$K_{et} = -\frac{2}{\pi\hbar}\int dE\Gamma_i \mathrm{Im}(G). \tag{4.83}$$

Now, Γ_i could be computed using the expression for the corresponding self-energy term. Keeping in mind that the "left electrode" includes a single state with the certain energy ϵ_i, one obtains

$$\begin{aligned}\Gamma_i &= \mathrm{Im}\left(\frac{|\tau_i|^2}{E - \epsilon_i + i\eta}\right)\\ &= \pi|\tau_i|^2\delta(E - \epsilon_i).\end{aligned} \tag{4.84}$$

Accordingly, the expression (4.83) for the transfer rate may be reduced to the form

$$K_{et} = -\frac{2}{\hbar}|\tau_i|^2 \mathrm{Im}[G(\epsilon_i)] \tag{4.85}$$

where τ_i represents the coupling between the donor and the molecule bridge. It must be stressed that within the chosen model, τ_i is the only term representing the relevant state coupling, which may be identified with the electronic transmission coefficient H_{DA} in the general expression (4.79) for K_{et}. This leaves us with the following expression for the Franck–Condon factor:

$$(FC) = -\frac{1}{\pi}\mathrm{Im}[G(\epsilon_i)]. \tag{4.86}$$

As discussed before, at weak coupling of the bridge to the leads, the electron–vibron interaction opens the set of metastable channels for the electron transport at the energies $E_n = \tilde{E}_0 + n\hbar\Omega$ ($n = 0, 1, 2 \ldots$), where $\tilde{E}_0$ is the energy of the bridge state with the polaronic shift included. The Green's function may be approximated as a weighted sum of contributions from these channels:

$$G(\epsilon_i) = \sum_{n=0}^{\infty} P(n)\left[\epsilon_i - \tilde{E}_0 - n\hbar\Omega + i\eta\right]^{-1} \tag{4.87}$$

where $\eta \rightarrow 0^+$, and the coefficients $P(n)$ are probabilities for the channels to appear given by Eq. (4.2).

Substituting Eq. (4.87) into Eq. (4.86), we get

$$(FC) = \sum_{n=0}^{\infty} P(n)\delta(\Delta F - n\hbar\Omega). \tag{4.88}$$

Here, $\Delta F = \epsilon_i - \tilde{E}_0 \equiv \epsilon_i - E_0 + \Lambda^2/(\hbar\Omega)$ is the exoergicity of the transfer reaction, that is, the free energy change originating from the nuclear displacements accompanied by the polarization fluctuations. The effect of the latter is inserted via the reorganization term $\Lambda^2/(\hbar\Omega)$ related to the polaron formation. The exoergicity in the transfer reaction takes on the part similar to that of the bias voltage in the electron transport through molecular junctions. It gives rise to the electron motion through the molecules. In the particular case when the voltage drops between the initial state (left electrode) and the molecular bridge, these two quantities are directly related by $|e|V = \Delta F$.

Usually, the long-range electron transfer is observed at moderately high (room) temperatures, so the low-temperature approximation (4.88) for the Franck–Condon factor cannot be employed. However, the expression (4.86) remains valid at finite temperatures if the expression for the Green's function is modified to include the thermal effects. It is shown [49] that within the high-temperature limit ($k_B T > \hbar\Omega$), the expression for the (FC) may be converted to the well-known form first proposed by Marcus:

$$(\mathrm{FC}) = \frac{1}{\sqrt{4\pi E_p k_B T}} \exp\left[-\frac{(\Delta F - E_p)^2}{4E_p k_B T}\right] \tag{4.89}$$

where $E_p = \Lambda^2/(\hbar\Omega)$ is the reorganization energy.

While studying the electron transfer reactions in practical macromolecules, one keeps in mind that both donor and acceptor subsystems in the standard donor-bridge-acceptor triad are usually complex structures including multiple sites coupled to the bridge. Correspondingly, the bridge has a set of entrances and a set of exits which an electron can employ. At different values of the tunnel energy, different sites of the donor and/or acceptor subsystems can give predominant contributions to the transfer. Consequently, an electron involved in the transfer arrives at the bridge and leaves from it via different entrances/exits, and it follows different pathways while on the bridge.

Also, nuclear vibrations in the environment could strongly affect the electron transmission, destroying the pathways and providing a transition to completely incoherent sequential hopping mechanism of the electron transfer. All this means that proper computation of the electron transmission factor H_{DA} for practical macromolecules is a very complicated task. The strong resemblance between the electron transfer reactions and the electron transport through molecules gives grounds to believe that studies of molecular conduction can provide important information concerning quantum dynamics of electron participating in the transfer reactions. One may expect that some intrinsic characteristics of the intramolecular electron transfer such as pathways of tunneling electrons and distinctive features of donor/acceptor coupling to the bridge could be obtained in experiments on the electron transport through molecules. For instance, it was recently suggested to characterize electron pathways in molecules using the inelastic electron tunneling spectroscopy, and other advances in this area are to be expected.

Chapter 5
Electronic Structure Calculations in Molecules

5.1 Kohn–Sham Equations

Electronic structure calculations are commonly recognized as the indispensable basis for studies of important observable properties in the variety of materials. These calculations are widely used in condensed matter physics and quantum chemistry, providing useful predictions for solids and solid surfaces nanostructures, molecules, and atoms. Correspondingly, a great number of computational methods and approaches were developed to carry out the electron structure calculations. Even a brief description of these methods is far beyond the scope of the present work. We restrict ourselves with a short introduction into Kohn–Sham density-functional theory (DFT) which presently is a top choice approach to electronic structure calculations in many-body systems including molecules and molecular clusters [377–379]. It is known that computational effort required in conventional quantum chemistry approaches exponentially grows as the number of involved electrons N increases. Therefore, these approaches, which give extremely accurate results for small molecules, require enormous computational costs when applied to larger molecules/molecular complexes. At the same time, within the DFT the computational costs increase significantly slower, being proportional to N^2 or N^3. Actually, this means that DFT can be successfully applied to molecules including hundreds of atoms, whereas the quantum chemistry approaches are practically useful for smaller molecules including a few tens of atoms at most. The comparably low computational costs combined with a reasonable accuracy of the results favor the DFT over elaborate quantum chemistry approaches and make this theory an indispensable tool in studies of large molecules and/or molecular complexes [380]. While the DFT-based computations may have difficulties to adequately capture excitations and transport spectra, these computations bring good and well-calibrated with precise bonding geometry results as far as total ground-state energies are concerned. The DFT is a mean field theory which suggests an efficient way to reduce the intractable many-body problem of interacting electrons subject to an external potential to a tractable problem of noninteracting electrons moving in an effective

N.A. Zimbovskaya, *Transport Properties of Molecular Junctions*, Springer Tracts in Modern Physics 254, DOI 10.1007/978-1-4614-8011-2_5,

potential. This effective potential includes the external potential and additional terms arising due to electron–electron interactions. The most important characteristics of DFT are the absence of many-body wave functions and the prominent part taken by the local electron density $n(\mathbf{r})$. In the case of a molecule/molecular cluster the considered many-body system includes N electrons and M nuclei. The system is described by the Hamiltonian:

$$H = \sum_{i=1}^{N} \left(-\frac{\hbar^2}{2m} \nabla_i^2 + V_{\text{nuc}}(\mathbf{r}_i) + \frac{1}{2} \sum_{j \neq i}^{N} \frac{e^2}{|\mathbf{r}_i - \mathbf{r}_j|} \right) + \frac{1}{2} \sum_{k=1}^{M} \sum_{l \neq k}^{M} \frac{Z_k Z_l}{|\mathbf{R}_k - \mathbf{R}_l|} \tag{5.1}$$

where m is the electron mass and $\mathbf{r}_i$ and $\mathbf{R}_k$ indicate positions of the electrons and nuclei, respectively; Z_k are the nuclei charges and the potential $V_{\text{nuc}}(\mathbf{r})$ is created by the nuclei.

Assuming that the nuclei positions are fixed (this corresponds to the Born–Oppenheimer approximation), the ground state of the electron system is described by a wave function $\psi(\mathbf{r}_1, \ldots, \mathbf{r}_N)$, which is a solution of the time-independent Schrodinger equation corresponding to the Hamiltonian given by Eq. (5.1). However, this equation could be solved with sufficient accuracy only for systems including relatively few electrons. The explosive increase in the number of electron configurations, which accompanies the increase in the number of electrons N, leads to practically unsurmountable computational difficulties. The DFT avoids these difficulties by introducing the electron density $n(\mathbf{r})$ as the key variable instead of the many-electron wave function ψ. The local electron density operator $\hat{n}(\mathbf{r})$ depends only on coordinates of electrons included into the system:

$$\hat{n}(\mathbf{r}) = \sum_{i=1}^{N} \delta(\mathbf{r} - \mathbf{r}_i). \tag{5.2}$$

Excluding from consideration the last term in the Hamiltonian (5.1) which describes nuclear interactions, one may rewrite the electron part of this Hamiltonian in the form

$$H_{\text{el}} = T + V + U \tag{5.3}$$

where T corresponds to the kinetic energy of electrons (the first term in the Hamiltonian (5.1)) and the term V describes the effect of the electrostatic potential of nuclei:

$$V = e \int d\mathbf{r} \hat{n}(\mathbf{r}) V_{\text{nuc}}(\mathbf{r}). \tag{5.4}$$

Finally, the term U corresponding to Coulomb interactions between the electrons is given by the expression:

$$U = \frac{1}{2} \int d\mathbf{r} \int d\mathbf{r}' \frac{e^2}{|\mathbf{r} - \mathbf{r}'|} \hat{n}(\mathbf{r}) \left[\hat{n}(\mathbf{r}') - \delta(\mathbf{r} - \mathbf{r}') \right]. \tag{5.5}$$

Here, the second term in the brackets is introduced to exclude the interaction of an electron with itself.

We start the derivation of basic DFT equations following the way suggested by Argaman and Makov [381] basing on thermodynamic considerations. First, we consider the grand canonical ensemble including electrons described by the Hamiltonian (5.3) which are kept in contact with a reservoir characterized by a certain finite temperature T and a chemical potential μ. We remark that the DFT mostly addresses the ground-state properties of the electron system and this is not consistent with a nonzero temperature. However, the consistency could be later restored by taking $T \to 0$ limit. The grand potential Ω for the considered system could be presented as follows:

$$\Omega = -k_{\mathrm{B}} T \ln \left\{ \mathrm{Trace} \left(\exp \left[\frac{H_{\mathrm{el}} - \mu \hat{N}}{k_{\mathrm{B}} T} \right] \right) \right\} \tag{5.6}$$

where $\hat{N}$ is the overall electron number operator and k_{B} is the Boltzmann's constant, as usual. In this expression, the trace indicates summation over all possible electron numbers N (if N is not fixed) and over all states of the Hamiltonian H_{ell} for each N. As follows from Eq. (5.6), the expectation value N of the electron number operator $\hat{N}$ is given by

$$N = -\frac{\partial \Omega}{\partial \mu}. \tag{5.7}$$

The free energy of the system $F(N, T)$ can be obtained by applying the Legendre transform:

$$F(N, T) = \Omega(\mu, T) + \mu N \tag{5.8}$$

where μ depends on the expectation number N determined by the relation (5.7). The important point is that inversion of the relationship (5.7), which allows to express μ in terms of N, requires a one-to-one correspondence between these variables. This correspondence is guaranteed by the convexity of the thermodynamic potential Ω which takes place in the absence of phase transitions in the system.

The above conclusions could be generalized in such a way that one may treat the local density of electrons $\hat{n}(\mathbf{r})$ instead of their total number. As follows from the expressions (5.3), (5.4), and (5.6), the thermodynamic potential Ω is a functional of the electrostatic potential $V_{\mathrm{nuc}}(\mathbf{r})$. In the considered system of electrons interacting with atomic nuclei, this potential is created by the latter. However, similar analysis could be carried out without specification of the nature and origin of this potential. It may be simply referred to as an external potential $v(\mathbf{r})$ affecting the electrons. The expectation value of the electron density $n(\mathbf{r}) \equiv \langle \hat{n}(\mathbf{r}) \rangle$ is given by the functional derivative of the grand potential with respect to the external potential:

$$n(\mathbf{r}) = \frac{\delta \Omega}{\delta v(\mathbf{r})}. \tag{5.9}$$

Now, one may introduce the Hohenberg–Kohn free energy defined by means of the functional Legendre transform [382]:

$$F_{\mathrm{HK}}[n(\mathbf{r})] = \Omega[v(\mathbf{r}) - \mu] - \int d\mathbf{r} \hat{n}(\mathbf{r})(v(\mathbf{r}) - \mu). \tag{5.10}$$

The grand potential Ω, which appears on the right-hand side, depends on the variable $v(\mathbf{r}) - \mu$ (the explicit temperature variable is omitted for brevity). The difference $v(\mathbf{r}) - \mu$ can be treated as a single functional variable corresponding to the given electron density $n(\mathbf{r})$. Introducing the free energy by the definition (5.10), one assumes that for each choice of $n(\mathbf{r})$, there exists the unique function $v(\mathbf{r}) - \mu$ which corresponds to the equilibrium density distribution given by the chosen $n(\mathbf{r})$. This assumption was first proven for a system characterized by a non-degenerate ground state, and it is known as the first Hohenberg–Kohn theorem [382]. Basing on this theorem, as well as on the general convexity of thermodynamic potentials, one may conclude that there exists a one-to-one relationship between the variables $v(\mathbf{r}) - \mu$ and $n(\mathbf{r})$.

To further explore the correlation between these variables, one may introduce the following function closely related to the grand potential:

$$\Omega_\mu(N, T) \equiv F(N, T) - \mu N. \tag{5.11}$$

This function is defined in such a way that it brings back the original grand potential given by Eq. (5.6) being minimized with respect to N. Since the chemical potential μ is a monotonous function of N, the minimum in Eq. (5.11) is unique. Now, we may generalize the expression for the Hohenberg–Kohn free energy by replacing the grand potential in Eq. (5.10) by the function determined by Eq. (5.11). Following this way, one defines the free energy functional:

$$\Omega_{v-\mu}[n(\mathbf{r})] = F_{\mathrm{HK}}[n(\mathbf{r})] + \int d\mathbf{r} n(\mathbf{r})(v(\mathbf{r}) - \mu). \tag{5.12}$$

In this expression, both $v(\mathbf{r})$ and $n(\mathbf{r})$ are treated as independent functional variables. Assuming that chemical potential and temperature are fixed, we may minimize the free energy functional with respect to $n(\mathbf{r})$. According to the definition of the grand potential function, the functional derivative of the left-hand side of Eq. (5.12) becomes zero, and we obtain the relation

$$\frac{\delta F_{\mathrm{HK}}}{\delta n(\mathbf{r})} = \mu - v(\mathbf{r}) \tag{5.13}$$

The free energy functional $\Omega_{v-\mu}[n(\mathbf{r})]$ is equal to the grand thermodynamic potential provided that $n(\mathbf{r})$ and $v(\mathbf{r})$ obey the relation (5.13), describing the one-to-one correspondence between this variables. So, this relation is similar to Eq. (5.7), determining the relationship between N and μ. The existence of a functional of the local electron density $n(\mathbf{r})$ which possesses the property described by Eq. (5.13) is claimed by the second Hohenberg–Kohn theorem [382], and it is the basic point of the DFT.

As it was mentioned above, most DFT applications are studies of ground-state properties. This means that the DFT-based analysis is usually referred to situations when the temperature is negligibly small. Therefore, it is important to remark that the previous analysis remains valid in the limit $T \to 0$. The single exception is the statement that N and μ retain a one-to-one correspondence within this limit. Actually, it was shown that at zero temperature, both $\Omega(N)$ and $F(N)$ become piecewise linear functions, and a whole linear segment in one of these functions may correspond to a single point with discontinuous derivative in the other [383]. However, this property of thermodynamic potentials does not bring significant problems hindering the development of the DFT as well as most of its applications.

To derive the basic DFT equations known as Kohn–Sham equations [377], one may insert into to electron Hamiltonian (5.3), a dimensionless parameter w characterizing the strength of electron–electron interactions. Accordingly, the interaction term U is to be replaced by wU. The parameter w varies between 0 (which corresponds to the system of noninteracting electrons) and 1 (which corresponds to the situation when the full interaction strength is manifested). The Hohenberg–Kohn free energy may be presented in the form

$$F_{\mathrm{HK}}[n] = F_0[n] + \int_0^1 \frac{\partial F_{\mathrm{HK}}[n]}{\partial w} dw \tag{5.14}$$

where $F_0[n]$ represents the system of noninteracting electrons. As follows from the relation between the thermodynamic potentials Ω and F, $\partial F/\partial w$ is equal to $\partial\Omega/\partial w$, and the latter equals the interaction energy U. The expression (5.14) may be rearranged in the following way:

$$F_{\mathrm{HK}}[n] = F_0[n] + E_{\mathrm{es}}[n] + E_{\mathrm{XC}}[n]. \tag{5.15}$$

Two last terms in this expression represent the interaction energy, both being functionals of the local electron density $n(\mathbf{r})$. The term $E_{\mathrm{es}}[n]$ is commonly approximated by the Hartree electrostatic energy:

$$E_{\mathrm{es}}[n] = \frac{e^2}{2} \int d\mathbf{r} \int d\mathbf{r}' \frac{n(\mathbf{r})n(\mathbf{r}')}{|\mathbf{r} - \mathbf{r}'|}. \tag{5.16}$$

The term $E_{XC}[n]$ is the exchange-correlation energy. Formally, this contribution could be defined as the difference between the actual value of the integral included in the expression (5.14) for the free energy and the Hartree energy given by Eq. (5.16), namely,

$$E_{XC}[n] = \frac{e^2}{2}\int_0^1 dw \int d\mathbf{r} \int d\mathbf{r}' \frac{n(\mathbf{r})}{|\mathbf{r}-\mathbf{r}'|}\rho_{XC}(\mathbf{r},\mathbf{r}',[n],w). \tag{5.17}$$

Using Eq. (5.5), one obtains

$$n(\mathbf{r})\rho_{XC}(\mathbf{r},\mathbf{r}',[n],w) = \langle \hat{n}(\mathbf{r})[\hat{n}(\mathbf{r}') - \delta(\mathbf{r}-\mathbf{r}')]\rangle_{n(\mathbf{r}),w} - n(\mathbf{r})n(\mathbf{r}').. \tag{5.18}$$

Here, the expectation value describes the density of electron pairs (where the electrons are located in the positions $\mathbf{r}$ and $\mathbf{r}'$) in the system with the interaction strength characterized by a certain value of the parameter w and with the electron density distribution $n(\mathbf{r})$. The function ρ_{XC} in the integrand of Eq. (5.17) describes the effect of interelectronic repulsions, that is, the fact that an electron present at the point $\mathbf{r}$ reduces the probability of finding another electron at the nearby point $\mathbf{r}'$. The exchange-correlation energy may be considered as the energy resulting from interaction between an electron and the corresponding exchange-correlation hole.

Obviously, one needs to employ different expressions for external potential for different interaction strengths. Only for the full strength interactions ($w = 1$), the external potential $v(\mathbf{r})$ suffices to reproduce the correct electron density distribution. In general case ($0 < w < 1$), the potential $v(\mathbf{r})$ should be replaced by a w-dependent effective potential $v_w(\mathbf{r})$. For a system of noninteracting electrons ($w = 0$), this effective potential $v_{\text{eff}}(\mathbf{r})$ is called the Kohn–Sham potential [377, 380]. Basing on the expression (5.15) for the free energy, one may find the relation between the original external potential $v(\mathbf{r})$ and the Kohn–Sham potential:

$$\mu - v_{\text{eff}}(\mathbf{r}) - e\varphi(\mathbf{r}) + v_{XC}(\mathbf{r}) = \mu - v(\mathbf{r}) \tag{5.19}$$

where $\varphi(\mathbf{r})$ is the electrostatic potential created by electrons:

$$\varphi(\mathbf{r}) = -\frac{e}{2}\int d(\mathbf{r}')\frac{n(\mathbf{r}')}{|\mathbf{r}-\mathbf{r}'|} \tag{5.20}$$

and the exchange-correlation potential $v_{XC}(\mathbf{r})$

$$v_{XC}(\mathbf{r}) = \frac{\delta E_{XC}}{\delta n(\mathbf{r})}. \tag{5.21}$$

The exchange-correlation potential could be found provided that the electron density distribution $n(\mathbf{r})$ is known, and the latter could be computed by solving the system of single-electron Schrodinger equations of the form

$$\left[-\frac{\hbar^2}{2m}\Delta + v_{\text{eff}}(\mathbf{r})\right]\varphi_i(\mathbf{r}) = \epsilon_i\varphi_i(\mathbf{r}). \tag{5.22}$$

These are the so-called Kohn–Sham equations [377], and their solutions $\varphi_i(\mathbf{r})$ are known as Kohn–Sham orbitals. The electron density $n(\mathbf{r})$ could be expressed in term of these wave functions, namely,

$$n(\mathbf{r}) = \sum_{i=1}^{N} \left|\varphi_i(\mathbf{r})\right|^2. \tag{5.23}$$

The set of equations (5.19)–(5.23) must be solved self-consistently. It provides a scheme for finding $n(\mathbf{r})$ and the ground-state energy for the system of N interacting electrons exposed to the external potential.

Now, we return back to the system of electrons interacting with each other and with atomic nuclei which is described by the Hamiltonian (5.1). Again, we assume that the nuclei positions are fixed, so the ground state of the system is described by the wave function $\psi(\mathbf{r}_1, \mathbf{r}_2, \dots \mathbf{r}_N)$. Here, we do not discuss the relationship between this wave function and Kohn–Sham orbitals. Nevertheless, we may write the following expression for the electron density distribution corresponding to the ground state of the considered system:

$$\begin{aligned} n(\mathbf{r}) &\equiv \sum_{i=1}^{N} \delta(\mathbf{r} - \mathbf{r}_i) \\ &= N \int d\mathbf{r}_2 \int d\mathbf{r}_3 \dots \int d\mathbf{r}_N \psi^*(\mathbf{r}, \mathbf{r}_2, \dots, \mathbf{r}_N)\psi(\mathbf{r}, \mathbf{r}_2, \dots, \mathbf{r}_N) \end{aligned} \tag{5.24}$$

which is easier to describe and analyze than the wave function ψ itself.

As shown by Hohenberg and Kohn [382], Eq. (5.24) could be reversed. This means that the wave function ψ associated with the ground state of the system is completely determined by the electron density, corresponding to this state. In other words, the function ψ is a functional of the electron density

$$\psi = \psi[n] \tag{5.25}$$

This remarkable result allows one to consider the ground-state energy of the electron system E as a functional of n. Using the expression (5.3) for the electron part of the Hamiltonian, one may write

$$\begin{aligned} E[n] &= \langle\psi[n]|T + V + U|\psi[n]\rangle \\ &\equiv T[n] + V[n] + U[n]. \end{aligned} \tag{5.26}$$

Again, the accurate expression for the electron density provides the minimum for the energy functional (5.26). So, a successful minimization of the energy functional yields the ground-state electron density which could be used to calculate other ground-state observables. These observables include vibrational modes [384–386], infrared and Raman spectra [385, 387], ionization energies, electron affinities, polarizabilities [388–390], and magnetic properties [151–154, 391, 392]. The task of finding a good approximation for the energy functional is greatly simplified by using a different separation for the energy functional introduced by Kohn and Sham [377], namely,

$$E[n] = T_0[n] + e \int d\mathbf{r} n(\mathbf{r}) V_{\text{nuc}}(\mathbf{r}) + E_{\text{es}}[n] + E_{\text{XC}}[n]. \tag{5.27}$$

Here, T_0 is the kinetic energy that a system with the electron density n would have in the absence of electron–electron interactions. The second and third terms give expressions for the electrostatic energy originating from the interactions of the electrons with the nuclei and from the Coulomb repulsion of electrons, respectively. As in the expression (5.14) for the free energy of the system of interacting electrons, the last term $E_{\text{XC}}[n]$ represents the exchange-correlation energy. As was already mentioned, the key point of the DFT is that the correct expression for the electron density must minimize the energy functional (5.27) and the minimum value of $E[n]$ should correspond to the ground-state energy of the electron system. Applying the variational principle, one may write

$$\begin{aligned}\frac{\delta E[n]}{\delta n} =& \frac{\delta T_0}{\delta n} + V_{\text{nuc}}(\mathbf{r}) \\ &+ \frac{e^2}{2} \int \frac{d\mathbf{r}' n(\mathbf{r}')}{|\mathbf{r} - \mathbf{r}'|} + \frac{\delta E_{\text{XC}}[n]}{\delta n} \\ =& \mu \end{aligned} \tag{5.28}$$

where μ is the Lagrange multiplier associated with the requirement that the number of electrons in the considered system remains constant. Looking for the electron density satisfying Eq. (5.28), we again arrive at the set of Kohn–Sham equations written in the form

$$\left\langle \delta\varphi_{i\sigma} \right| - \frac{\hbar^2}{2m} \nabla^2 + v_{\text{eff}}(\mathbf{r}) - \epsilon_i \left| \varphi_{i\sigma} \right\rangle = 0. \tag{5.29}$$

Now, the potential v_{eff} inserted into these equations has the form

$$v_{\text{eff}}(\mathbf{r}) = v_{\text{nuc}} + \frac{\delta E_{\text{XC}}}{\delta n} - e\varphi(\mathbf{r}) \tag{5.30}$$

and the electron density is expressed in terms of Kohn–Sham orbitals $\varphi_{i\sigma}$ by Eq. (5.23).

In practical calculations, the Kohn–Sham orbitals are expressed in terms of a set of orthonormalized basis functions. The latter are usually presented as products of the orbital functions $f_j(\mathbf{r})$ and spinors χ_σ. Employing these basis functions, one obtains

$$\varphi_{i\sigma}(\mathbf{r}) = \sum_{j,\sigma'} C^{i\sigma}_{j\sigma'} f_j(\mathbf{r})\chi_{\sigma'}. \tag{5.31}$$

Once this ansatz is introduced, one varies the expansion coefficients $C^{i\sigma}_{j\sigma'}$ rather than the value of each wave function at each point in space which leads to a secular equation. One aspect of the approximations to the DFT is that some integrals required to solve equations (5.29) cannot be computed in a closed form. Therefore, some numerical work is always required to solve these equations. Many algorithms for the solution of the above equations are developed, and the particular algorithm that one uses rather strongly depends on the form of the spatial parts of basis functions. Several sets of the orbital functions had been and still are employed in practical computations. The simplest among them is the plane-wave basis [393,394], which is used when one needs delocalized Kohn–Sham orbitals. Also, there exist commonly used localized sets of spatial basis functions such as augmented plane waves [395], linearized muffin-tin orbitals [396], and Gaussian orbitals [387, 397]. Presently, Gaussian orbitals are often applied for building up basis sets. From the beginning, it was realized that integrals of the Gaussian orbitals have simple analytic forms which help in construction of secular equations and the following electron structure calculations. As the research moved on to involve the systems of large size, two other aspects of Gaussian orbitals had been recognized as useful. First, these functions are localized in space, so as the number of electrons in the system (N) increases, the overlap matrix becomes sparse. While general basis sets lead to $O(N^2)$ pairs of basis functions products which give nonzero matrix elements of the overlap matrix, the employment of Gaussian orbitals reduces this number to $O(N)$. Secondly, the computational problems one faces within a Gaussian orbital framework are easily parallelized [248]. At present, there exists an extensive literature where the computational methods using the Gaussian orbitals as basis functions are discussed in detail [248, 384, 398–401].

Beyond the Born–Oppenheimer approximation, nuclei coordinates $\mathbf{R}_1, \mathbf{R}_2, \ldots, \mathbf{R}_M$ for atoms included in the considered system could be treated as variables of the DFT energy functional, as well. Therefore, one can perform the atomic structure optimization by minimizing the energy functional with respect to the atomic coordinates. Usually, the energy functional has several local minima where the energy gradients

$$\mathbf{g}_\alpha = \frac{\partial E}{\partial \mathbf{R}_\alpha} \tag{5.32}$$

become zero for all atomic coordinates $\mathbf{R}_\alpha$. Here, index α is labeling atoms included in the considered system. Each minimum corresponds to a stable or at least metastable atomic structure. The gradients $\mathbf{g}_\alpha$ determine forces acting upon atoms which vanish if the optimized molecular geometry is achieved. Once the molecular geometry is optimized, and the wave functions are self-consistently computed, one may calculate the relevant matrix elements and the observables of interest using these wave functions as the basis set.

All terms in Eq. (5.27) with the exception of $E_{XC}[n]$ may be exactly evaluated. At the same time, the exact expression for the exchange-correlation energy is known only for the free-electron gas. For other systems (including molecules) the functional $E_{XC}[n]$ must be approximated. For practical applications, the DFT must be implemented with a suitable approximation for the exchange-correlation energy.

5.2 Approximations for the Exchange-Correlation Energy Term

Several approximations for this term were suggested and employed in DFT-based computations [402–406]. The simplest among them is the local-density approximation (LDA). This approximation was first suggested to describe systems with slowly varying electron density whose properties (to a certain degree) resemble those typical for a homogeneous gas of interacting electrons. For such systems the exchange-correlation energy could be approximated by the expression

$$E_{XC} = \int d\mathbf{r} n(\mathbf{r}) \epsilon_{XC}(n(\mathbf{r})) \tag{5.33}$$

where $\epsilon_{XC}(n)$ has the same form as the exchange-correlation energy per electron in a uniform electron gas with the density n. This quantity could be computed using Monte Carlo technique for the relevant electron densities, and it can be accurately calculated within the high electron density limit. In practical computations concerning electron systems with moderately high densities, $\epsilon_{XC}(n)$ may be estimated using various interpolation formulas. For example, this quantity may be approximated by the expression [407]:

$$\epsilon_{XC}(n) = -\frac{0.458}{r_s} - 0.0666 X\left(\frac{r_s}{11.4}\right). \tag{5.34}$$

Here, $r_s = (3/4\pi n)^{-1/3}$ is the Wigner–Seitz radius expressed in units of the Bohr radius, and the function $X(x)$ has the form

$$X(x) = \frac{1}{2}(1 + x^3)\ln(1 + x^{-1}) - x^2 + \frac{1}{2}x - \frac{1}{2}. \tag{5.35}$$

Within the LDA, the exchange-correlation potential $v_{XC}(\mathbf{r})$ is described by a simple formula

$$v_{XC}(\mathbf{r}) = \frac{d}{dn}(n\epsilon_{XC}(n))_{n=n(\mathbf{r})}. \tag{5.36}$$

Despite its simplicity, the LDA in the form given by Eq. (5.33) with the corresponding approximation for ϵ_{XC} brings surprisingly good results for many atomic, molecular, and crystalline interacting electron systems, even though in some of these systems the density of electrons varies rather fast. The LDA may be generalized by removing the specific form of $\epsilon_{\mathrm{XC}}(n)$ and replacing it by function $g(n)$ whose form is originally undetermined but holding on the general idea that the functional E_{XC} solely depends on the electron density:

$$E_{\mathrm{XC}}^{\mathrm{LDA}}[n] = \int g(n)n(\mathbf{r})d\mathbf{r}. \tag{5.37}$$

Approximate models aiming at the description and analysis of many-electron system which use the local electron density in the manner similar to DFT were known much earlier than the latter theory was developed. The best-known among these predecessors of the DFT is the Thomas–Fermi model suggested in 1927 to approximate the distribution of electrons in atoms. The DFT could be reduced to this model if one omits from consideration the exchange-correlation energy and then computes the free energy of noninteracting electrons $F_0[n]$ treating $n(\mathbf{r})$ as the density of a homogeneous electron gas. This results in the following approximation:

$$F_0[n] \approx \frac{3}{10}(3\pi^2)^{2/3} \int d\mathbf{r} n(\mathbf{r})^{5/3}. \tag{5.38}$$

The advance made by introducing the DFT is twofold. First, it shows that the electron density distribution indeed contains all necessary information about the considered electron system, thus giving justification to calculation schemes based on this assumption. Secondly, working within the DFT approach, one may ceaselessly improve the approximations used to describe E_{XC}, thus increasing the accuracy of the DFT computations in whole.

A natural and often implemented improvement of LDA is the local spin-density approximation (LSDA). The employment of this approximation is motivated by the fact that exchange-correlation interactions are very different for electrons with parallel and antiparallel spins. In consequence, better approximation for E_{XC} could be developed if one introduces separate densities for spin-up and spin-down electrons. Therefore, the latter could be used as independent functional variables $n_\uparrow(\mathbf{r})$ and $n_\downarrow(\mathbf{r})$. Within the local spin-density approximation (LSDA), the exchange-correlation term depends on the electron spin densities $n_\downarrow(\mathbf{r})$ and $n_\uparrow(\mathbf{r})$:

$$E_{\mathrm{XC}}^{\mathrm{LSDA}}[n_\uparrow, n_\downarrow] = \int g(n_\downarrow, n_\uparrow)n(\mathbf{r})d\mathbf{r} \tag{5.39}$$

where $n(\mathbf{r}) = n_\downarrow(\mathbf{r}) + n_\uparrow(\mathbf{r})$ is the total electron density. A very good approximation for $g(n_\downarrow, n_\uparrow)$ had been suggested basing on quantum Monte Carlo simulations in jellium [408].

Approximations for E_{XC} could be further improved using the generalized gradient model where the exchange-correlation functional is supposed to depend on the electron spin densities $n_\downarrow(\mathbf{r})$ and $n_\uparrow(\mathbf{r})$ and on their gradients calculated at the same coordinates as these densities themselves. This approximation suggested by Perdew and coworkers [403] and by Becke [405] is known as the generalized gradient approximation (GGA). Within this approximation, E_{XC} could be presented in the form

$$E_{XC}^{GGA} = \int g(n_\downarrow, n_\uparrow, \nabla n_\downarrow, \nabla n_\uparrow) n(\mathbf{r}) d\mathbf{r}. \tag{5.40}$$

The "semilocal" GGA functionals have demonstrated useful improvement over LSDA-based results being applied to atoms, molecules, and solids. On the whole, in comparison with LSDA, GGA reveals a tendency to bring more accurate results for total energies, atomization energies barriers, and structural energy differences. Typically, GGA favors inhomogeneity of the electron density distribution more than LSDA does [408].

To carry out practical GGA-based calculations, one needs to find a suitable approximation for the function $g(n_\downarrow, n_\uparrow, \nabla n_\downarrow, \nabla n_\uparrow)$. One may start the corresponding analysis from the exact expression for the exchange-correlation energy:

$$E_{XC}[n_\downarrow, n_\uparrow] = \frac{e^2}{2} \int_0^1 dw \int d\mathbf{r} \int d\mathbf{r}' \frac{n_{XC}(\mathbf{r}, \mathbf{r}', [n_\downarrow, n_\uparrow], w)}{|\mathbf{r} - \mathbf{r}'|}. \tag{5.41}$$

This expression is similar to the previous one given by Eq. (5.17). The difference between the two originates from the fact that in Eq. (5.41), the electron densities for the spin-up and spin-down electrons are treated as separate variables, whereas the exchange-correlation energy described by Eq. (5.17) solely depends on the total electron density $n(\mathbf{r})$. The factor $n_{XC}(\mathbf{r}, \mathbf{r}', [n_\downarrow, n_\uparrow], w)$ in the integrand of Eq. (5.41) represents the density of the exchange-correlation hole surrounding an electron at the position $\mathbf{r}$. This hole density is taken at $\mathbf{r}$ for a certain value of the parameter w indicating the strength of electron–electron interactions. The exchange-correlation hole density introduced by Eq. (5.41) is the sum of two terms:

$$n_{XC}(\mathbf{r}, \mathbf{r}', [n_\downarrow, n_\uparrow], w) = n_X(\mathbf{r}, \mathbf{r}', [n_\downarrow, n_\uparrow]) + n_C(\mathbf{r}, \mathbf{r}', [n_\downarrow, n_\uparrow], w). \tag{5.42}$$

Here, the term n_X is the exchange hole density which does not depend on the interaction strength. Therefore, the whole effect of the electron–electron interactions on n_{XC} is included into the correlation term n_C. It was proven [409] that the exchange and correlation contributions to n_{XC} obey the following equations:

$$\int d\mathbf{r}' n_X(\mathbf{r}, \mathbf{r}', [n_\downarrow, n_\uparrow]) = -1, \tag{5.43}$$

$$\int d\mathbf{r}' n_c(\mathbf{r}, \mathbf{r}', [n_\downarrow, n_\uparrow], w) = 0, \tag{5.44}$$

$$n_X(\mathbf{r},\mathbf{r}',[n_\downarrow,n_\uparrow]) \le 0. \tag{5.45}$$

These equations could be derived basing on a sum rule for the exchange-correlation holes. This rule requires that every exchange-correlation hole surrounds a single electron, so

$$\int d\mathbf{r}' n_{XC}(\mathbf{r},\mathbf{r}',[n_\downarrow,n'_\uparrow],w) = -1. \tag{5.46}$$

Keeping in mind that this sum rule must be satisfied for all possible values of the interaction parameter w and presenting n_{XC} in the form (5.42), one concludes that integration of the w-dependent term n_C over the space variables must return zero result in agreement with Eq. (5.44). Comparing Eqs. (5.42)–(5.44), one immediately arrives at Eq. (5.46).

The GGA expression for exchange-correlation energy was derived basing upon the constraints (5.43)–(5.45) on the system-averaged hole [410]. The same expression was obtained in a different way based on the constraints on the exchange-correlation energy itself [404]. The resulting GGA for the function $g(n_\downarrow,n_\uparrow,\nabla n_\downarrow,\nabla n_\uparrow)$ could be written in the form

$$g(n_\downarrow,n_\uparrow,\nabla n_\downarrow,\nabla n_\uparrow) = g(n_\downarrow,n_\uparrow)F(r_s,P,s). \tag{5.47}$$

Here, $g(n_\downarrow,n_\uparrow)$ corresponds to the LSDA [see Eq. (5.39)], and the enhancement factor depends on Wigner–Seitz local radius r_s, the relative spin polarization $P = (n_\uparrow - n_\downarrow)/(n_\uparrow + n_\downarrow)$, and the dimensionless density gradient $s = |\nabla n|/2(3\pi^2 n)^{1/3}n$. When $s = 0$, the enhancement factor $F(r_s,P,s) = 1$, and GGA becomes reduced to LSDA. A good analytic approximation for the function $F(r_s,P,s)$ is presented in [403].

Further improvement could be achieved if one assumes that the integrand in Eq. (5.40) depends on the second derivatives (Laplacians) of the electron spin-oriented densities. Such approximation is referred to as meta-GGA [408]. Within this approximation,

$$E_{XC}^{MGGA}[n_\downarrow,n_\uparrow] = \int g(n_\downarrow,n_\uparrow,\nabla n_\downarrow,\nabla n_\uparrow,\nabla^2 n_\downarrow,\nabla^2 n_\uparrow,\tau_\downarrow,\tau_\uparrow)n(\mathbf{r})d\mathbf{r}. \tag{5.48}$$

Besides the gradients and Laplacians of the spin-oriented densities, the integrand in Eq. (5.48) depends on the Kohn–Sham orbital kinetic energy densities τ_σ given by the expressions

$$\tau_\sigma(\mathbf{r}) = \frac{\hbar^2}{2m}\sum\left|\nabla\varphi_{i\sigma}(\mathbf{r})\right|^2. \tag{5.49}$$

The latter appear in the Taylor expansion of the exchange hole density near $|\mathbf{r} - \mathbf{r}'| = 0$. These kinetic terms are implicit functionals of electron densities, and this makes them important parameters which must not be neglected.

A nonempirical meta-GGA was suggested by Tao, Perdew, Staroverov, and Scuseria (TPSS) [411]. The TPSS is constructed by using constraints on the exchange-correlation energy without introducing additional empirical parameters. It was suggested [408] that one could reconstruct TPSS basing on the constraints on the exchange-correlation holes given by Eqs. (5.43)–(5.45). Numerical tests of TPSS [412, 413] showed that this approach satisfactory worked on the meta-GGA level, producing a functional that is competitive with known semiempirical approximations. One notices that the described sequence of approximations including LSDA, GGA, and meta-GGA has an important property. Every next term in this sequence gives more opportunities for analysis of nonlocal effects than the previous one. One may construct further approximations adding extra terms to this sequence, thus approaching the limit where the effects originating from long-range interactions (such as van der Waals attraction) could be adequately accounted for. However, every next step in the process of improvement of the approximation for E_{XC} by better representation of nonlocal features of the electron density distribution leads to a significant lengthening and complication of computations. Another approach to the problem of long-range interactions is briefly discussed in the next section.

Presently, the GGA in the form suggested by Perdew, Burke, and Ernzerhof [404] (also called PBE-GGA) is widely used by the physics community. This model does not include adjustable parameters. This makes calculations more accurate but, at the same time, lengthens them. Other approaches result in other approximations for E_{XC}, including empirical adjustable parameters which are also used in DFT-based calculations. For instance, in the chemistry community a popular approximation is one combining the exchange part suggested by Becke [405] with the correlation contribution in the form proposed by Lee, Yang, and Parr [406]. This approximation is known as BLYP functional. Even better recognized is another hybrid functional (so-called B3LYP) which combines the Becke exchange contribution with the expression for the energy following from the Hartree–Fock theory. The B3LYP functional includes three adjustable parameters indicating how much of the exchange described by the Hartree–Fock expression is mixed in.

5.3 Long-Range Interactions in the DFT

As mentioned before, the DFT meets with serious difficulties being applied to the systems where long-range interactions play a significant part. For instance, it is very difficult to reproduce within DFT van der Waals forces, which (among other things) are important in many biological molecules. In principle, DFT can yield a good approximation for the ground-state energy including contributions from long-range interactions. However, practical and commonly used approximations for the exchange-correlation functional such as LSDA and/or GGA fail to capture essential effects of long-range interactions.

A strategy which allows to remedy this difficulty needs to be suggested. A natural starting point for the building up of such a strategy is to divide the whole interaction energy in two parts corresponding to the short-range (U_{sr}) and long-range (U_{lr}) interactions, respectively,

$$U(r) = U_{\mathrm{sr}}(r) + U_{\mathrm{lr}}(r) \tag{5.50}$$

where $r = |\mathbf{r} - \mathbf{r}'|$. As suggested by Kohn and cowokers [414], we may rewrite the electron Hamiltonian as a function of parameter w which "turns on" the long-range interactions, namely,

$$H(w) = T + V_w + U_{\mathrm{sr}} + wU_{\mathrm{lr}}. \tag{5.51}$$

As usual, in this expression, T represents the kinetic term, and the external potential V_w is chosen in such a way that the ground-state electron density remains unchanged as the parameter w varies between 0 and 1. When $w = 0$ the long-range interactions disappear and electron–electron interactions become entirely short range. In the opposite limit $w = 1$ the potential V_w coincides with the external potential. The ground-state energy of the system described by the Hamiltonian (5.51) also depends on w. At $w = 0$ it takes on the value $E(0)$ which corresponds to the system characterized by the absence of the long-range interactions. The real physical system has the ground-state energy $E = E(1)$. The relation between these energies is given by [414]

$$\begin{aligned} E =& E(0) + \int_0^1 dw \frac{dE}{dw} \\ =& E(0) + e \int d\mathbf{r}[v(\mathbf{r}) - v_0(\mathbf{r})]n(\mathbf{r}) \\ &+ \frac{1}{2} \int d\mathbf{r} \int d\mathbf{r}' U_{\mathrm{lr}}(|\mathbf{r} - \mathbf{r}'|) \\ &\times \left[\int_0^1 \langle \hat{n}(\mathbf{r})\hat{n}(\mathbf{r}') \rangle_w dw - n(\mathbf{r})\delta(\mathbf{r} - \mathbf{r}') \right] \end{aligned} \tag{5.52}$$

where $v(\mathbf{r})$ is the external potential in the real system, $v_0(\mathbf{r})$ is the potential in the system with the long-range interactions switched off, and $\langle \ldots \rangle_w$ indicates the expectation value in the ground state computed for a certain value of the parameter w.

The energy $E(0)$ could be computed employing DFT. The result may be presented in the form similar to (5.27)

$$\begin{aligned} E(0) = & T_0[n] + e \int d\mathbf{r} n(\mathbf{r}) v_0(\mathbf{r}) \\ &+ \frac{1}{2} \int d\mathbf{r} d\mathbf{r}' U_{\mathrm{sr}}(|\mathbf{r} - \mathbf{r}'|) n(\mathbf{r}) n(\mathbf{r}') + E_{\mathrm{XC}}^{\mathrm{sr}}[n]. \end{aligned} \tag{5.53}$$

The expression for the interaction term is given by Eq. (5.16), provided that $U_{\rm sr} = e^2/|\mathbf{r}-\mathbf{r}'|$ represents the Coulomb interactions between electrons. Also, it is worthwhile to stress that the exchange-correlation term in Eq. (5.53) should be computed disregarding the long-range interactions. Combining Eqs. (5.52) and (5.53), one obtains

$$
\begin{aligned}
E = T_0[n] &+ e\int d\mathbf{r}v(\mathbf{r})n(\mathbf{r}) \\
&+ \frac{1}{2}\int d\mathbf{r}d\mathbf{r}'U(|\mathbf{r}-\mathbf{r}'|)n(\mathbf{r})n(\mathbf{r}') \\
&+ E^{\rm sr}_{\rm XC}[n] - \frac{1}{2}NU_{\rm lr}(0) + \Delta E[n].
\end{aligned} \tag{5.54}
$$

Here, N is the total number of electrons in the considered system. The term $\Delta E[n]$ corresponds to the contribution to the ground-state energy originating from the long-range interactions. It has the form

$$
\begin{aligned}
\Delta E[n] =& \frac{1}{2}\int d\mathbf{r}d\mathbf{r}'U_{\rm lr}(|\mathbf{r}-\mathbf{r}'|) \\
&\times \int_0^1 dw \big\langle[\hat{n}(\mathbf{r})-n(\mathbf{r})][\hat{n}(\mathbf{r}')-n(\mathbf{r}')]\big\rangle_w .
\end{aligned} \tag{5.55}
$$

To proceed, one may employ the fluctuation-dissipation theorem [93]. By means of this theorem one can express the integrand in the integral over w in Eq. (5.55) in terms of the retarded linear susceptibility χ, namely,

$$
\big\langle[\hat{n}(\mathbf{r})-n(\mathbf{r})][\hat{n}(\mathbf{r}')-n(\mathbf{r}')]\big\rangle_w = -\frac{1}{\pi}\int_0^\infty d\omega \mathrm{Im}\chi(\mathbf{r},\mathbf{r}'\omega,w). \tag{5.56}
$$

As usual, the successibility χ determines linear response of the electron density in the considered system to a time-dependent perturbing potential $\delta v = v_1(\mathbf{r},\omega)\exp(-i\omega t)$. The electron density $\delta n = n_1(\mathbf{r},\omega,w)\exp(-i\omega t)$ responding to the perturbation is related to the latter by the equation

$$
n_1(\mathbf{r},\omega,w) = \int d\mathbf{r}'\chi(\mathbf{r},\mathbf{r}',\omega,w)v_1(\mathbf{r}',\omega) \tag{5.57}
$$

which defines the susceptibility. Substituting the expression (5.56) into Eq. (5.55), we get

$$
\begin{aligned}
\Delta E[n] =& -\frac{1}{2\pi}\int d\mathbf{r}d\mathbf{r}'U_{\rm lr}(|\mathbf{r}-\mathbf{r}'|) \\
&\times \int_0^1 dw \int_0^\infty d\omega \mathrm{Im}\chi(\mathbf{r},\mathbf{r}',\omega,w).
\end{aligned} \tag{5.58}
$$

So, if the susceptibility is known, one may straightforwardly compute $\Delta E[n]$ for any specific form of the long-range interaction. However, finding and appropriate approximation for $\chi(\mathbf{r},\mathbf{r}',\omega,w)$ is an intricated and very nontrivial task. It was shown [414, 415] that the susceptibility χ obeys the following integral equation:

$$\chi(\mathbf{r},\mathbf{r}'\omega,w) = \chi_0(\mathbf{r},\mathbf{r}'',\omega) + \int d\mathbf{r}''d\mathbf{r}'''\chi_0(\mathbf{r},\mathbf{r}'',\omega) \times \left[U(|\mathbf{r}''-\mathbf{r}'''|) + f_{\mathrm{XC}}(\mathbf{r}'',\mathbf{r}''',\omega,w)\right]\chi(\mathbf{r}''',\mathbf{r}',\omega,w). \quad (5.59)$$

In this equation, $\chi_0(\mathbf{r},\mathbf{r}',\omega)$ is the response function for the noninteracting system of electrons, and the term f_{XC} describes the effects of exchange and correlation. Regretfully, practical applications of this equation are quite limited. The self-consistent solution of this equation can be found only for systems possessing very high symmetries such as spherical atoms. For other systems, solving of this equation is computationary forbidding.

Another and less cumbersome method of calculation of the response function $\chi(\mathbf{r},\mathbf{r}',\omega,w)$ requires introduction of time-dependent variables [414]. The susceptibility may be treated as the Fourier transform of time-dependent function $\chi(\mathbf{r},\mathbf{r}',t,w)$:

$$\chi(\mathbf{r},\mathbf{r}',\omega,w) = \int dt\chi(\mathbf{r},\mathbf{r}',t,w)\exp(i\omega t). \quad (5.60)$$

This time-dependent function determines the response of the electron density $n_1(\mathbf{r},t,w)$ to the perturbation $v_1(\mathbf{r},t)$. It is defined by the standard relation

$$n_1(\mathbf{r},t,w) = \int d\mathbf{r}' \int dt'\chi(\mathbf{r},\mathbf{r}',t-t',w)v_1(\mathbf{r}',t'). \quad (5.61)$$

Substituting the expression (5.60) into Eq. (5.58), one obtains

$$\Delta E[n] = -\frac{1}{2\pi}\int d\mathbf{r}d\mathbf{r}'U_{\mathrm{lr}}(|\mathbf{r}-\mathbf{r}'|) \times \int_0^1 dw \int_0^\infty \frac{dt}{t}\chi(\mathbf{r},\mathbf{r}',t,w). \quad (5.62)$$

The computational procedure aiming at finding a suitable approximation for the time-dependent response function $\chi(\mathbf{r},\mathbf{r}',t,w)$ was described in [414]. It includes solving Schrodinger equations for noninteracting electrons moving in the time-dependent mean field. The potential corresponding to this field includes terms describing exchange and correlation effects. The suggested procedure was successfully applied to calculate the contribution to the ground-state energy of a pair of spherically symmetric atoms which originate from the van der Waals interactions between them. Despite this and some other partial successes we must repeat that in its presently existing form, DFT is rather poorly fitted for studies of the effects of long-range interactions in practical systems.

5.4 Excitation Energies

An important task is to generalize the DFT in such a way that it will provide means to describe excited electronic states of the considered system along with its ground state. We start the analysis of the DFT applicability to studies of excitation energies from presenting a few rigorously justified results. Considering Kohn–Sham equations (5.22) one notices that the eigenvalues ϵ_i could be treated as Lagrange factors which appear due to the requirement that the Kohn–Sham orbitals must be orthogonal. Therefore, the physical meaning of these eigenvalues is not immediately obvious. Within the Hartree–Fock theory, this physical meaning is clarified by using Koopman's theorem [379]. The theorem states that

$$\epsilon_i^{\mathrm{HF}} = E(n_1, \ldots, n_i, \ldots, n_m) - E(n_1, \ldots, n_{i-1}, \ldots, n_m). \tag{5.63}$$

Here, ϵ_i^{HF} is the Hartree–Fock eigenvalue, and $E(n_1, \ldots, n_i, \ldots, n_m)$ is the total energy of the system of electrons with occupation numbers $n_1, n_2, \ldots, n_m$. The meaning of this theorem is that when an electron is taken out of the system, all remaining orbitals do not change. Then ϵ_i^{HF} could be identified with the energy required to remove the electron from the ith orbital. In general, this theorem is not valid within the density-functional formalism where one finds that [416]

$$\epsilon_i(n_1, n_2, \ldots, n_m) = \frac{\partial E}{\partial n_i}. \tag{5.64}$$

Using this expression, one may present the difference in energies given by Eq. (5.63) in the form

$$\begin{aligned} &E(n_1, \ldots, n_i, \ldots, n_m) - E(n_1, \ldots, n_i - 1, \ldots, n_m) \\ &= \int_0^1 dn \epsilon_i(n_1, \ldots, n_i + n - 1, \ldots, n_m). \end{aligned} \tag{5.65}$$

Within the DFT, this expression could be justified only if the occupation numbers on the left-hand side of Eq. (5.65) correspond to the ground states of the systems, respectively, including N ($N = \sum n_i$) and $N - 1$ electrons. This means that ϵ_i should refer to the highest occupied state in the system of N electrons. It was shown [383] that within the exact DFT formalism, the energy for the highest occupied orbital obeys the equation

$$\epsilon_m(n_1, \ldots, n_m = n) = \epsilon_m(n_1, n_2, \ldots, n_m = 1) \tag{5.66}$$

for an arbitrary value of the occupation number n belonging to the interval $0<n<1$. Substituting this expression into Eq. (5.65), one can easily compute the integral over "n" which gives the result

$$E(n_1, \ldots, n_m = 1) - E(n_1, \ldots, n_m = 0) = \epsilon_m(n_1, \ldots, n_m = 1). \tag{5.67}$$

Therefore, the physical meaning of the Kohn–Sham highest eigenvalue associated with an occupied state is clarified within the DFT formalism. It gives the ionization energy of the system which could be computed as the difference between the ground-state energies of the systems with N and $N - 1$ electrons. As for the remaining Kohn–Sham eigenvalues, they cannot be easily identified with physically meaningful quantities.

Presently, there are basically three classes of methods for calculating excitation energies using the DFT. First, there are computational schemes based on the Dyson equation. It was shown by Sham and Kohn [417] that the self-energy term which enters the Dyson equation is a functional of the electron density distribution. Exploring the properties of this functional, they concluded that it had a short range and became negligible when $|\mathbf{r} - \mathbf{r}'|$ took on values beyond a certain range. They suggested an approximation for the self-energy term, relating the latter to the relevant characteristics of a homogeneous electron gas. The suggested approach was further developed in several works (see the review article [379] and references therein) and some interesting results we obtained. However, the approach based on the Dyson equation is more appropriate for systems with slowly varying electron density than for atoms and molecules. Considering the latter one might choose another class of methods where the DFT is more or less straightforwardly used to compute the total energies of both ground state and excited state. The difference between the two is assumed to approximate the excitation energy. As was already mentioned, these calculations lack justification. Nevertheless such schemes were (and still are) employed in several works. In many cases this brings estimations for the excitation energies which do not contradict experimental data.

In practical calculations of energy differences, the Slater's transition-state computational scheme [389] is often used. Within this method, one starts from Eq. (5.65) where the integral over "n" is replaced by its average and this average is approximated by the integrand taken at the middle of the interval:

$$E(n_1, \ldots, n_i, \ldots, n_m) - E(n_1, \ldots, n_i - 1, \ldots, n_m) \approx \epsilon_i(n_i - 0.5). \quad (5.68)$$

To proceed, one assumes that the eigenvalues are almost independent of a change of occupation numbers (if this change is small compared to the total number of electrons in the system) provided that the approximation chosen for the E_{XC} is appropriate. Accepting this assumption, one may use the eigenvalue for the occupation number n_i in Eq. (5.68). Then this equation is reduced to the Koopman's theorem, and the eigenvalues become directly related to the excitation energies. Although not properly justified, this approach is used in some electron structure calculations even now. Another and better justified method of calculation of excited states was suggested by Theophilou [418]. Using this method one analyzes M lowest energy states of the system including N electrons by considering a subspace S of the Fock space which is spanned by the M state vectors of lowest energy. Denoting the lowest eigenstates of the electron Hamiltonian H_{el} as $\psi_1, \psi_2, \ldots, \psi_M$, one may write

$$\text{Trace}_s(\hat{H}_{\text{el}}) = \sum_{i=1}^{M} \langle \psi_i | \hat{H}_{\text{el}} | \psi_i \rangle = \sum_{i=1}^{M} E_i \tag{5.69}$$

where the trace is taken over the subspace S and E_i are the energies corresponding to the states ψ_i. It can be proven that $\text{Trace}_s(\hat{H}_{\text{el}})$ is a functional of electron density operator:

$$\text{Trace}_s(\hat{H}_{\text{el}}) \equiv E_s[\hat{n}]. \tag{5.70}$$

Theophilou had shown that the functional $E_s[\hat{n}]$ reaches its minimum when the subspace S coincides with the space spanned by the exact eigenstates. This minimum principle could be applied to find a set of single particle equations and the corresponding approximations for their wave functions u_i. The first step in determining the excited state energies is to find the ground-state energy of the system E_1. Then one may consider the electron density expressed in terms of the wave functions u_i, namely,

$$n(\mathbf{r}) = \sum_{i=1}^{N-1} |u_i(\mathbf{r})|^2 + \frac{1}{2}|u_N(\mathbf{r})|^2 + \frac{1}{2}|u_{N+1}(\mathbf{r})|^2. \tag{5.71}$$

Using this expression for $n(\mathbf{r})$ in minimizing the functional $E_s[\hat{n}]$, one may find the energy $E_1 + E_2$ which allows to find the energy of the first excited state. In principle, this procedure may be repeated several times to find the energies of as many excited states as desired. However, for sufficiently large systems with numerous low-lying states, this method becomes very time consuming, and this put limitations on its applicability. Also, the minimum principle was employed by Perdew and Levy [419]. They showed that the exchange-correlation energy must be a nonmonotonic function if one considers a tight-bound atom. Similar lack of monotony may be expected in molecules. It was suggested that extrema of E_{XC} could indicate the excited states of the system.

Finally, we briefly discuss a recently proposed variational formalism aiming at the computation of the wave functions for the excited states of a molecule [420]. In developing this formalism, a constraint of orthogonality between many-electron Slater determinants that are constructed from the ground state and the considered excited state is applied. The calculation starts from a single Slater determinant constructed basing on N Kohn–Sham orbitals and representing the ground-state wave function for the system including N electrons:

$$\psi = A(\varphi_1, \varphi_2, \ldots, \varphi_N) \tag{5.72}$$

where A is the antisymmetrizing operator. The wave functions for single-electron excitations may be arranged using $N - 1$ original occupied orbitals φ_i $(i \neq h)$ and an unoccupied orbital φ_p:

$$\Phi = A(\varphi_1, \varphi_2, \ldots, \varphi_{h-1}, \varphi_p, \varphi_{h+1}, \ldots, \varphi_N) \tag{5.73}$$

where the subscripts h and p are referring to the hole and particle states. In these expressions, the occupied orbitals are defined up to an arbitrary transformation, so the optimal hole state is required to lie in the space spanned by the occupied orbitals. Now, the electron density is given by $n_{\text{ex}} = n_g - n_h + n_p$ where the contribution n_g is the ground-state density determined by Eq. (5.23), $n_h = |\varphi_h|^2$ and $n_p = |\varphi_p|^2$. The excited state Hamiltonian depends on n_{ex}, and one may approximate it as

$$H_{\text{ex}} = H(n_{\text{ex}}) \tag{5.74}$$

To make further progress, one introduces a perturbative Hamiltonian of the form

$$\Delta H = \alpha(H_{\text{ex}} - H_g) \tag{5.75}$$

where α is a tuning parameter and H_g is the ground-state Hamiltonian. Single particle orbitals corresponding to the excited state may be perturbatively improved. For example,

$$\varphi'_k = \varphi_k + \alpha \sum_{j>N;k\neq h} \frac{\langle \varphi_j|\Delta H|\varphi_k\rangle}{\epsilon_j - \epsilon_h} \varphi_j \tag{5.76}$$

where $\epsilon_{j,h}$ are single particle energies associated with the corresponding orbitals in the Kohn–Sham Schrodinger-like equations (5.22). Orthonormalizing these perturbatively improved orbitals, one obtains the set of orthonormal and α-dependent functions $\{\varphi''_j\}$, which could be used to make up a new Slater determinant. The resulting determinant depends on the tuning parameter α, and it is orthogonal to the ground-state wave function given by Eq. (5.72). The electron density becomes expressed in terms of the functions φ''_j:

$$n'' = \sum_j |\varphi''_{j\neq h}|^2 + |\varphi''_p|^2. \tag{5.77}$$

Then one may compute all relevant energies basing on the electron density given by Eq. (5.77). These energies depend on the parameter α, and the next step in the calculations is the minimization of the energy functional with respect to this tuning parameter. Once the best α is found, this provides one with a new way of estimating an excited-state single-electron Hamiltonian H_{ex} which gives a new ΔH. Given the improved ΔH, one can return to Eq. (5.75) to further improve the estimation of single particle orbitals. The process can be iterated until self-consistency is established.

5.5 Time-Dependent DFT

Further development of the density-functional theory includes its time-dependent versions. Time-dependent density-functional theory (TDDFT) is used to study the effects of time-dependent potentials on the properties of many-body systems including molecules as well as solids. It is especially useful in studying excitation energies, frequency-dependent response properties, and photoabsorption spectra of such systems. Conceptual foundations of TDDFT and DFT are similar. The starting point for both theories is the statement that there exists a one-to-one correspondence between the wave function and the electron density. In the case of TDDFT both quantities depend on time. Basing on the "equivalence" of the electron density and the wave function, one may introduce the effective potential of a system of noninteracting quasiparticles which returns the same electron density as that inherent to the actual interacting system. Then one may introduce a system of Schrodinger-like equations using the effective potential. This system is the analog of Kohn–Sham equations used as an important part of the DFT formalism. However, within the TDDFT building up such a system is much more complicated task. The reason is that the time-dependent effective potential at any given instant is determined by the values taken by the electron density at all previous times. Obviously, this difficulty does not occur when one is dealing with the time-independent versions of DFT.

The formal foundation of the time-dependent DFT is the Runge–Gross theorem [421] which is the analog of the Hohenberg–Kohn theorem [382] for time-dependent systems. This theorem states that for a certain initial many-body wave function, there exists a unique one-to-one correspondence between the time-dependent external potential acting on the considered electron system and its time-dependent electron density. This implies that all properties of the system can be determined from the knowledge of electron density distribution. Basing on the Runge–Gross theorem one may derive the system of Kohn–Sham equations for an electron system subject to a time-dependent external potential $V_{\text{ext}}(t)$. The Hamiltonian for such a system could be presented in the form

$$H(t) = T + V_{\text{ext}}(t) + U \tag{5.78}$$

where, as before, T represents the kinetic energy operator and U describes electron–electron interactions. In general, the potential $V_{\text{ext}}(t)$ includes a time-independent part arising due to interactions between electrons and atomic nuclei [see Eq. (5.1)]. A nontrivial time dependence of V_{ext} originates from external fields (usually electric and/or magnetic) affecting the electrons. The many-body wave function ψ evolves according to the time-dependent Schrodinger equation

$$H\psi = i\hbar\frac{\partial\psi}{\partial t} \tag{5.79}$$

and it satisfies a single initial condition

$$|\psi(0)\rangle = |\psi\rangle. \tag{5.80}$$

Now, one may follow the Kohn–Sham approach and replace the system of interacting electrons described by the Hamiltonian (5.78) by a system of noninteracting quasiparticles exposed to the time-dependent potential $V_{\text{eff}}(\mathbf{r},t)$. This system consists of N equations (N being the total number of electrons), and it has the form

$$\begin{aligned} i\hbar\frac{\partial\varphi_i(\mathbf{r},t)}{\partial t} &= \left(-\frac{\hbar^2}{2m}\Delta + V_{\text{eff}}(\mathbf{r},t)\right)\varphi_i(\mathbf{r},t), \\ \varphi_i(\mathbf{r},0) &= \varphi_i(\mathbf{r}). \end{aligned} \tag{5.81}$$

Here, the potential $V_{\text{eff}}(\mathbf{r},t)$ could be presented in the form similar to that given by Eq. (5.19), namely,

$$V_{\text{eff}}(\mathbf{r},t) = V_{\text{ext}}(\mathbf{r},t) - e\varphi(\mathbf{r},t) + V_{\text{XC}}(\mathbf{r},t). \tag{5.82}$$

Again, the potential $V_{\text{eff}}(\mathbf{r},t)$ is decomposed to extract the external potential, the time-dependent term describing Coulomb interactions of the quasiparticles, and the exchange-correlation term. The many-body wave function corresponding to the introduced system of noninteracting quasiparticles can be simply represented as Slater determinant constructed using single particle wave functions $\varphi_i(\mathbf{r},t)$. Accordingly, the time-dependent electron density distribution is given by

$$n(\mathbf{r},t) = \sum_{i=1}^{N} |\varphi_i(\mathbf{r},t)|^2. \tag{5.83}$$

The density $n(\mathbf{r},t)$ determined by this expression is equal to the electron density corresponding to the original system of interacting electrons.

To proceed, one needs to find the expression for the exchange-correlation contribution to the Kohn–Sham effective potential. The problem of determining suitable approximations for $V_{\text{XC}}(\mathbf{r},t)$ is a challenging one. In their original work [421], Runge and Gross suggested to define this potential using an action-based argument. They started from the Dirac action $A[\psi]$ given by the expression

$$A[\psi] = \int dt \Big\langle \psi(t) \Big| H - \frac{i\hbar\partial}{\partial t} \Big| \psi(t) \Big\rangle. \tag{5.84}$$

The Dirac action is a functional of the wave function ψ. Varying the wave function to minimize $A[\psi]$, one may get the time-dependent Schrodinger equation which corresponds to the minimum action. At the same time, it is known that there exists a one-to-one relationship between the many-body wave function ψ and the electron density distribution. So, one may treat the Dirac action as a functional of the electron density

$$A[\psi[n]] \equiv A[n]. \tag{5.85}$$

Runge and Gross derived a formal expression for the exchange-correlation component of the action by functional differentiation of the expression (5.84). Since the exchange-correlation part of the Dirac action determines the exchange-correlation term in the Kohn–Sham effective potential $V_{\mathrm{XC}}(\mathbf{r}, t)$, they did obtain the expression for the latter. However, later it was demonstrated that an approach basing on the Dirac action brings paradoxical results inconsistent with the causality of response functions. The density response function χ (which is defined as the functional derivative of the density with respect to the external potential) should obey the causality principle. This means that a change in the potential occurring at a certain instant cannot affect the electron density values corresponding to earlier times. The response functions generated by the Dirac action are symmetric in time, so they lack the required causal structure. Another approach which does not suffer from the lack of proper causal structure of the resulting response functions was suggested by R. van Leeuwen [422]. Within this approach, the Kohn–Sham equations are formally identified with a stationary point of an action functional defined within the Keldysh formalism of complex-time path integration. Suitable approximations for the time-dependent Kohn–Sham effective potential are still being explored [423], and further progress in these studies is expected as well as needed.

When the external time-dependent perturbation is so weak that it is unable to completely destroy the ground state of the considered system, one may restrict oneself with the analysis of linear response of the system to the external disturbance. This is a much simpler task. Within the linear response theory variations of the system solely depend on the ground-state wave function. This gives the opportunity to use some results obtained within the time-independent DFT to derive the expressions for the response functions. In the presence of a small time-dependent perturbation $\delta V_{\mathrm{ext}}(t)$, the Hamiltonian of the electron system takes on the form

$$\begin{aligned} H'(t) &= H_{\mathrm{el}} + \delta V_{\mathrm{ext}}(t) \\ &= H_{\mathrm{KS}}[n] - e\delta\varphi(\mathbf{r}, t) + \delta V_{\mathrm{XC}}([n], t) + \delta V_{\mathrm{ext}}(t) \end{aligned} \tag{5.86}$$

where $H_{\mathrm{KS}}[n]$ is the Hamiltonian appearing in the Kohn–Sham equations (5.22) which correspond to the electron system undisturbed by the time-dependent perturbation and the remaining terms represent the time-dependent corrections to the original effective potential which appear due to the external perturbation. The linear response of the electron density to the perturbation is given by

$$\delta n(\mathbf{r}, t) = \int d\mathbf{r}' \int dt' \chi_{\mathrm{KS}}(\mathbf{r}, t, \mathbf{r}', t') \delta V_{\mathrm{eff}}(\mathbf{r}', t'). \tag{5.87}$$

Here, the response function $\chi_{KS}(\mathbf{r},t,\mathbf{r}',t')$ is defined by the above equation. Within the linear response domain the variations of the Hartree term $-e\delta\varphi(\mathbf{r},)$ and the exchange-correlation term $V_{XC}(\mathbf{r},t)$ may be expanded in powers of the density variation δn. The linear in δn terms have the form

$$-e\delta\varphi = \frac{e^2}{2}\int \frac{d\mathbf{r}'}{|\mathbf{r}-\mathbf{r}'|}\delta n(\mathbf{r}',t), \tag{5.88}$$

$$\delta V_{XC} = \int d\mathbf{r}' \int dt' f_{XC}(\mathbf{r},t,\mathbf{r}',t')\delta n(\mathbf{r}',t'). \tag{5.89}$$

Also, the linear response of the electron density is related to the external perturbation as follows:

$$\delta n(\mathbf{r},t) = \int d\mathbf{r}' \int dt' \chi(\mathbf{r},t,\mathbf{r}',t')\delta V_{ext}(\mathbf{r}',t'). \tag{5.90}$$

Inserting the expressions (5.88) and (5.89) into Eq. (5.87) and comparing the result with Eq. (5.90), one obtains the integral equation establishing the relationship between $\chi(\mathbf{r},t,\mathbf{r}',t')$ and $\chi_{KS}(\mathbf{r},t,\mathbf{r}',t')$, namely,

$$\begin{aligned}\chi_{KS}(\mathbf{r},t,\mathbf{r}',t') =& \chi(\mathbf{r},t,\mathbf{r}',t') - \int d\mathbf{r}'' \int dt'' \chi_{KS}(\mathbf{r},t,\mathbf{r}'',t'') \\ &\times \left[\frac{e^2}{2}\frac{1}{|\mathbf{r}'-\mathbf{r}''|} + f_{XC}(\mathbf{r}'',t'',\mathbf{r}',t')\right].\end{aligned} \tag{5.91}$$

Within the TDDFT, excitation energies of the electron system are corresponding to the poles of the linear response function. Therefore, these energies could be found using Eq. (5.91).

In the following three sections we consider a few applications of the stationary DFT which are related to transport characteristics of some molecules. Also, we briefly touch upon some time-dependent DFT-based calculations in Sect. 5.8 while discussing light-harvesting molecular complexes.

5.6 Dielectric Response of Fullerenes

Reduction in the spatial dimensions of nanoparticles results in a number of interesting properties such as the widening of the bandgap, reduction of the melting temperature, and increased magnetic moments. The fundamental understanding of the size dependence of properties of nanoparticles, i.e., at what size the quantum size effects set in, is important for tailoring these systems for possible nanotechnological applications. Studies along these lines continue to be frontline research in nanoscience [424–427], and fullerenes are among the objects of these

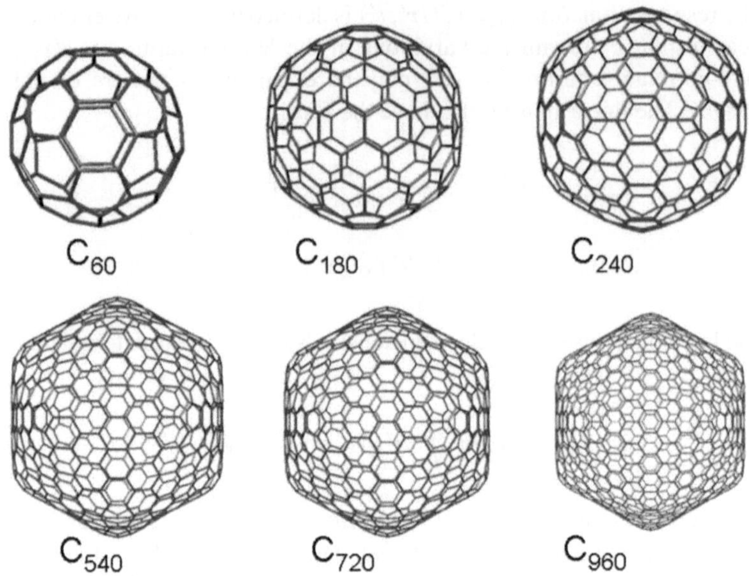

Fig. 5.1 Optimized geometries for some fullerenes whose polarizabilities are calculated [390]. Reprinted with the permission of the APS

studies. Fullerenes are hollow cages formed by sp^2-bonded carbon atoms. They are finite analogs of graphene with 12 pentagonal defects that introduce curvature and make them closed structures. Unlike solid nanoparticles, however, fullerene molecules have all the atoms on the surface, whereas solid spherical nanoparticles have roughly $N^{1/3}$ atoms on the surface. This feature results in rapid increase of their volume as the number of atoms (i.e., its size) increases. As dipole polarizability is proportional to volume, fullerenes are most suitable for investigating quantum size effects manifested in polarizability.

In the last decade, several studies of the polarizability of fullerenes were carried out (see, e.g., [390, 427, 428]). The DFT was successfully adopted to compute the first-order electric response of carbon fullerenes of various sizes from C_{60} to C_{2160} by calculating their static dipole polarizabilities [390,429]. In the work [390], polarizabilities of carbon fullerenes with optimized geometries shown in Fig. 5.1 were computed using the sum-over-states (SOS) method and the random phase approximation within the linear theory. The fullerene structures were held rigid, so the ionic (vibronic) contributions to dipole polarizabilities were ignored.

The working SOS expression for the xy component of polarizability tensor can be written as

$$\alpha_{xy} = 2 \sum_{m}^{\text{unocc}} \sum_{n}^{\text{occ}} \frac{\langle \varphi_m | \hat{x} | \varphi_n \rangle \langle \varphi_m | \hat{y} | \varphi_n \rangle}{|\epsilon_m - \epsilon_n|}. \tag{5.92}$$

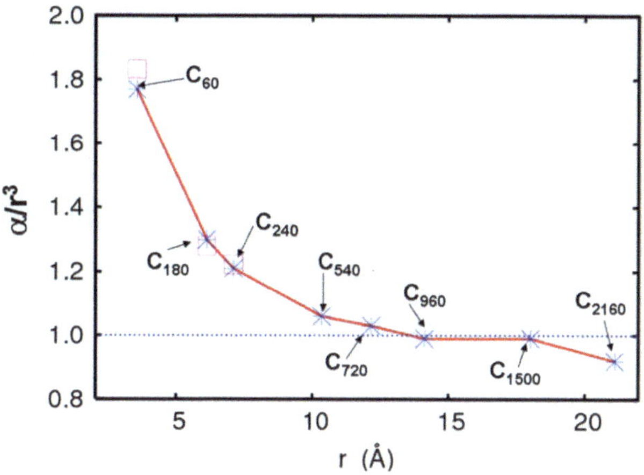

Fig. 5.2 The ratio of polarizability, α^{RPA}, and volume, r^3, as a function of fullerene radius r. The *squares* represent polarizability obtained by the finite-field method [390]. Reprinted with the permission of the APS

Here, φ_i are single particle orbitals, and ϵ_i is the single particle energy of the ith orbital. The $\alpha_{\mathrm{SOS}} = \frac{1}{3}(\alpha_{xx} + \alpha_{yy} + \alpha_{zz})$ was calculated according to Eq. (5.92) once the self-consistent solution of the Kohn–Sham equations for each fullerene was obtained [390]. The computational efforts involved in determining the large number of transition dipole moments could be reduced by making use of group theoretical techniques. The α_{SOS} determined using Eq. (5.92) gives the unscreened polarizability since the field-induced polarization effects (local field effects)are missed in the calculations. Consequently, the SOS overestimates the experimentally accessible screened polarizability. For the smallest fullerene (C_{60}), α_{SOS} is roughly 3.6 times larger than the screened polarizability [429]. This overestimation slightly increases (to a factor $\sim$ 4) for C_{180}, C_{240}, and C_{540}. The screened polarizability can be deduced from SOS using the linear response theory [430] and the random phase approximation (RPA) by means of the following expression:

$$\alpha^{\mathrm{RPA}} = \left[1 + \frac{\alpha_{\mathrm{SOS}}}{R^3}\right]^{-1} \alpha_{\mathrm{SOS}}. \tag{5.93}$$

Here, R is the fullerene radius [430]. For fullerenes larger than C_{60} that are polyhedral in structure, the average radius r can be used instead of R.

One receives even better approximation using an effective radius $r + \delta$ where the factor δ accounts for the π-electron cloud around the bare nuclear skeleton. Its typical values are 1.2–1.3 Å [427, 430]. The dependence of the screened polarizability of the fullerene radius is shown in Fig. 5.2. As mentioned above, within the classical electrostatic theory, the polarizability is proportional to the volume, so α/r^3 must remain constant. Comparing the computed results for the fullerenes with the classical estimate (the latter is represented by the horizontal line),

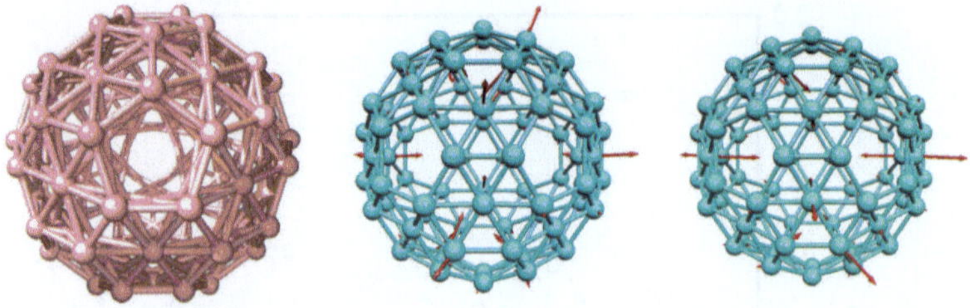

Fig. 5.3 *Left panel*: optimized geometry of a B_{80} cage. *Middle* and *right panels*: unstable vibrational modes of an icosahedral B_{80} cage [432]. Reprinted with the permission of the APS

one may see that the quantum size effects significantly contribute to polarizabilities of smaller fullerenes and cease to be important for larger carbon fullerenes including 2,160 or more atoms [390]. Similar conclusions concerning the part taken by the quantum size effects were made in the earlier work [427] with the exception that the minimum number of atoms which allows to neglect quantum effects in the dipole polarizability was estimated as 3,840.

In the case of carbon fullerenes, vibrational contributions to the polarizability are insignificant, and one may neglect them [431]. However, the situation can be different in other cagelike structures. Recent work by Szwacki et al. [432] added boron to the list of elements that can form fullerene-like hollow cage structures. Several studies have reported stable clusters. It was shown that the B_{80} fullerene is a stable hollow cage. The basic structure of this cluster is similar to that of C_{60} with 12 pentagonal and 20 hexagonal rings. The B_{80} cluster has an additional boron atom at the center of each hexagon (as shown in Fig. 5.3). It is known that the boron analog of C_{60} is not a stable structure. However, these additional 20 boron atoms at the centers of the hexagonal rings stabilize the B_{60} to form a stable B_{80} fullerene with a binding energy of 5.77 eV per atom. The structure of a B_{80} cluster was suggested by Szwacki et al. to maintain the same icosahedral point group symmetry as in the C_{60} fullerene.

Vibrational stability and electronic properties of B_{80} fullerenes including its static dipole polarizability were studied by Baruah and coworkers within the DFT, assuming that the Kohn–Sham orbitals had been expressed as linear combinations of Gaussian orbitals [386]. The calculations showed that the icosahedral B_{80} cluster was vibrationally unstable with two sets of imaginary frequencies. One of the two sets of modes is fourfold degenerate, and the other one is a threefold degenerate mode. The vibrational motions of atoms in these two modes are depicted in Fig. 5.3. These modes are generated by the out-of-plane motions of the boron atoms capping hexagons. Full unconstrained relaxation of B_{80} cluster demonstrated that the cluster had T_h symmetry. This result is consistent with another recent study, which reports the T_h symmetry of B_{80} clusters [433]. Also, it was shown that the T_h cluster was vibrationally stable. As mentioned before, the C_{60} fullerene can be built from the graphene sheet by introducing pentagonal rings. These pentagonal rings are defects that introduce curvature in the planar sheet, and 12 pentagons make the structure closed. The C_{60} is the smallest fullerene in which the pentagonal rings are isolated. In analogy with C_{60}, the structure of the B_{80} fullerene can also be thought to be derived from the triangular boron sheet. However, its structure is also compatible with the hole doped triangular sheet. As in C_{60}, the 12 pentagonal rings in B_{80} are isolated. The bonding in B_{80} can also be explained using the three-center bonding [434].

The dipole polarizability of B_{80} was computed using the finite-field method [386]. Within this method, the total energies and/or dipole moments are self-consistently calculated for various values of the applied electric field. Then one may build the polarizability tensor from well-converged total energies and/or dipole moments given by the expressions [387]

$$\mu_\beta = -\frac{\partial E}{\partial G_\beta},$$

$$\alpha_{\beta\gamma} = -\frac{\partial^2 E}{\partial G_\beta \partial G_\gamma} \tag{5.94}$$

where $\beta, \gamma = \{x, y\}$, μ_β, and G_β are the components of the dipole moment and the external electric field, respectively, and E is the total energy.

Due to the quasispherical symmetry of the B_{80} cluster, the off-diagonal elements of the polarizability tensor are zero. The mean polarizability obtained by the finite-field method is 149 Å^3. The mean polarizability of C_{60} fullerene determined using the same set of approximations is 82 Å^3. The larger polarizability of the B_{80} cluster is principally due to its larger volume. Using the radius of the outermost atoms, the volume of the B_{80} cluster can be estimated to be roughly 1.74 times that of the C_{60} fullerene. Using 1.74 as a scaling factor, the polarizability of the C_{60} fullerene can be employed to estimate the polarizability of the B_{80} cluster. This rough estimate is 143 Å^3, in good agreement with the polarizability obtained by the finite-field method. The vibrational frequencies determined in the analysis of the vibrational stability of the B_{80} cluster also can be used to compute the vibrational contribution to the dipole polarizability. In general, the vibrational polarizability is the second largest contribution to the total polarizability tensor. For the case of ionic and hydrogen-bonded systems, the vibrational contribution may be comparable to or even larger than the electronic contribution. The vibrational contribution to the polarizability tensor within the double harmonic approximation [431] is given as

$$\alpha_{i,j}^{\text{vib}} = \sum_m Z_{i,m} \Omega_m^{-2} Z_{j,m}^T. \tag{5.95}$$

Here, Ω_m is the frequency of the mth vibrational mode, and $Z_{i,m}$ is the effective charge tensor (see [431] for more details). The vibrational contribution to polarizability is 5.5 Å^3, which is much smaller than the electronic contribution but significantly larger than that observed in carbon fullerenes, and it cannot be neglected. A noticeable vibrational contribution to the dipole polarizability of B_{80} fullerenes is related to the higher reactivity of this boron fullerene as compared to that of the carbon fullerene C_{60}. It makes the B_{80} cluster a potentially interesting candidate as an electron receptor in molecular complexes where long-range electron transfer reactions are used for practical purposes such as light harvesting.

5.7 Density-Functional-Based Investigations of Molecular Magnets

As discussed above, magnetic molecules, that is, the molecules with nonzero global net spin, attract a significant interest of the research community. This interest mostly originates from the potentialities of these molecules as building blocks for nanoelectronic and spintronic devices. Single-molecule magnets (SMMs) are three-dimensional arrays of identical high-spin molecules. Each molecule has a large effective spin with a large magnetic anisotropy barrier (MAB) and responds to an external magnetic field as a nanoscale, single-domain magnetic particle. Usually, magnetic molecules are composed of a few transition metal atoms which are locked at their lattice sites by a host consisting of organic molecules and ligands. The most extensively studied SMM is Mn_{12} [292]. A single molecule of Mn_{12} (Fig. 5.4)

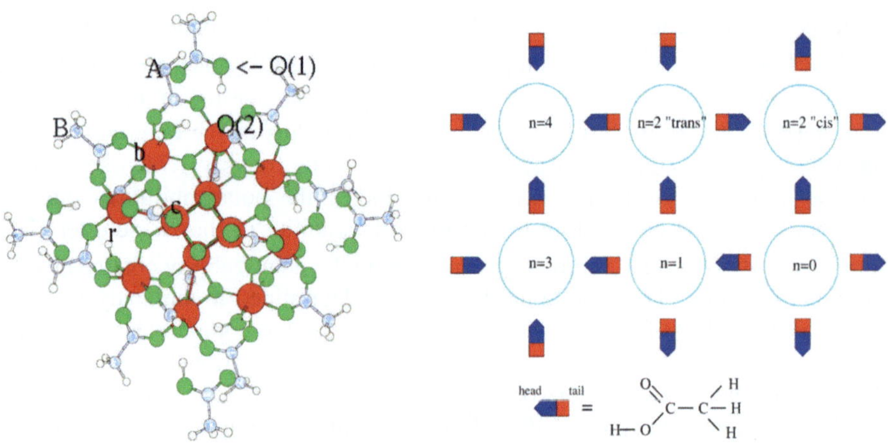

Fig. 5.4 *Left panel*: a single Mn_{12} molecule surrounded by four HAc solvent molecules, $[Mn_{12}O_{12}(CH_3COO)_8(HCOO)_8(H_2O)_4]4(CH_3COOH)$. The sizes of atoms decrease in the order of Mn, O, C, and H. The four inner cubane Mn_{4+} ions ($S = 3/2$) are ferromagnetically coupled to each other, and the eight outer crown Mn_{3+} ($S = 2$) ions also are ferromagnetically coupled. The Mn spins in the crown are antiparallel to those in the cubane, which results in the total effective ground-state spin of $S = 10$. The easy axis of the molecule is normal to the page. Two inequivalent acetates are labeled as "A" and "B." The hydrogen bond is formed between oxygen atoms $O(1)$ and $O(2)$. *Right panel*: six distinct configurations of a Mn_{12} molecule with four HAc solvent molecules. The *circles* represent Mn_{12} molecules, n is the number of HAc molecules whose heads point into a Mn_{12} molecule. From the top leftmost, clockwise, the symmetry of each configuration is S_4, C_2, C_1, S_4, C_1, and C_1 (after [443]). Reprinted with the permission of the APS

has an effective ground-state spin of $S = 10$ with a magnetic easy axis along the crystal c axis and has S_4 symmetry. Low-temperature magnetization hysteresis loop measurements on Mn_{12} revealed steps at quantized magnetic fields, in which macroscopic quantum tunneling (MQT) occurred between spin-up states and spin-down states through the magnetic anisotropy barrier [292,435].

As discussed in [392], both the MQT and the MAB in Mn- and Fe-based magnetic molecules are mostly determined by the second-order spin–orbit interactions. Matrix elements of the interaction energy associated with the spin–orbit interactions could be presented in the form [388,392]

$$U_{j\sigma;k,\sigma'} = \frac{1}{i}\Sigma_\gamma \langle f_j|V_\gamma|f_k\rangle\langle\chi_\sigma|S_\gamma|\chi_{\sigma'}\rangle \tag{5.96}$$

where the summation is carried out over the Cartesian coordinates ($\gamma = \{x, y, z\}$), f_j and χ_σ are the orbital functions and spinors used to build up the basis set for the expansions of Kohn–Sham orbitals [see Eq. (5.31)], and the operator V_x is defined as follows:

$$\langle f_j|V_x|f_k\rangle = \frac{1}{2c^2}\left\langle f_j\left|\frac{d\Phi}{dy}\frac{d}{dz} - \frac{d\Phi}{dz}\frac{d}{dy}\right|f_k\right\rangle \tag{5.97}$$

Here, c is the speed of light; $\Phi(\mathbf{r})$ is the total Coulomb potential acting upon an electron which moves in the electric field created by the nuclei and the remaining electrons included into the considered molecule. The matrix elements for the operators V_y and V_z could be found by cyclic permutations of the coordinate labels in the above equations. The spin–orbit coupling was shown to bring the second-order correction to the total energy [392]:

$$E_S = \sum_{\sigma,\sigma'}\sum_{\gamma,\gamma'} A_{\gamma\gamma'}^{\sigma\sigma'} S_\gamma^{\sigma\sigma'} S_{\gamma'}^{\sigma'\sigma} \tag{5.98}$$

where the matrix elements $A_{\gamma\gamma'}^{\sigma\sigma'}$ are computed using the collection of Kohn–Sham spin orbitals $\{\varphi_{i\sigma}\}$:

$$A_{\gamma\gamma'}^{\sigma\sigma'} = -\sum_{jk}\frac{\langle f_j|V_\gamma|f_k\rangle\langle f_k|V_{\gamma'}|f_j\rangle}{\epsilon_{j\sigma} - \epsilon_{k\sigma'}} \tag{5.99}$$

and

$$S_\gamma^{\sigma\sigma'} = \langle\chi_\sigma|S_\gamma|\chi_{\sigma'}\rangle. \tag{5.100}$$

By diagonalizing the matrix A, one may transform the Hamiltonian (5.98) and reduce it to the expression which agrees with Eqs. (3.49) and (3.50), namely,

$$H_s = DS_z^2 + E(S_x^2 - S_y^2) \tag{5.101}$$

where S_z is the spin operator projected onto the z axis which is identified with the magnetic easy axis and D and E are the uniaxial and transverse second-order magnetic anisotropy parameters caused by spin–orbit coupling. High-field, high-frequency electron paramagnetic resonance (EPR) experiments on the SMM Mn_{12} provided that $(-D) = 55\,\mathrm{K}$ [436, 437]. The ground-state characteristics of Mn_{12} and similar magnetic molecules where some Mn ions were substituted by Fe ions were computed within the DFT. Basing on these computations, the constant D which is related to the height of the energy barrier (the latter is approximately ten times greater in magnitude than $|D|$) was calculated, and its value for Mn_{12} indicated a very good agreement between the theory and the experiment [152, 392]. Also, low-lying excited states and their effect on the magnetic anisotropy barrier in Mn_{12} molecules were studied [151]. Considering the S_4 symmetry of a Mn_{12} molecule (without solvent molecules), we know that $E = 0$. Quantum tunneling of magnetic moment is then allowed for pairs of states that satisfy a certain selection rule. The two states must have S_z-eigenvalues differing by integer multiples of 4. However, tunnelings between states which violate this selection rule were clearly observed in magnetic tunneling measurements [292, 435]. To resolve this puzzle, three different theories have been proposed. Chudnovsky and Garanin [438, 439] proposed that dislocations in single crystals could produce a broad continuous distribution in E. Detailed x-ray diffraction studies by Cornia et al. [440] suggested that disorder caused by acetic-acid (CH_3COOH) solvent molecules could break the S_4 symmetry and provide nonzero locally varying E values. Pederson et al. suggested that the discrepancy could be erased if one takes into account the effect of spin–orbit–vibron coupling occurring in the molecule [397]. They showed that for isotropically impure samples, there were two different vibrational effects that caused coupling between M and $M \pm 2$ states and that the symmetry breaking would be random assuming the isotopic substitution. Magnetic quantum tunneling [441] and high-field, high-frequency EPR experiments [442] showed the results which were not in quantitative accord with all these theories.

To further explore the origin of the symmetry break which causes otherwise forbidden transitions, the effects of both positional disorder induced by acetic-acid solvent molecules and orientational disorder induced by methyl groups in the Mn_{12}-acetate molecule were analyzed [443]. Density-functional-based calculations showed that only hydrogen-bonded solvent molecules contributed to the transverse anisotropy among all six possible configurations shown in Fig. 5.4 (right panel). The maximum value of the transverse anisotropy was estimated as $E = 0.016\,\mathrm{K}$. The nonzero value of this parameter indicated the symmetry break which made possible the transitions observed in experiments. Also, this value agrees with the experimental results reported in [441, 442]. As for the effect of the orientational disorder, it was proven to be negligible.

Another example of SMM is a dimerized Mn_4 $[Mn_4O_3Cl_4(O_2CCH_3CH_2)_3(C_5H_5N)_3]_2$ molecule (see Fig. 5.5). The Mn_4 molecules exhibit a special behavior different from that typical for Mn_{12} and Fe_8 magnetic molecules. In particular, these molecules exhibit different tunneling behavior [444]. This includes quantum tunneling prior to magnetic-field reversal and an absence of quantum tunneling at

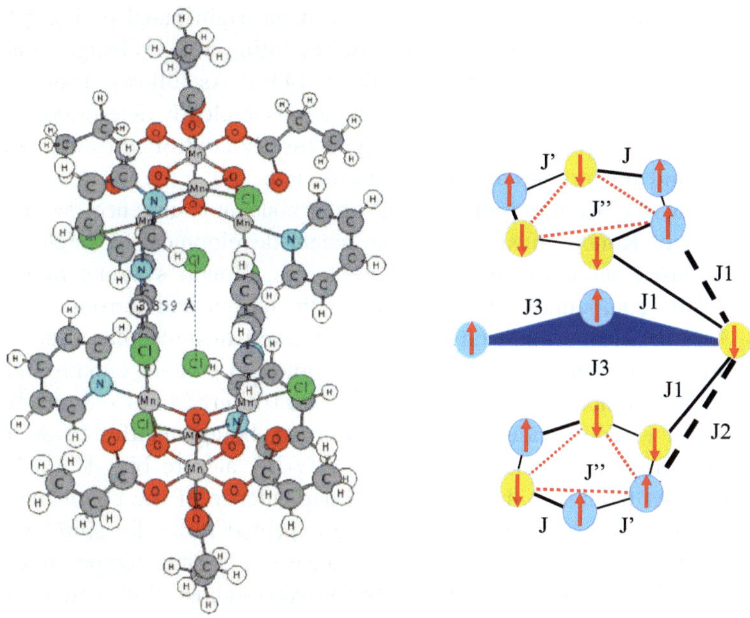

Fig. 5.5 *Left panel*: Mn_4 dimer geometry. The dimer is formed by inversion of the threefold symmetric monomer. Each monomer has a magnetic core consisting of three ferromagnetically coupled Mn_{4+} ($S = 3/2$) ions leading to a total spin of $S = 9/2$. *Right panel*: the 15 magnetic vanadium atoms of the $K_6[V_{15}As_6O_{42}(H_2O)]$ molecule. They form two hexagonal layers and an inner triangular layer of vanadium atoms sandwiched within. The *arrows* show the lowest energy spin configuration found from DFT calculations [446]. The six exchange parameters used in the Heisenberg Hamiltonian are shown as *lines*. Reprinted with the permission of the APS

zero field in contrast to Mn_{12} and Fe_8. To understand the basis for this qualitative deviation, both the magnetomolecular anisotropy and the intermolecular exchange interaction in the Mn_4 dimer were calculated using density-functional theory [445]. The results confirm that there exists an appreciable *antiferromagnetic* exchange interaction between the monomers and that tunneling fields in this dimer are strongly influenced by the presence of the monomer-monomer exchange interaction. This interaction produces a bias field that encourages monomeric magnetic-moment reversal below zero field and prevents two monomers from simultaneously flipping their magnetic moments at zero field. It was determined that the origin of the exchange interaction was not dominated by either kinetic or exchange-correlation terms and that the total "exchange" interaction was, in fact, an order of magnitude smaller than the kinetic contribution. On the contrary, in the case of Mn_{12} and Fe_8, the exchange interactions are large enough to make the spin–orbit and anisotropy to predominate in the controlling of the behavior of these systems.

A quintessential example of a molecular magnet where the spin–orbit and anisotropy are insignificant and their part is reduced to that of a tiny perturbation to the low-energy spin excitations is the molecule $K_6[V_{15}As_6O_{42}(H_2O)]$ which is commonly referred to as V_{15}. The arrangement of magnetic vanadium atoms in the V_{15} molecule is schematically shown in the right panel of Fig. 5.5. DFT-based calculations of the exchange parameters fitting V_{15} exchange interactions were carried out by Kortus and collaborators [446]. As follows from studies of various SMMs, it is still left to future investigations to clearly determine how many of the molecular magnets have their behavior mostly controlled by the spin–orbit interactions and anisotropies and how many are not.

As known, the Kondo problem has undergone a resurgence, with atomic-scale studies of Kondo physics. With the recent development of single-molecule transistors based on individual small molecules, Kondo systems now include organometallic compounds and fullerenes with normal and ferromagnetic leads [27, 28, 31, 146, 303]. Kondo physics was observed in molecular nanostructures incorporating transition metal complexes designed to contain unpaired electrons [447, 448]. As a function of gate voltage V_G, transitions from Coulomb blockade conduction to Kondo conduction were observed, being manifested as a strong peak in the differential conductance, $g \equiv \partial I/\partial V$, at zero bias $V = 0$ in one charge state. At fixed V_G, the temperature dependence of the conductance peak amplitudes and widths agreed well with the expected forms for spin-$1/2$ Kondo resonances. Observed in single-molecule transistors, Kondo temperatures are $\gtrsim 50K$, comparable to those in purely metallic Kondo systems. However, it was found that $T_K(V_G)$ was strongly inconsistent with the simple model of Kondo physics as seen in semiconductor devices. Contrary to the predictions following from this model, the measured Kondo temperature only slightly changed as the gate voltage varied. In some samples T_K increased as the gate voltage shifted away from the nearest charge degeneracy [447]. To explain this deviation, the electronic structure calculations were performed.

These calculations reveal an additional energy scale that is often small in semiconductor dots but in the molecular systems is comparable energetically to the inferred energies Γ, U_0, μ (where Γ is describing the coupling of the molecular complex to the leads and U_0, μ are the charging energy and the electrochemical potential for the leads, respectively) and the expected single particle level spacing. In both Co and Cu complexes studied in the experiments of [447], intramolecular exchange was strong. In the Co case, to a good approximation ligand field-effect split, the d states into one doubly degenerate, one nearly doubly degenerate, and one singly degenerate sets of states. The complex had a total spin of $S = 3/2$, as was confirmed by EPR. A minority spin half-occupied d state was at the Fermi level with two nearly degenerate d states 1.1 eV above the Fermi level and the remaining occupied minority d electron state 0.5 eV below the Fermi level. Exchange splittings pulled the majority d band significantly down, with the highest occupied majority d state 1.8 eV below the Fermi level. The fermiology was complicated by an additional delocalized unpolarized twofold degenerate carbon $2p$ state lying 0.05 eV below the Fermi level. Also, there was a delocalized doubly degenerate molecular state 0.85 eV below the Fermi level. The Cu complex had a total spin of $S = 1/2$, with the Fermi level unchanged in comparison to Co. This means that the minority spin d states were pulled to lower energies by approximately 1.1 eV. The structure calculations also confirmed significant delocalization of the majority spin states, extending into the ligands, consistent with large Γ values for these systems. The occasional observation of Kondo resonances with similar T_K values on both sides of charge degeneracy points further indicated that intramolecular exchange cannot be neglected in these devices. All these results demonstrated that correlated states involving single-molecule transistors could exhibit rich effects not seen in their semiconductor quantum dot counterparts.

Finally, we remark that there exist some questions concerning the applicability of the density-functional theory to studies of multireference systems such as Mn_{12} and other large magnetic molecules. These questions were and still are discussed, and presently several extensions and generalizations of the DFT are suggested. However, a multitude of calculations shows that density-functional-based Slater determinants constructed from Kohn–Sham orbitals provide a good starting point to explore many-electron wave functions, and numerous researchers have adopted it to address the multireference characteristics of a many-spin problem.

5.8 Photoexcitation of the Light-Harvesting Carotenoid-Porphyrin-C_{60} Molecular Triad

The quest for efficient ways to exploit renewable energy sources has become one of the key problems of our times. The idea of a solid-state device capable of transforming solar light energy into electric potential energy is quite old, and enormous progress has been made in optimizing traditional junction-based devices with respect to their thermodynamical efficiency and production cost [449]. Recently, entirely new classes of materials have been proposed as building blocks for next-generation solar cells. Among these nanostructured semiconductors, organic-inorganic hybrid and supramolecular assemblies appear to be the most promising [450]. Biological light-harvesting systems, found mostly in plants and bacteria, show an impressive efficiency for solar energy conversion. The efficiency of the light-harvesting process measured as the ratio of the amount of energy converted to another form to the amount of energy absorbed is more than 90 % for the natural light-harvesting systems. The solar energy conversion devices based mostly on semiconductors have a much smaller efficiency. Therefore, biomimetic materials are currently widely studied to achieve efficient photovoltaic devices.

In the class of supramolecular assemblies, the donor–acceptor dyads and triads are objects of careful experimental study. They are in fact considered an ideal mimic for the primary photosynthetic process, which basically consists of a photoexcitation followed by a charge transfer between the component units. The same process can be exploited in photoconversion units like dye-sensitized (or Gratzel) solar cells [451, 452]. The search for ideal components of light-harvesting dyads has led to porphyrin-C_{60} assemblies. This bioinspired molecular photovoltaic was first synthesized by Liddell and coworkers [453, 454], and it contains three components: a diaryl base porphyrin, a β-carotenoid molecule, and a pyrrole-C_{60} molecule.

The system is composed of 207 atoms with 632 valence electrons. Its structure corresponds to the formula $C_{132}H_{68}N_6O$. The pyrrole ligand breaks the icosahedral symmetry of the C_{60} molecule and also connects it to the porphyrin. The porphyrin and the carotene are connected through an amide connector. The base porphyrin functions as a chromophore, the carotenoid as an electron donor, and the C_{60} as an electron acceptor. The porphyrin molecule absorbs light in the red and blue regions of the spectrum. The change in absorption due to the presence of the hydrogens is small, so that the main absorption peaks remain in the blue and red regions.

The C_{60} molecule is known to accelerate photoinduced charge separation. It has a particularly deep triply degenerate lowest unoccupied molecular orbital (LUMO) at approximately -4.3 eV versus vacuum, capable of accommodating six electrons from neighboring molecules. Moreover, the highly conjugated nature of its bonds, together with a cage radius of 3.45 Å, stabilizes the extra charges in a very efficient way, so that little repulsive forces are felt by the excess electrons. A further step towards solar cells based on these components consisted of transforming the donor–acceptor unit into a donor-bridge-acceptor unit. This change has the main purpose of obtaining a better charge separation and a longer lifetime of the final (excited) state. While many experimental properties of porphyrin-C_{60}-based dyads and triads are known, less is known from an *ab initio* theoretical point of view. The main reason is that the large size of the systems makes traditional accurate quantum chemistry methods (such as configuration interaction) impractical or infeasible. Less accurate but computationally efficient methods such as density-functional theory have successfully been employed to describe the ground state of such compounds [420, 455]. Also, their excited-state properties were analyzed [420, 456]. Results concerning photoabsorption spectra demonstrating the best agreement between the theory and experiment were obtained by using a time-dependent version of density-functional theory. As discussed before, the time-dependent DFT provides a rigorous and, in principle, exact method that allows for the calculation of excited states through an extension of DFT to the domain of time-dependent external potentials [422]. Its efficiency was demonstrated by the first TDDFT calculation of the electronic structure and absorption spectrum of the carotenoid-diaryl-porphyrin-C_{60} triad (abbreviated hereafter as triad or C-P-C_{60}).

The optimized geometry of the triad, which is shown in Fig. 5.6b, was obtained in [455]. The results of computation of the ground-state characteristics showed that no significant hybridization of the molecular orbitals belonging to three subunits (namely, pyrrole-C_{60}, β-carotenoid with the amide and the diaryl porphyrin) appeared upon the triad formation. This is a reasonable result for the distances between the subunits are too large for pronounced hybridization to occur. The states are mostly localized on the parent subunits. Thus, the highest occupied state (HOMO) is localized on the β-carotenoid, while the lowest unoccupied orbital (LUMO) is localized on the C_{60}. Comparing the values of the HOMO and LUMO energies of the C-P-C_{60} complex with those of the isolated moieties shows that the energy of the HOMO in the triad and in the isolated β-carotenoid differ by only 40 meV as well as the energy of the LUMO in the triad and in the isolated pyrrole-C_{60}. In some other cases, the correspondence between the localization of orbitals

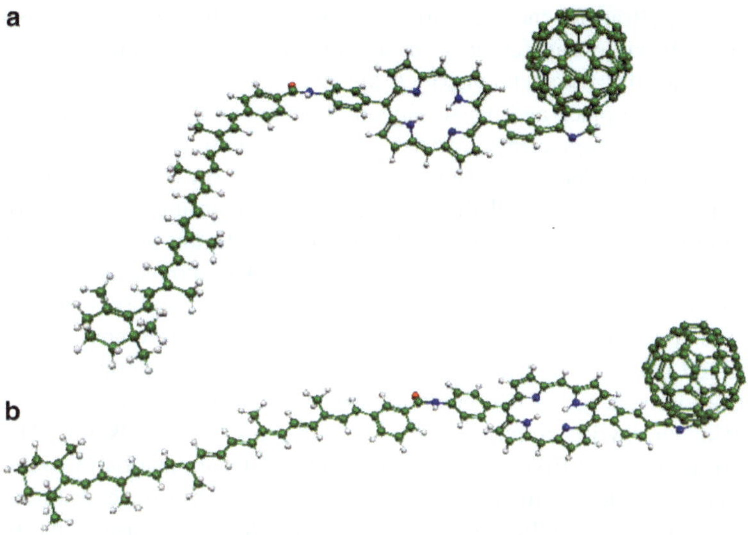

Fig. 5.6 The geometries of a carotenoid-porphyrin-C_{60} triad molecule [455]. Reprinted with the permission of the AIP

remains, but some degeneracies are lifted [456]. For example, the triply degenerate C_{60} LUMO is split into three separate levels in the C-P-C_{60}. Finally, some orbitals present a mixed character, such as LUMO+5, LUMO+6, and LUMO+7 with mixed P-C_{60} character or HOMO-6 with mixed C-P character. These orbitals however do not appear to have a major role in the photoexcitation process. From the above results, one concludes that the electronic structure of the ground state of the triad is well approximated by the sum of the electronic structures of the isolated molecules. This conclusion implies that it is possible to design triad-based photoconversion devices by covalently joining preoptimized molecular components.

The optical absorption spectra for the triad were computed using TDDFT in the way described in [423]. Starting from the ground-state Kohn–Sham orbitals, the system was instantaneously perturbed with a weak external electric potential of magnitude k_0 along the principal Cartesian direction (i.e., by applying the external potential $\delta v_{\text{ext}}(\mathbf{r}, t) = -\mathbf{k}_0 \cdot \mathbf{r}\delta(t)$). The magnitude of the perturbation was kept small in order to keep the dipolar response linear. From the Fourier transform of the real-time response to this perturbation, the dynamical polarizability and finally the dipole strength function, which is proportional to the photoabsorption cross section, were obtained by averaging the polarizability over the three Cartesian directions [456].

The obtained dependence of the photoabsorption cross section of the triad on the external photon energy revealed a reasonably good agreement with the experimental data of [457]. Also, the main features of the absorption spectra were better described within the time-dependent DFT than applying the Fermi golden rule to Kohn–Sham eigenvalues and eigenfunctions, as it was carried out in the earlier work [455]. At the same time, the time-dependent DFT-based calculations reported in [456] do not yield the charge-transfer energies which are the relevant figures of merit in studies of photoinduced charge-transfer reactions in large molecules including the considered C-P-C_{60}. The charge-transfer energy is the energy that relates to the removal of an electron from one place and its subsequent attachment at another place inside the molecule. In long-range electron transfer reactions, these positions are situated far apart. Baruah and Pederson showed that the transfer energy (C) required to move an electron between two identical alkali atoms is mostly a ground-state property [420], which could be simply expressed in terms of the relevant ionization energy (I) and electron affinity (A). For two well-separated alkali atoms, the charge-transfer energy is given by $I - A - 1/R$ with R being the separation between the atoms. The $1/R$ stabilization is simply due to the classical Coulomb interaction between two charged particles, and it immediately falls out by considering the difference of the Coulomb energy of two neutral atoms compared to two oppositely charged atoms. This energy is correct to $O(R^{-4})$ for the case of two alkalies. The lowest correction term arises from the polarization-induced stabilization of each ion that occurs due to the electric field of the counterion. This energy depends linearly upon the sum of the polarizabilities of the counterions. As in the case of the Coulomb interaction, monopole-induced- dipole stabilization as well as other electrostatic interactions between the separated ions are extractable from ground-state density-functional calculations.

In computing the ground-state characteristics of a system consisting of separated cation and anion, one meets some difficulties. If one attempts to self-consistently determine the energy of the well separated cation and anion using a standard ground-state density-functional theory, the system quickly finds the real ground state consisting of two neutral atoms. Part of the problem is that the standard DFT iterative method reorders occupation numbers to account for Fermi-level misalignments. The misalignment present for the cation–anion pair has a physical meaning, as it correctly identifies the fact that the electronic configuration of the charge-separated state does not coincide with that of the ground state. A computational method that allows for minimization of the functional within the constraint of orthogonality to the ground state does not have this shortcoming.

Such approach is described in the previous sections of the present chapter. Being applied to the triad, this approach brings the value 2.46 eV for the excitation energy which agrees with the estimate $I - A - 1/R$ based on the carotenoid ionization energy, the fullerene C_{60} affinity, and the fullerene-carotenoid separation in the triad [420]. When the triad is bathed in a radiation similar to that of sunlight, a large number of dipole-allowed transitions between various electronic states and vibronic states can occur. Recently, the main approximations for the transition probabilities were computed assuming the temperature $T = 0$ [420], and these probabilities were used to obtain the risetime of the charge-separated states in the triad by carrying out a kinetic Monte Carlo simulation. It was found that the incident solar radiation fails to excite the triad to any of the large dipole states. This situation is similar to that which occurs when the experiments are done using toluene as solvent. Experimentally, charge transfer is not observed when the triad is solvated in toluene (dipole moment $\mu = 0.375D$) but is observed in 2-MTHF

($\mu = 1.47D$) and benzonitrile ($\mu = 4.18D$) [453]. Carbonera et al. have estimated from electrochemical measurements that the energy of the charge-separated state in 2-MTHF is 1.24 eV above the ground state [454]. This is clearly an effect of the solvent polarization. The polar solvent encourages charge transfer by lowering the energy of the relevant states. If the polarization is ignored, then the large dipole states associated with the fullerene are situated above the porphyrin excited states. Also, the energy ordering of these states does not match that of the experiment without accounting for polarization of the solvents.

The energy shift due to polarization depends on the molecular volume. Assuming that the triad occupied the volume of $100 \times 35 \times 35\,(\text{a.u.})^3$, it was found that the transfer energy corresponding to the transition between the lowest states with a hole on the carotenoid and an extra electron on the fullerene was reduced to the value close to those observed in experiments. However, it was also found that an average of about 62 s should be taken to reach the target states [420]. The situation could be radically improved if one assumes that some counterions are present in the ambience. They should create an extra electric field further reducing the transfer energies and shortening the transfer times, so that the transfer reactions could develop fast enough for a useful solar cell.

5.9 Specifics of the Electron Structure Calculations for MMM Junctions

A significant challenge in modeling of nanoscale devices originates from differences in time, length, and energy scales, characterizing different parts of such a device which is a very inhomogeneous system. Considering a MMM junction as a typical nanodevice, one realizes that electronic structure calculations for such a system combining elements of a microscale (leads) with that of a nanoscale (the molecular linker) should be carried out using a combination of computational methods ranging from those appropriate for solids to those based on quantum chemistry. Also, it is necessary to find a way to mix and match band structure techniques employed to describe properties of the molecular bridge and the leads. This mixing should be carried out with precaution since the combining of different computational approaches can bring significant inaccuracies in the results. Nevertheless, such mixing procedure may appear to be the only practical way to perform electron structure calculations on highly inhomogeneous systems including MMM junctions. To emphasize the degree of the inhomogeneity in the metal–molecular junctions, we refer to the system consisting of aniline molecules adsorbed by a surface of lowly doped silicon layer [458]. It was shown that in this system the depletion width takes on values about 10^{-6} m which is too large to treat this depletion region atomistically. At the same time, electronic properties at the molecule-silicon interface are controlled by the bondings of the molecules to the silicon surface, charge transfer, and band alignment at the interface. The relevant region spans just a few nanometers. So, in carrying out electron structure computations on this and other similar systems, one must combine short-ranged chemical correlations that need *ab initio* theories with long-ranged band structure and, possibly, dipolar correlations that merit another approaches.

Early band structure calculations on metal–molecule junctions were usually performed, employing semiempirical theories (see, e.g., [57]). Also, DFT-based approaches were applied to describe such systems assuming a jellium model for the electrodes [459,460]. In later works these simplifications were discarded. Full-scale methods for computing the electronic structure, electronic transport characteristics, and forces acting on the atoms included in consideration were developed within the DFT to be applied to atomic-scale systems attached to semi-infinite electrodes such as MMM junctions. A strong advantage of the DFT-based approaches is their ability to catch many-body correlation effects which are strong in small nanostructures. Despite the difficulties in capturing excitations and transport spectra, the DFT gives satisfactory results well calibrated with precise bonding geometries and band structure as far as total ground-state energies go.

However, as discussed in previous sections, the DFT lacks the knowledge of the exact form of the exchange-correlation potential. For this reason, several approximations for the exchange-correlation potential were suggested and used by solid-state physicists to analyze the electronic structure of solids. Regretfully, the most commonly adopted approximations such as LSDA and GGA do not include a self-interaction correction (SIC) which arises due to the tendency of an electron to act on itself. The LSDA–GGA approximations lack the SIC. Therefore, the corresponding model potentials remain orbital independent. The SIC omission leads to underestimation of bandgaps and overestimation of Coulomb costs in solids. In the case of MMM junctions and similar systems, the underestimation of the bandgaps for the electrodes results in underrating of the heights of barriers for the electron current flow through the system. As discussed before, these barrier heights may be evaluated as differences between the Fermi energies (chemical potentials) of the electrodes and the closest conducting HOMO/LOMO level on the bridge. So, the underrating of the barrier height automatically brings the overrating of the current and causes discrepancies between theoretically evaluated and experimentally observed transport characteristics.

Quantum chemists use to circumvent difficulties arising due to the omission of the SIC by employing hybrid DFT approximations such as B3LYP model functional combining a LDA-type expression (capturing correlation part of the exchange-correlation potential) with Hartree–Fock term describing the exchange. The point of the B3LYP approximation is that the self-interaction errors unavoidable while using LSDA or GGA expressions for the exchange-correlation contribution to the effective potential should be canceled within this scheme. Several other hybrid approximations were suggested such as one proposed by Aulbur and coworkers [461] which is referred to as GW approach. This approach starts from the LDA and then introduces correlations through an energy-dependent self-energy matrix. In principle, the GW should be able to bring exact results, but in practical applications, it remains reduced to some model expressions. Also, one could mention an approach known as LDA+U. This is a modification of the LDA by adding an extra orbital-dependent potential U. The form of this extra term is usually determined by employing models or semiempirical fits. In the quantum chemistry community, several configuration approaches are being employed. Within these approaches, one accounts for correlations by taking into consideration a set of many-electron configurations including as many elements as practically possible.

It is worthwhile to mention a semiempirical approach based on the extended Huckel theory (EHT) [462, 463]. Originally developed to perform electron structure calculations on conjugated molecules, this approach was later extended to describe band structures of various systems including crystalline solids. Within the EHT model, the bonding chemistry is captured by relating off-diagonal terms in the Hamiltonian with the overlap between the corresponding basis sets multiplied by their average parametrized on-site energies. This approach resembles the nonorthogonal tight-binding version of DFT [464] with included empirical parameters chosen to fit the basis set rather than the Hamiltonian. Once fitted to a benchmark structure (e.g., a bulk silicon or graphite), the model parameters could be used to compute band structure in modifications of these materials subjected to strain and reconstruction. Obviously, since the EHT hopping integrals are proportional to overlaps between basis sets, one only needs to recompute these overlaps for the considered modification of the benchmark structure without reparametrization of the basis sets themselves. Self-consistent charging and polarization may be included into this model by employing several methods as described by Ghosh [5]. EHT model may be suitable to compute the electronic structures of the molecular bridges which are still amenable to DFT-based studies, despite the errors in the bandgap estimations. However, this approach could be very useful in studies of the contact band structure which determines the energy-dependent couplings of the electrodes to the bridge levels and controls the magnitude of the current flowing through the system.

Electron structure calculations for strongly inhomogeneous MMM junctions are simplified by the fact that the only quantity communicating the coupling between the dissimilar regions (a left/right electrode and a molecular linker) remaining in equilibrium is the interfacial Green's function $g_{L,R}(E)$ (see Sect. 2.1). If this quantity is known, the interfacial transmission which characterizes transport properties could be determined. This means that one may carry out separate electronic structure calculations for the electrodes and the molecular bridge provided that one can match the results for the interfacial Green's function obtained from these calculations over a sufficiently large energy range. The computational methods applied to describe the electron structure of these dissimilar regions may be entirely different. For instance, one can use EHT with a semiempirical basis set to carry out the relevant computations for the contacts and the DFT formalism for the molecular bridge-related computations. Then one could formally match $g_{L,R}(\mathbf{r}, \mathbf{r}', E)$ computed for the corresponding electrode with atomistic self-energy determined by the atomistic surface cluster Hamiltonian. These functions are to be matched at the interface separating the electrode from the extended molecule. The match is possible to achieve provided that nonlocal correlations remain insignificant and could be

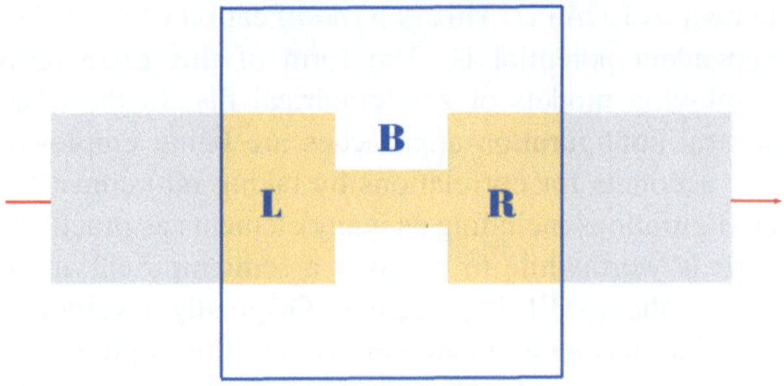

Fig. 5.7 Schematics of the finite part of a system which determines the transport properties. The Hamiltonian matrix elements inside the *L* and *R* regions take on bulk electrode values. The external region is not directly relevant for the calculations. The direction of electron transport is denoted by the *arrow*

ignored. The boundary between the regions must be judiciously chosen to justify this assumption. The matching of the continuum (related to the electrodes) and atomistic (related to the molecular linker) representations for the relevant Green's function could be reached by expanding the basis functions for an electrode Φ_m in terms of the basis set for the molecule $\{\psi_\alpha\}$, namely,

$$\Phi_m(\mathbf{r}) = \sum_\alpha a_{m\alpha}\psi_\alpha(\mathbf{r}). \tag{5.102}$$

Using this expansion, one may derive the transformation rule for the $g_{L,R}(\mathbf{r}, \mathbf{r}', E)$. The "continuous" Green's function computed adopting the electron structure of the left/right electrode is related to the "atomistic" Green's function as follows:

$$(g_{L,R}^{\text{con}})_{\alpha\beta} = \sum_{mn} a_{m\alpha}^* a_{n\beta}(g_{L,R}^{\text{atom}})_{mn}. \tag{5.103}$$

The values of overlaps $a_{m\alpha}$ between two basis sets determine the energy range over which the matching works.

Similar approaches could be developed for application to biased metal–molecular junctions and their transport properties [465,466]. For illustration, we describe here a scattering state approach suggested by Brandbyge, Mozos, and coworkers [465]. These authors consider a system consisting of two semi-infinite electrodes (left and right) coupled via a contact region (bridge). As it is typical for MMM junctions, the direct coupling between the electrodes is neglected, so all matrix elements of the Hamiltonian and overlap integrals between orbitals on atoms belonging to different electrodes take on zero values. The Hamiltonian and the density matrix are assumed to be converging to the bulk values within the electrodes. All these quantities accept values different from their values on the bridge and in the regions of contact between the bridge and the electrodes.

Transport properties of the system are determined by a finite part of the latter including the L–B–R region as shown in Fig. 5.7. The retarded Green's function of the L–B–R system could be found in a usual way by inverting the matrix

$$\begin{pmatrix} H_L+\Sigma_L & \tau_L & 0 \\ \tau_L^\dagger & H_B & \tau_R \\ 0 & \tau_R^\dagger & H_R+\Sigma_R \end{pmatrix} \tag{5.104}$$

where $H_{L,R}$ and H_B are the Hamiltonian matrices for the left and right electrodes and the bridge, respectively, and the matrices $\tau_{L,R}$ describe the coupling of the electrodes to the bridge. The matrix elements of the Hamiltonians $H_{L,R}$ could be determined from separate electron structure calculations carried out for the bulk of the left and right electrodes. Also, from these calculations one could determine the self-energy terms $\Sigma_{L,R}$ corresponding to the continuous interface Green's function $g_{L,R}^{\text{con}}$. The remaining parts of the matrix given by Eq. (5.104) should be computed in a different manner.

In biased systems, these terms are determined by the nonequilibrium electron density. To find the latter, one may employ the scattering state approach. The scattering states starting on the left electrode may be generated from the unperturbed states of the uncoupled bulk left electrode Φ_l^0 using the retarded Green's function which corresponds to the bridge coupled to this electrode:

$$\Phi_l(\mathbf{r}) - \Phi_l^0(\mathbf{r}) + \int d\mathbf{r}' G^r(\mathbf{r},\mathbf{r}')\tau_L(\mathbf{r}')\Phi_l^0(\mathbf{r}'). \tag{5.105}$$

Similar expressions could be written for the wave functions $\Phi_r(\mathbf{r})$ corresponding to the right electrode. Neglecting the direct interaction between the electrodes, one may present the electron density matrix from the incoming scattering states in the form

$$\begin{aligned} \rho(\mathbf{r},\mathbf{r}') = & \sum_l \Phi_l(\mathbf{r})\Phi_l^*(\mathbf{r}') f_L(\epsilon_l - \mu_L) \\ & + \sum_r \Phi_r(\mathbf{r})\Phi_r^*(\mathbf{r}') f_R(\epsilon_r - \mu_R). \end{aligned} \tag{5.106}$$

Here, the summation is carried out over all states associated with the left (first term) and right (second term) electrodes and $f_{L,R}$ are the Fermi distribution functions for these electrodes with the electrochemical potentials $\mu_{L,R}$. This density matrix describes only those bridge states which are coupled to the continuum of electrodes states. However, these states are most important in determining transport properties of the system. Introducing a set of basis functions for the molecular bridge, one may expand the wave functions $\Phi_{l,r}(\mathbf{r})$ in the "atomistic" basis functions $\psi_\alpha(\mathbf{r})$ (see Eq. (5.102)). Substituting these expansions into Eq. (5.105), one obtains

$$a_{l\alpha} = a_{l\alpha}^0 + \sum_\nu \left[G(\epsilon_l + i\eta)\tau_L\right]_{\alpha\nu} a_{l\nu}^0 \tag{5.107}$$

where the coefficients $a^0_{l\nu}$ are corresponding to the expansion of the unperturbed orbital Φ^0_l in basis functions ψ_α. Similar expressions could be written for the coefficients $a_{r\alpha}$. Matrix elements of the density matrix within the atomistic basis equal:

$$\begin{aligned}\rho_{\alpha\beta} =& \sum_l a_{l\alpha} a^*_{l\beta} f_L(E - \mu_L) \\ &+ \sum_r a_{r\alpha} a^*_{r\beta} f_R(E - \mu_R).\end{aligned} \tag{5.108}$$

Introducing the spectral function for the left electrode $A^L(E)$,

$$A^L_{\alpha\beta}(E) = \pi \sum_l a_{l\alpha} a^*_{l\beta} \delta(E - \epsilon_l) \tag{5.109}$$

and the similar function for the right electrode $A^R(\epsilon)$, one may rewrite Eq. (5.108) as

$$\rho_{\alpha\beta} = \frac{1}{\pi} \int_{-\infty}^{\infty} dE \left\{ A^L_{\alpha\beta}(E) f_L(E - \mu_L) + A^R_{\alpha\beta}(E) f_R(E - \mu_R) \right\}. \tag{5.110}$$

Employing Eq. (5.107), we can express electron density matrix elements in terms of "unperturbed" coefficients $a^0_{l\alpha}$. Since the important part of the density matrix is related to the scattering region, and the first terms in these equations become zero within this region, the expressions for $\rho_{\alpha\beta}$ within the scattering region take on the form

$$\rho^L_{\alpha\beta}(E) = \left\{ G(E) \text{Im}[\tau g^L(E) \tau^\dagger] G^\dagger(E) \right\}_{\alpha\beta}. \tag{5.111}$$

Her, $g^L(E)$ is the unperturbed Green's function for the left contact

$$g^L(E)_{\alpha\beta} = \sum_l \frac{a^0_{l\alpha} a^{0*}_{l\beta}}{E - \epsilon_l + i\eta}. \tag{5.112}$$

From the expression (5.111), one may identify the retarded self-energy $\Sigma^L(E)$ and its imaginary part $\Gamma^L(E)$.

The density matrix of electron system departing from the equilibrium state is given by Eq. (2.29). In this expression, the integrand is the lesser Green's function $G^<(E)$ which is expressed in terms of the retarded and advanced Green's functions via the Keldysh equation (2.35). Computing matrix elements of the relevant Green's functions and the matrices $\Gamma^{L,R}(E)$ determined by Eq. (5.111) within the basis $\{\psi_\alpha(\mathbf{r})\}$ and using the relation (5.102) between the "continuous" and "atomistic" interfacial Green's functions, one may find matrix elements of the electron density matrix including the information concerning the nonequilibrium occupations. Also, in the case of ballistic transport through the system one may use the obtained results for the Green's functions and self-energy terms to calculate the electron current through the system.

A somewhat different computational procedure was employed by Arnold, Weigend, Everts, and Koentopp [466, 467]. They suggested that the self-energy terms $\Sigma^{r,a}$ which enter the Dyson equations of the form (2.23) for the electron Green's functions corresponding to the molecular bridge are known and given by simple approximations. For instance, considering a system including a molecule coupled to a single-electron reservoir (electrode), they approximated $\Sigma(\mathbf{r}, \mathbf{r}', t, t')$ as follows [466]:

$$\Sigma(\mathbf{r}, \mathbf{r}', t, t') \approx \big[\delta\epsilon(\mathbf{r}) + i\gamma(\mathbf{r})\big]\delta(\mathbf{r} - \mathbf{r}')\delta(t - t'). \tag{5.113}$$

Here, the local absorption rates $\gamma(\mathbf{r})$ and energy shifts $\delta\epsilon(\mathbf{r})$ are allowed to accept nonzero values only on the surface S separating the molecule from the reservoir which corresponds to absorbing boundary conditions. These functions as well as the profile of the interface S should be constructed in accordance with specific properties of the considered metal–molecular junction. The approximation (5.113) may be justified assuming that characteristic times for the processes of electron transfer from the reservoir to the molecule and back are very short. A properly constructed expression for the self-energies may be used to compute the relevant Green's functions and density matrix elements employing an iterative self-consistent method. This technique was used to compute the electron transmission and current through benzenedithiol and 9,10-bis((2′-para-mercaptophenyl9,10-bis(2′-para-mercaptophenyl)-ethinyl)-anthracene molecules sandwiched between gold contacts [466,467].

We remark that the approach described here, as well as other scattering theory-based approaches, is a single particle in nature. Therefore, it disregards many-body effects such as the Coulomb blockade in the transport process. Accordingly, within these approaches, the current through the junction is presented by the simple Landauer formula [see Eqs. (2.54) and (2.56)]. Nevertheless, good results showing agreement between theoretical computations and experimental data were obtained in several works. For instance, such agreement was demonstrated for scanning tunneling spectra of a fullerene C_{60} molecule placed on silicon [468]. In computing of these spectra, the silicon electrode band structure was captured using the EHT formalism, while the molecule itself and its bonding with the silicon electrode were described employing the LDA version of DFT. The two approaches were matched at the silicon-molecule interface using the basis mixing and transformation discussed before. Positions of the computed spectral peaks as well as their magnitudes demonstrate a rather good agreement with the corresponding experimental results, as shown in Fig. 5.8.

As discussed before, electron–phonon interactions may significantly affect transport properties of metal–molecule junctions and similar systems. The simplest way to account for these interactions and their dynamical implications is to cast

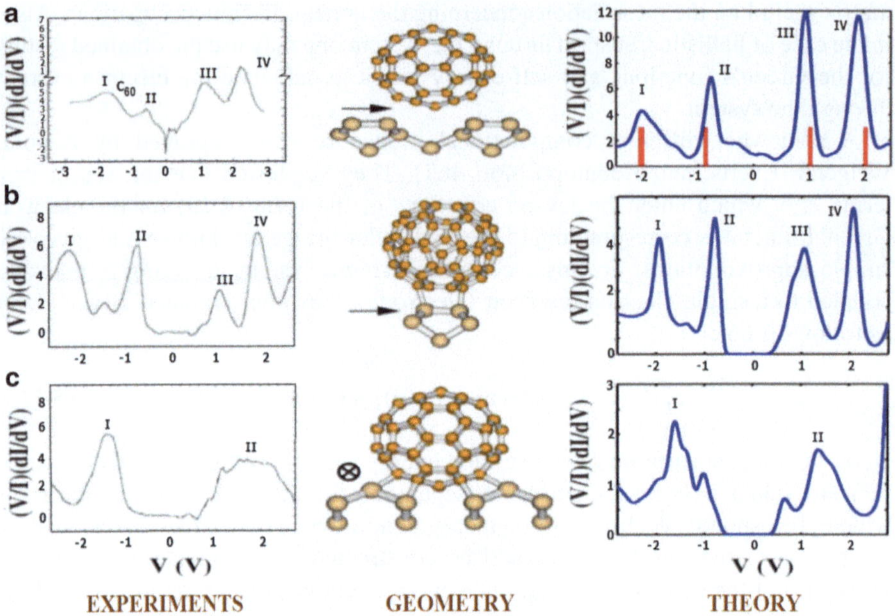

Fig. 5.8 Theoretical and experimental scanning tunneling spectra of the fullerene C_{60} bounded on Si. Observed spectra show a good correlation with first-principle-based theoretical results. Reprinted from [468] with the permission of APS

the calculations within the framework of scattering theory. This could be done adopting the Tersoff–Hamann approach [190, 191] which had been developed for studies of electron transport through MMM junctions in STM configurations. Within this approach, the current is dominated by the phonon-modified electron density of states at the STM tip and by occupation of the tip and substrate electronic states. Calculations based on this approach (where the required density of states was obtained using NEGF-DFT formalism) were carried out by Lorente, Persson, and coworkers [469, 470], and the results were successfully used to interpret early experimental data reported in [163] and some other works. Beyond the STM configuration, various extensions of Landauer methodology [65] are being employed. The electron–phonon interactions could be taken into account by using perturbation expansions about the Landauer expression for the elastic transmission [104, 205, 206]. Certainly, these are low-order calculations. Nevertheless, good agreement with the experiment was reported for off-resonance tunneling properties of MMM junctions including various molecular bridges. So, the perturbative scattering theory-based approximation provides a useful tool for exploring of inelastic tunneling despite its insufficient theoretical foundation. Presently, there exist several versions of perturbative scattering approximations.

Within the version suggested by Troisi and Ratner [104, 207, 208], one starts from the expression for the elastic electron transmission given by Eq. (2.66). Then

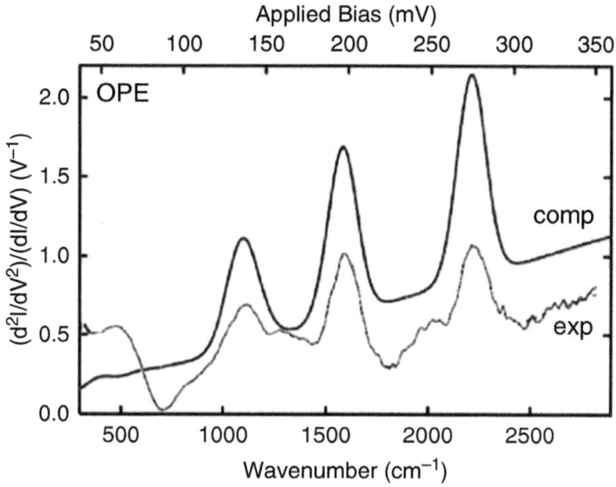

Fig. 5.9 IETS spectrum of phenylene ethynylene trimer computed theoretically [104] in comparison with the experimental results reported in [106]. The linewidths are arbitrarily chosen to better fit the experimental data but the intensities (reprinted by the areas under the *curves*) are computed. Reprinted with the permission of the ACS

one assumes that the Green's functions included into this expression depend on the nuclear configuration and expands them in normal mode coordinates, keeping only zero and first-order terms in these expansions. The electronic and vibrational structure needed as an input may be obtained from DFT-based calculations. These calculations should be carried out for particular models for the bridge. The simplest model assigns to the primary set only molecular modes, but more extended analysis could be developed [104]. One optimizes the structure and performs vibrational analysis on the isolated molecule taking into account the coupling terms which appear as a result of the Green's function expansions in the normal mode coordinates. Then the optimized structure and normal modes may be directly translated into the geometry of the junction where the molecule serves as a linker. This approach provides a rather simple way to analyze the characteristics of inelastic tunneling, and it was successfully applied to several organic molecules. The resulting IETS spectra consist of series of peaks whose positions and intensities agree with the experiments. An example is presented in Fig. 5.9. However, the shape of the spectral peaks cannot be obtained from the low-order calculations. In Fig. 5.9 the lineshapes are arbitrarily chosen (keeping the already computed intensities to provide the best possible fits with the experimental results). It should be emphasized that the described perturbational approach remains justified only within the Landauer–Imry regime, that is, far away from electron resonances.

Since the Green's functions matrix elements are expanded in modes, the obtained results have the form of sums of contributions from these modes. Therefore, the obtained results (e.g., IETS spectra) may be classified according to the symmetries of the molecular bridge within the junction. Analyzing these contributions one may single out particular predominating pathways for the electron transport. A combined study of the pathway aspect and the symmetry aspect leads to a set of propensity rules. The latter takes on the part similar to that of selection rules known in vibrational spectroscopy. The propensity rules are important because they can help to reach better understanding of both pathways for electron transport through molecular bridges and possible geometries of the molecule within the junction. Finally, we remark that electron–phonon interactions could be (and actually are) analyzed using the DFT-based electron structure calculations and NEGF-based transport studies. The effects of electron–phonon couplings are usually treated at self-consistent Born approximation level. Several interesting results are obtained which are reviewed in [26]. The results of [104] and many other works show that DFT-based calculations are often unable to correctly describe the profile of the electron transmission through MMM junctions. Within the standard DFT, the transmission peaks are often given broader and smoother shapes than those agreeing with observed results concerning electron conductance and IETS. Also, the position of the peaks sometimes disagrees with experimental data. It was argued [471] that the root of these discrepancies is the omission of an extra exchange-correlation corrections in the Landauer expression for the current through a MMM junction. These corrections may be small, but they bring derivative discontinuities in the following approximations, thus causing significant changes in the shape of the transmission peaks as well as in their positions. This analysis shows that various aspects of electron structure calculations including the effects of electron–phonon interactions need further analysis which presently is being carried out.

Chapter 6
Nanoelectronic Applications of Molecular Junctions

6.1 Field-Effect Transistors

It is a common knowledge that the first functioning transistor was invented in the late 1940s by Bardeen, Brattain, and Shockley, and this invention had marked the starting point for the microelectronic revolution. The metal-oxide-silicon field-effect transistors (MOSFET) appeared in the 1960s, and they dominated the development of microelectronics in the following forty years. Computer industry and digital communication systems give two examples of MOSFET applications. One of the most important characteristics of the progress in microelectronics is the process of miniaturization of electronic devices. As early as 1965, Gordon Moore has made his famous prediction that the number of transistors placed on a single silicon chip would double every one and one half year [472]. This prediction became the roadmap of the semiconductor industry. The increase in the number of transistors situated on a sole chip requires their miniaturization. Actually, the size of MOSFET produced by the microelectronic industry during the last thirty years has shrunken a factor of more than one hundred, and presently it reaches a few tens of nanometer [472]. However, there are grounds to expect increasing difficulties slowing down the MOSFET miniaturization beyond the 10 nm mode. These problems mostly originate from the fact that a nanometer-sized MOSFET no longer behaves as a device with a long channel, that is, a device where the electrostatics and the current flow in the channel between the source and drain are effectively controlled by the gate. When the channel length becomes too short, the gate ability to control the channel deteriorates, and the so-called short-channel effects appear. These effects increase the MOSFET "off" current and render the current dependence of the bias voltage, so the device performance worsens. To improve the electrostatic control of the channel by the gate, new device structures are being explored, where the gate electrode is wrapped around the channel region. These devices are called multi-gate MOSFETs, and their size could be made smaller than that of usual planar devices. Functional multi-gate transistors made in a silicon nanowire with the diameter of 3 nm were reported in 2006 [473].

N.A. Zimbovskaya, *Transport Properties of Molecular Junctions*, Springer Tracts in Modern Physics 254, DOI 10.1007/978-1-4614-8011-2_6,

The quality of the electrostatic gate control over the channel can be estimated introducing the "natural length" of a device $\tilde{\lambda}$, given by the expression [474]

$$\tilde{\lambda} = \sqrt{\frac{\epsilon_{\text{si}}}{n\epsilon_{\text{ox}}} t_{\text{si}} t_{\text{ox}}} \tag{6.1}$$

where n is the number of gates and ϵ_{si}, t_{si} and ϵ_{ox}, t_{ox} represent the permittivities and thicknesses of the silicon film and the gate insulator, respectively. It had been shown that short-channel effects would become negligible in the devices with channel length L seven to eight times smaller than the "natural length" for all gate configurations. Satisfactory performances of MOSFETs were reported at $L \sim 4\tilde{\lambda}$ [475].

Device miniaturization requires reduction of the parameter $\tilde{\lambda}$. As follows from the estimate given by Eq. (6.1), one may reduce the value of this parameter increasing the number of gates and/or decreasing the film thickness t_{ox}. Also, one can choose a dielectric material with a high dielectric constant as the gate insulator. These ways were and still are followed aiming at the improvement of miniaturized MOSFETs performances. However, there exist natural reasons preventing the reduction of $\tilde{\lambda}$ beyond certain limits. First, thin gate insulator films permit quantum tunneling of electrons through them, and the tunneling probabilities exponentially increase as the insulator thickness decreases. This puts the limit on the oxide-film thickness. For silicon dioxide which is commonly used in MOSFETs manufacturing, the limit thickness was established to take on values about 1–2 nm assuming that the device operates at the voltage close to 1 V, and its channel length is about 20 nm [476]. Secondly, as the dielectric constant of the gate insulator increases, its bandgap reveals a tendency to decrease, so the band lineup at the silicon–insulator interface may become quite asymmetric. Obviously, this affects the device performance. Also, any insulating material used as the gate insulator must possess an ability to provide a low defect density at the contact with the silicon, comparable to that typical for Si–SiO_2 interfaces. This puts constraints on the choice of suitable dielectric materials.

There are other factors putting limits on the MOSFET miniaturization which are not accounted for in the estimate (6.1) for the natural length of the device. This estimate is derived within the classical approach to the electron transport. This approach remains appropriate for the analysis of the MOSFET properties when the cross-sectional areas of the devices stay above 10 nm. In the devices with smaller cross-sectional areas, quantum confinement of charge carriers (electrons/holes) in the direction perpendicular to the current flow introduces significant changes in the electric properties of silicon [477]. Besides, the charge carrier confinement may profoundly influence their spatial distribution, scattering, and transport mechanisms [478]. Also, the short length of the device (within the sub-10 nm range) provides opportunities for other quantum effects to appear. These include charge carrier tunneling through body-to-drain junction and direct source-to-drain tunneling. All above listed and other important factors hindering the MOSFET miniaturization beyond certain limit are thoroughly discussed by Xue and Ratner [479].

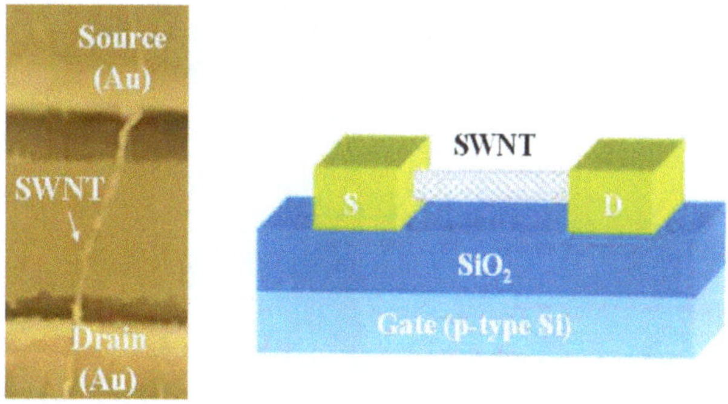

Fig. 6.1 Atomic force microscope image of a typical CNTFET sample [484] (*left panel*) and schematic diagram of CNTFET structure (*right panel*). Reprinted with the permission of AIP

Molecular-based electronic devices are being heavily explored in the recent decade. Such devices show potentialities which provide grounds to hope that someday they may overcome the size limitations of the current silicon-based electronics technology. We start the description of already existing molecular electronic devices from their general definition. Following Xue and Ratner [479]we define a molecular electronic device as a system partially or wholly consisting of molecules or other similar entities (such as carbon nanotubes/nanowires and quantum dots), which can perform useful functions in electronic apparatus through atomic-scale control. This rather cumbersome definition is chosen for it allows to include into consideration such entities as carbon nanotubes/nanowires, which are presently used in manufacturing of nanodevices more often than molecules themselves. An important common feature of all these entities is that they are fabricated using bottom-up methods, and this emphasizes their similarity to the molecules.

Carbon nanotubes (CNT) field-effect transistors fabricated with a CNT taking part of the channel were first demonstrated in the 1990s [480,481]. In these devices, a CNT was placed on an oxidized silicon layer between two metal electrodes as shown in Fig. 6.1. The silicon substrate situated underneath the insulating layer took the part of the gate. Later, more sophisticated device layouts and fabrication methods were developed to manufacture CNTFETs. In addition to their small size, CNTs possess some important features which make them attractive as parts of electronic nanodevices. The CNTs are composed by wrapping graphene sheets. They consist of one (single-wall) or a few (multi-wall) layers of carbon atoms in a honeycomb lattice arrangement. The bonding of carbon atoms in a graphene sheet is strong, which leads to high mechanical and thermal stability of CNTs. Their electron properties could vary within a broad range. Specific CNTs may be insulating, semiconducting, or metallic like depending on their chirality. Conducting CNTs could carry current densities of the order of 10^9 A/cm^2. The CNTs are quasi-one-dimensional entities, and this implies a reduced phase space for carriers scattering by impurities and lattice vibrations and favorably influences their electrostatic behavior. The probability of back scattering by defects and acoustic phonons in CNTFETs is lower than in planar silicon devices [482,483]. Also, it was demonstrated that CNTFETs can have

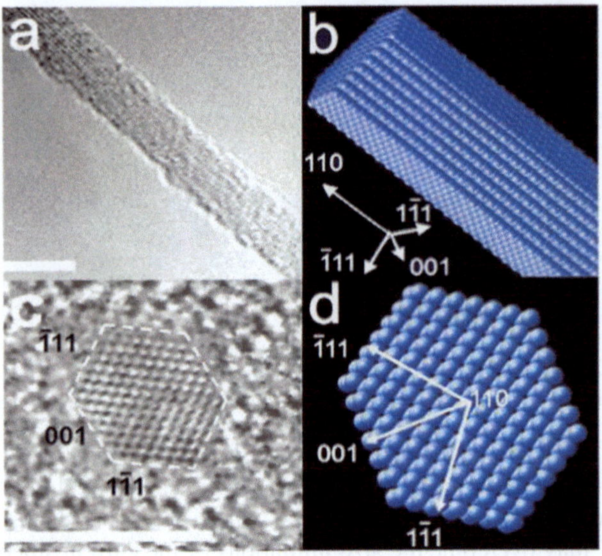

Fig. 6.2 The transmission electron microscopy (TEM) image (**a**) and the schematic diagram (**b**) of a 3.8 nm silicon nanowire grown along the $\langle 1\,1\,0 \rangle$ direction; the high-resolution TEM image (**c**) and the schematic diagram (**d**) of the wire cross-section [493]. Reprinted with the permission of the ACS

lower switching delays than MOSFETs with similar on–off ratios [482, 484] and it was suggested that high-frequency performance of these devices may be better than that of silicon-based FETs [485].

In contrast to MOSFETs, the source and drain contacts in carbon nanotube-based transistors are usually made out of metals. Typical metals used for this purpose include Ni, Ti, Al, and Au. The metal/CNT junctions form Shottky barriers, and the barriers may significantly affect transport characteristics of the devices [486, 487]. Both heights and shapes of the barriers may have important effect on transport characteristics. The properties of the barriers may be favorably changed by chemical treatment of the contact/CNT interfaces [488] and/or by a suitable choice of metal electrodes [479]. Doping in CNTs is controlled by mechanisms which significantly differ from those typical for semiconductors. It may be chemically introduced by various means. To dope a CNT, one may expose it to alkali metals, insert fullerene molecules into the tube interior, or to perform the CNT surface functionalization with molecules/polymers to provide the charge-transfer doping. Using specific chemical treatment, the doping type could be converted between p-type and n-type [489]. Also, doping in carbon nanotubes might be introduced by means of electrostatic gating and the charge transfer between the contacts and the tube [490]. It was shown that the CNT doping could originate from curvature-induced redistribution of the electric charge, which shifts the position of Fermi level inside the bandgap [491, 492].

Nanowires (see Figs. 6.2 and 6.3) represent another kind of quasi-one-dimensional nanostructures which could take on the part of active elements in

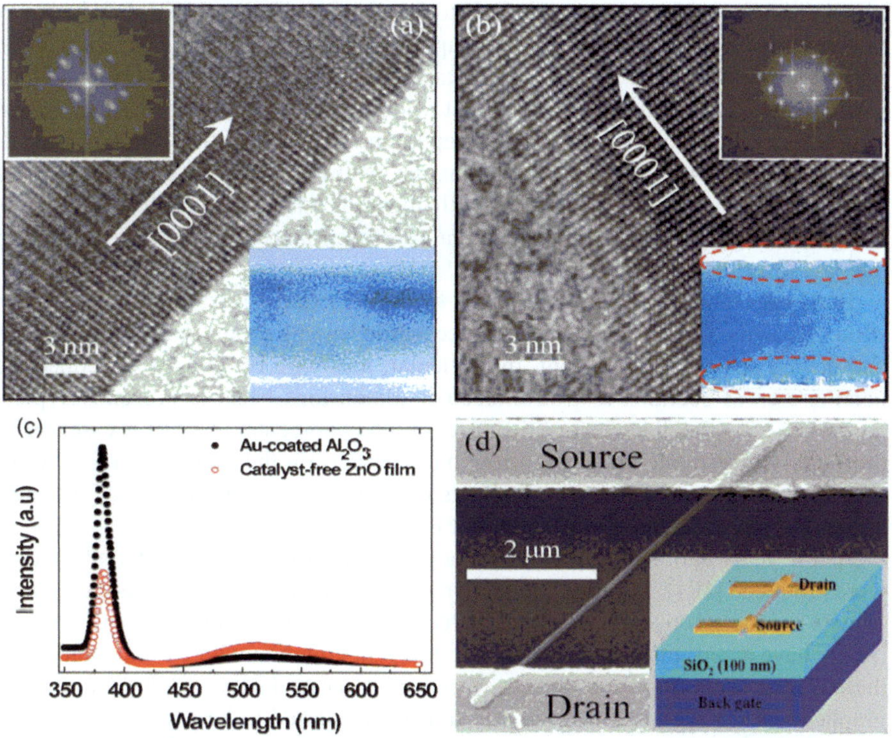

Fig. 6.3 High-resolution TEM images of ZnO nanowires grown on an Au-coated sapphire substrate (**a**) and on an Au-catalyst-free ZnO film (**b**); emission spectra of ZnO nanowires on the two substrates (**c**), and a field emission scanning electron microscopy image of a single ZnO nanowire connected between source and drain electrodes in FET device (**d**). *Insets* show computed Fourier transform patterns (*upper*) and low magnification TEM images [499]. Reprinted with the permission of the AIP

FETs and other nanoelectronic devices. Lieber group pioneered the vapor-liquid-solid phase growth process, which has led to the fabrication of single-crystal silicon nanowires [479]. Size distribution of nanowires obtained in this manner was determined by the size of catalyst nanoclusters used in the growth process. Later, new techniques were reported which allowed to transform silicon nanowires into metallic silicide ones [493]. Single-crystal metallic silicides are known to have high-conductivity and high-failure current. At the same time, they are capable of forming atomically sharp metal-semiconductor heterostructures with silicon wires of similar diameters. These properties open up possibilities for manufacturing of ultra-dense nanosystems, which integrate the active device area and highly conducting interconnects from a single nanowire building block. Silicon wires used as active elements in these nanosystems may be n-type and/or p-type doped. The doping may be chemically introduced into the wires by inserting the corresponding dopants during the growing process.

Silicon is not the sole material used to fabricate nanowires [494]. For instance, in recent decade significant effort was applied to the growth and characterization of ZnO nanowires due to their remarkable properties. These properties include a wide range of conductivities, a direct wide bandgap, radiation hardness, high transparency, piezoelectricity, room-temperature ferromagnetism, and chemical-sensing effects [495, 496]. The zinc oxide nanowires can be grown either by vapor-phase or aqueous solution growth methods [497]. Field-effect transistors using ZnO nanowire as the channel area have been fabricated and investigated in the past several years [497–500].

Both nanotube- and nanowire-based field-effect transistors have been demonstrated showing good quality performance, while leaving room for further materials and device design improvement. In particular, nanowire FETs need to reduce the nanowire cross sections to true molecular dimensions because presently typical nanowires have the diameters of the order of ten nanometers. Also, it is essential to thoroughly investigate the surface-roughness effects on the transport properties of nanowire-based transistors. These effects may profoundly alter their performance, as was demonstrated for ZnO nanowire FETs [497].

A few years ago, single sheets of graphene with linear dimensions of microns were produced [501]. Being a sole known material with atomic thickness, graphene reveals unique electron properties. Charge carriers move in the graphene sheet as if they have a zero mass and constant velocity. In common with carbon nanotubes, there is little scattering from defects. Very soon after its discovery, potentialities for graphene use in high-speed electronic transistors started to be explored. An ideal graphene sheet is characterized by a zero bandgap, and this limits its usefulness for manufacturing electronic devices. However, for narrow sheets or nanoribbons, the electron structure could be modified by adding various atoms or atomic groups at the edges. Transport properties of graphene nanoribbons including the possibilities of their nanoelectronic applications are being intensely studied (see, e.g., [502, 503]).

Turning to metal–molecule–metal junctions themselves, we remark that their potentialities as building blocks for nanoelectronic devices were intensively explored in the past two decades. Extensive studies of MMMs transport properties are largely inspired by these potentialities. Experimental explorations of the usefulness of molecules in electronics significantly advanced after the introduction of scanning tunneling microscopy (STM) and the use of break junctions. These methods make possible for molecules to span gaps of a nanometer or even less in between the metal electrodes, thus creating MMM junctions. Fabrication of FETs on molecule scale may be difficult due to the lack of the field effect in devices whose length has the order of a few nanometers. This happens because the gate control requires the arrangement of the gate in close proximity to the molecule while avoiding overlap with the source and drain electrodes. One can overcome this difficulty by inserting the device in an electrolyte. Then the gate voltage would mostly fall across the double layers at the electrode–electrolyte interfaces, and field effects could appear. This method was successfully used to observe a strong field effect in the MMM junction consisting of a 2.3 nm long perylene tetracarboxylic diimide molecule attached to gold leads [504] as shown in Fig. 6.4. Another way to achieve

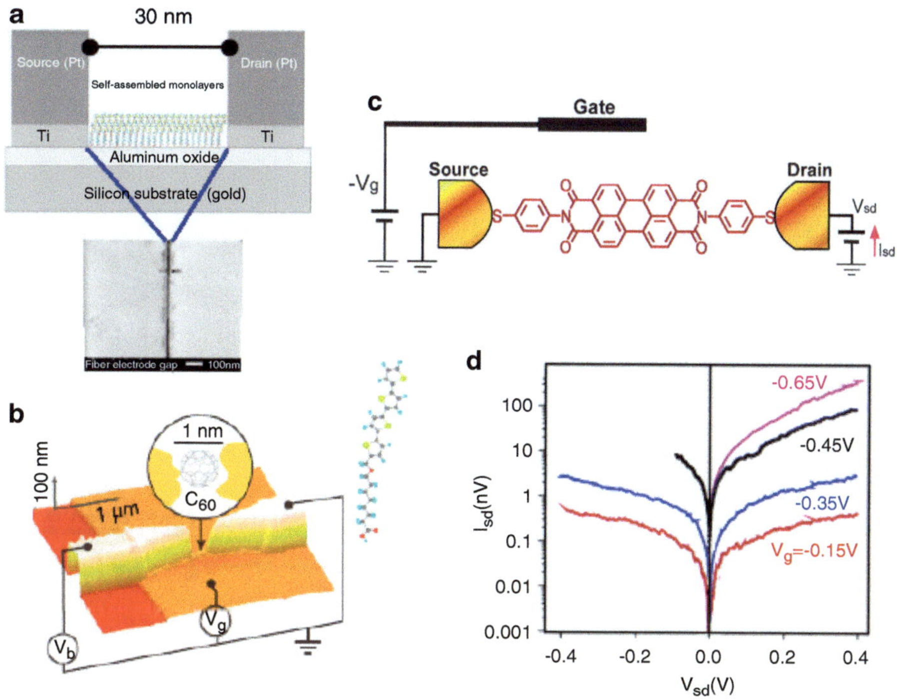

Fig. 6.4 Schematics of single-molecule transistors including a self-assembled monolayer (**a**), a single C_{60} fullerene (**b**), and of a single-molecule transistor with an electrochemical gate (wire reference electrode in $0.1NaClO_4$) where the gate and bias voltage are controlled with a bipotentiostate not shown in the figure (**c**). The current I_{sd} vs. bias voltage V_{sd} characteristics for the molecular transistor (**c**) at various gate voltages (**d**). At low gate voltages the curves in linear scale show the linear $I_{sd} - V_{sd}$ behavior [504]. Reprinted with the permission of the ACS

a pronounced field effect is to put charge species in close proximity to molecules. For instance, modification of current–voltage characteristics for a STM junction was attained by introducing a charge-transfer complex, whose electron acceptor was bonded to the molecular bridge, and the donor came from the ambient liquid [505]. Also, strong field effect was observed in the STM junction, where styrene-derived molecules were grown on a silicon (1 0 0) surface [169]. In this case, the effect originated from charged dangling bond states on the Si surface. The electric field produced by these bond states shifted the molecular energy levels.

Recently, new demonstrations of single-molecule FET had been reported by Song and coworkers [506] and by Ballmann and Weber [507]. In their experiments benzenedithiol molecules were bridged across nanogap between gold electrodes formed by electromigration break-junction method. The gate control was improved by using Al for the gate electrode and ultrathin Al_2O_3 oxide layer for the gate dielectric. This gate configuration was often used to study gate modulation of single-molecule conductance in various π-conjugated molecules (see, e.g., [9, 27, 31, 279, 447]). Gapless nonlinear current–voltage characteristics observed

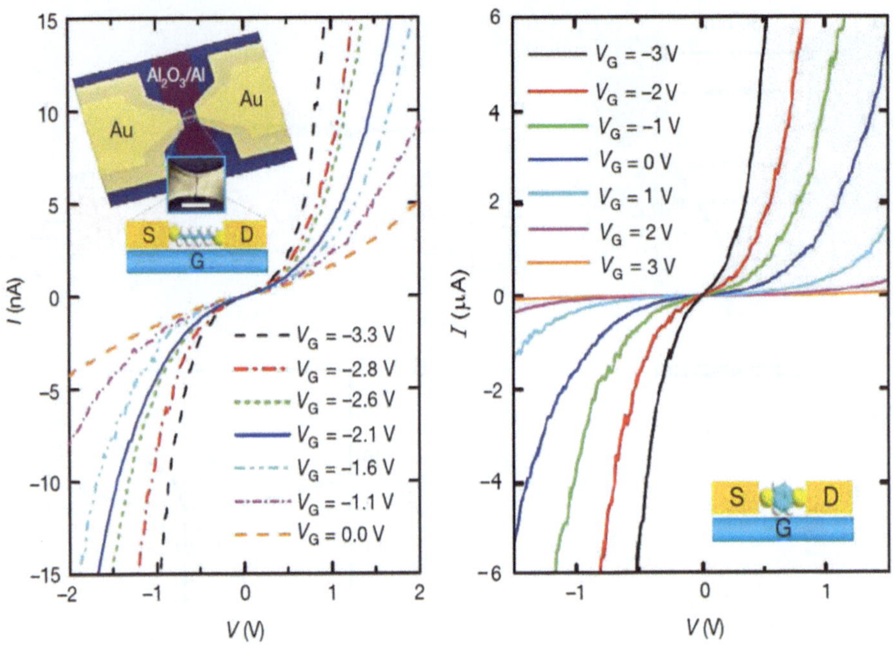

Fig. 6.5 Gate-controlled transport characteristics in *Au-ODT-Au* (*left*) and *Au-BDT-Au* (*right*) molecular junctions [506]. The current–voltage curves are measured at $T = 4.2\,\mathrm{K}$ for different values of the gate voltage V_g. *Inset*, the device structure and schematics. Scale bar, 100 nm. Reprinted with the permission of the NPG

in the experiment suggested that predominating transport mechanism was coherent electron tunneling via HOMO level of the benzenedithiol bridge. Overall results illustrated in Fig. 6.5 supported the gate modulation of single-molecule conductance occurring via electrostatic tuning of molecular orbital levels.

However, while the single-molecule FET was proven to work, the device yield was rather low (about 10 %). The issue of low yield remains critical for single-molecule FETs. To improve the yield one needs to provide formation of well-defined metal–molecule contacts with strong coupling, good controls of gate-to-molecule distance, and mitigation of electrode screening. All these difficulties are hard to overcome. Perhaps, these challenges could become surmountable by employing few layers of graphene packs as source and drain electrodes [508]. The inherent two-dimensionality of graphene sheets may appear to be useful to create sub-nanometer gaps separating molecular bridges from the gate as well as to reduce screening effects.

Measurements of the current–voltage characteristics for MMM junctions often bring contradictory results. The conductance per molecule may vary by several orders in magnitude for MMMs with identical molecular linkers. The discrepancies could originate from the presence of additional tunneling gaps, from uncertainties in the number of measured molecules, or from poorly defined surface areas in the junctions [178]. One may observe convergence and improvement in the reproducibility in the measurements of current–voltage characteristics of MMM junctions comparing, for example, the results obtained by Lortscher et al. [10] and Martin et al. [509] for junctions including benzenedithiol molecule attached to gold contacts through thiol groups. Nevertheless, at present single-MMM junctions are not sufficiently reliable, stable, and reproducible for technological applications. Self-assembled monolayers (SAM) may offer a more promising route to the fabrication of reliable devices. Metal-SAM-metal nanojunctions, where SAM of alkanethiol molecules is placed in between electrodes about 45 nm in diameter, had been fabricated several years ago [510, 511]. The junction electrodes diameters could be enhanced up to tens of micrometer [512], which makes such large-area junctions more suitable for device fabrication. Various kinds of molecules may be used to form a SAM. For instance, it was recently suggested to employ tungsten-polyoxometalate molecules for this purpose. Transport properties and possible applications of metal-SAM-metal junctions made out using these molecules were studied in the recent work [513]. The present situation in the field of diverse metal-SAM-metal junctions fabrication is described by Dasari and coworkers in their recent work [514].

6.2 Single-Molecule Diodes

A diode is known to be a two-terminal electronic device which facilitates current flow in one (forward) bias voltage polarity and suppresses it under the opposite (reversed) bias direction. Molecular design for diodes was first suggested by Aviram and Ratner nearly four decades ago [1]. A schematics of the suggested device is presented in Fig. 6.6. In the proposed system, the electrodes are bridged with a structure consisting of donor and acceptor molecules connected by σ-bond. This σ link serves as an insulating barrier separating the donor and acceptor segments from

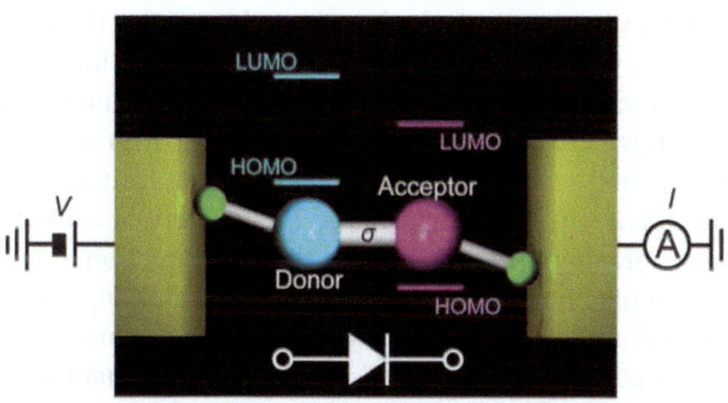

Fig. 6.6 Schematic model of Aviram–Ratner molecular diode [1] composed of a donor-σ-acceptor structure attached to source and drain electrodes (after [631]). Reprinted with the permission of the AIP

each other. Due to the presence of this barrier, molecular orbitals are localized on the donor and acceptor parts of the bridge. Because of the asymmetric structure, the HOMOs and LUMOs of the considered D-σ-A combination align with the electrode Fermi level more easily in one bias voltage polarity than in the other one thereby causing current rectification. This mechanism is far from being unique. Rectifying properties of MMM junctions may arise from several factors. The issue is discussed in Sect. 3.2. The plurality of mechanisms controlling the current rectification in metal–molecular junctions, and similar systems complicate the interpretation of their rectifying properties which were repeatedly observed in experiments [84,515].

First single-molecule diodes had been fabricated using lithographically defined break junctions [515, 516]. In these devices, molecular bridges were characterized by a donor–acceptor structure as in the Aviram–Ratner molecular diode. Also, the electronic states of the donor and acceptor units were localized. In this case, the localization was achieved by reducing the overlap of the donor and acceptor π-orbitals using the large torsion created by steric hindrance of methylene at D–A link. MMM junctions were built by attaching the bridge structures to gold leads by means of dithiol anchoring groups. The described junctions demonstrated rectifying behavior with the forward and reversed currents ratio of the order of 10. In this particular case, rectification was believed to occur due to the difference in heights of the conduction steps at positive and negative bias voltage polarities. The difference originated from electrostatic shifts of occupied and unoccupied molecular levels under the applied bias. Although DFT-based calculations provided a reasonable explanation of transport properties of the considered single-molecule diodes, these calculations did not take into account the effects of random molecule orientations. These effects appeared to be significant, and they led to junction-to-junction variations of transport characteristics and to difficulties in the determining of the appropriate bias voltage polarity. In later experiments the orientation of molecular linkers was better controlled to improve the molecular diodes performances. For instance, recent break-junction experiments [517, 518] incorporated two different protection groups to control the orientations of diode molecules. In these experiments, a self-assembled monolayer of dipyrimidinyl-diphenyl molecules was formed on a gold substrate. The molecules were attached to the substrate by deprotecting the thiol groups at their diphenyl sides. Then the protecting groups at the dipyrimidinyl sides of the molecules were also removed allowing formation of molecular junctions by creating S–Au bonds between the dipyrimidinyl ends of bridging molecules and a gold tip. This method regulated the orientation of molecular linkers. The latter were aligned in such a way that dipyrimidinyl units were heading towards the tip, whereby enabling unambiguous definition of the bias direction.

The single-molecule conductance was determined applying the repeated break-junction method, which was subsequently used as a reference for manufacturing single-molecule junctions. The recorded current–voltage characteristics were noticeably asymmetric so the considered MMM junctions worked as molecular diodes. The rectification rate could be estimated by introducing the parameter $\kappa(V)$ defined by the expression [518]:

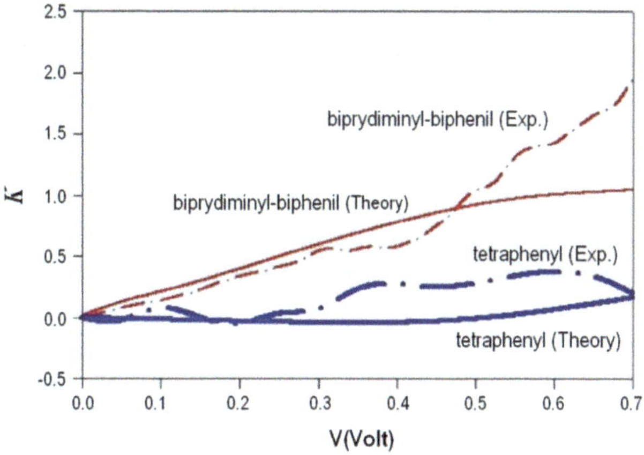

Fig. 6.7 κ-V plots of bipyrimidinyl-biphenyl single-molecular junctions. The *solid* (*red*) *line* represents NEGF results, and the *dash-dot line* (*red*) summarizes the experimental results. Results for tetraphenyl are also presented by the *thicker lines* (*blue*) as references. Each plot of experimental results is an average of individual junctions (after [518]). Reprinted with the permission of the ACS

$$\kappa(V) = \left[\frac{dI}{dV}(V) - \frac{dI}{dV}(-V)\right] \bigg/ \frac{dI}{dV}(0). \tag{6.2}$$

For MMM junctions demonstrating symmetric I–V characteristics $\kappa(V) = 0$. Nonzero values accepted by this parameter indicate that the current rectification occurs in the considered system. The dependencies of the rectification rate of the bias voltage for MMM junctions with dipyrimidinyl-diphenyl and tetraphenyl molecular linkers bridging gold electrodes are presented in Fig. 6.7. It is clear that the first kind of MMM junction provides the current rectification, whereas the second kind does not. Recent experimental reported in [517, 518] show that it is technically available to form MMM junctions functioning like diodes with well-defined molecule orientations.

Regretfully, the rectification ratio remains low ($\sim$5–10). To enhance this ratio one needs to improve the architecture of the junction by enhancing the difference in coupling strengths between the molecule and the electrodes (the asymmetric coupling of the molecule to the leads is recognized as an important factor controlling the rectification of electric current through the system) and by tuning the molecule electron structure. While employing molecular bridges consisting of donor and acceptor units separated by an insulating barrier, one way is to reduce the energy gap between the donor HOMO and the acceptor LUMO. This should help to utilize delocalized resonance states under electrostatic crossing of these two levels.

6.3 Quantum Dots and Carbon Nanotubes as Switches and Memory Elements

One of the fundamental achievements today is providing storage and processing of ever-increasing amounts of information. Up to now, the semiconductor memory industry continuously improved the storage density while simultaneously reducing the cost per information by scaling down the feature size. In 2008, the feature size for Flash memories has reached 45 nm. However, further reduction of memory features requires new technological approaches. Molecule-scale devices sensitivity to small values of the added or removed charge makes them attractive candidates for designing of miniaturized memory cells. Especially, self-organized quantum dots based on such materials as $GaAs$, $InAs$, InP provide several important advantages for new generations of nanostructures. Among these advantages is the possibility to simultaneously form billions of self-organized dots in a single technological step, which allows for massive parallel production within the bottom-up approach. Every single dot can store just a few or even a single charge carrier with a retention time which could be very long: potentially up to several years at room temperature. Assuming that a single dot represents one information bit, their enormous area density (about 10^{11} cm^2) gives means to store 10^{12} information bits per square inch. Also, the short characteristic time of the carriers capturing process makes the write time in such memories very short, which is an important prerequisite of fast-operating memories. A memory cell concept where the quantum dots were embedded in a p–n or p–i–n diode structure was suggested by Geller and coworkers [519]. The dots included in the self-organized system can be charged with electrons or holes representing 0 and/or 1 of the information bit. An emission potential barrier keeps the charge on the dot preventing its escape, and a capture potential barrier protects an empty (uncharged) dot from capturing unwanted charge carriers. The important advantage of this and similar systems is the possibility of tuning the heights of the barriers by the applied bias. Therefore, the capturing barrier could be eliminated during the write process making it easier to put charge carriers on the quantum dot states. To erase the information stored in the dot, one may suppress the emission barrier by the external bias voltage. The charge carrier relaxation time into the dot states may be as short as a few picoseconds at room temperature, which provides a high speed of recording/erasing information to/from the dots [520, 521].

The short write time is an important characteristic of memories. Another important property is their ability to store information for a long time. In quantum dots the information is coded by means of the electric charge on the dot which comes there with the captured charge carrier. Therefore, the storage time characterizes the charge carrier storage in the dots. Emission of electrons and holes from various kinds of quantum dots was thoroughly studied during the last decade. It was established that the controlling mechanism for the charge carrier emission from self-organized quantum dots exposed to an external electric field was the phonon-assisted tunneling. The thermal phonon-assisted emission rate for charge carriers is given by [522]

$$P = \gamma T^2 \sigma_\infty \exp\left[-\frac{E_a}{kT}\right] \tag{6.3}$$

where γ is a constant independent of temperature, σ_∞ is the apparent capture cross-section corresponding to the high-temperature limit $T \rightarrow \infty$, and E_a is the activation energy. So, the transmission rate P and the storage time $\tau = P^{-1}$ at a certain temperature strongly depend on the thermal activation energy E_a. This energy takes on different values for electrons and holes. Also, it depends on the material used to create quantum dots, the surrounding matrix, and the applied bias voltage. The greater this energy becomes, the longer is the storage time. Presently, the reported storage times vary from a few nanoseconds [523] to a few seconds [524] at the room temperature.

To further enhance the storage time in self-organized quantum dots, one needs to increase the charge carrier localization energy, which is closely related to the activation energy E_a. Charge carriers (especially holes) localization energies for the most promising material combinations for the quantum dots and the surrounding matrix such as $GaAs_xSb_{1-x}/GaAS$ and $In_xGa_{1-x}Sb/GaAs$, were theoretically estimated using eight-band $\mathbf{k} \cdot \mathbf{p}$ theory [525, 526]. It was shown that the hole localization energy could be increased from 350 up to 853 meV as the antimony content increases from 50 % to 100 %. Even more impressing results were obtained for $In_xGa_{1-x}Sb/GaAs$ quantum dots. The localization energy for holes was estimated as 989 meV for the indium content of 100 %. Basing on these theoretical estimates, one may expect the holes storage time in $InSb/GaAs$ quantum dots to be longer than an hour. It was conjectured that by replacing $GaAs$ matrix with the $AlAs$ matrix, one could enhance the storage time in $GaSb$ quantum dots up to years [519, 524], which shows the potential and advantages of using self-organized quantum dots to design nanomemories.

Another kind of molecule-scale entities which could be used in manufacturing nanomemory cells are CNTs. The intrinsic structural, thermal, and mechanical stability of CNTs gives grounds to confer a very long lifetime to these memory devices. Also, CNTs are known to be radiation resistant, so one may expect nanotube-based memory elements to remain robust in various ambiences [527]. Several types of CNT-based memory elements were suggested so far. These include memory cells based on CNT field-effect transistors, nanoelectromechanical systems, and systems using electromigration.

Both volatile and nonvolatile memory elements can be designed from a CNTFET functioning [528]. In these devices, the storage mediator is usually a charge stored near the nanotube employed as the FET channel. This charge causes a shift in the FET threshold potential, that is, the potential which indicates the opening of the nanotube channel for the charge carriers to flow. Due to the high carrier mobilities typical for CNTs, the information can be stored in a few electron configurations. The information reading may be accomplished by measuring current–voltage characteristics, and it could be written using a voltage applied to the gates. First CNTFET-based memory elements were proposed in 2002 by Fuhrer and coworkers

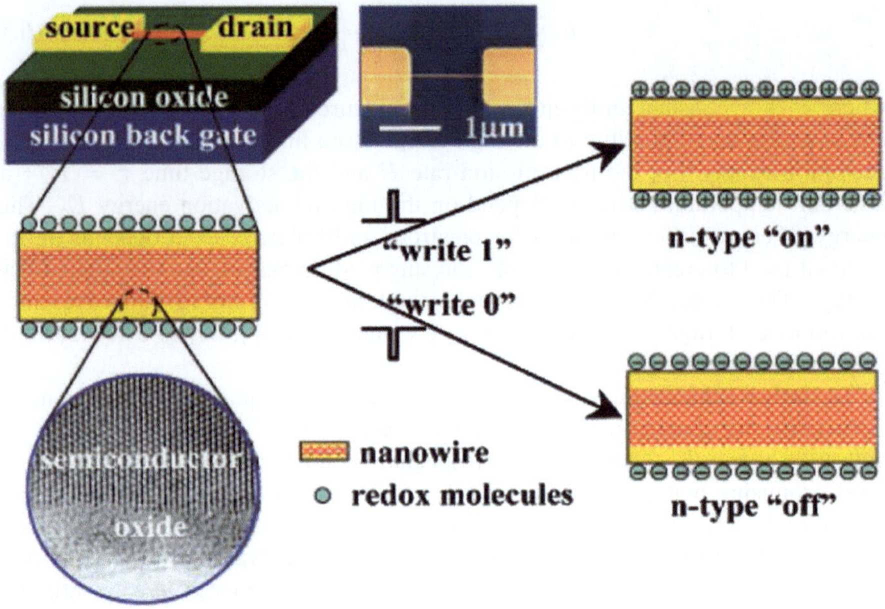

Fig. 6.8 Schematics of nanowire-based nonvolatile memory cell. The device consists of a semiconductor nanowire configured as a FET with the oxide surface functionalized with redox-active molecules. The *top middle inset* shows a SEM image of the device, and the *lower circular inset* shows a TEM image of the nanowire used to build up the device. Positive and negative charges are injected into and stored in the redox molecules with an applied bias voltage pulse. In a n-type nanowire, the positive charges create an "on" or logic "1" state, while the negative charges produce an "off" or logic "0" state [533]. Reprinted with the permission of the ACS

[529] and Radosavljevic and coworkers [530]. Even then these memory elements demonstrated the charge-storage time up to a few days at room temperature [531].

Soon after the publication of the first works [529–531] proposing CNTFET-based memories, it was confirmed that the shift of the threshold potential in these devices indeed was caused by an electric charge accumulated in the nanotube vicinity. This charge originated from the trapping of charge carriers in the relevant regions. There are various kinds of trapping centers which played important parts in functioning of early memory CNT-based devices. The trapping centers could be defects in the SiO_2 and/or molecules (e.g., water or alcohol) absorbed on the dielectric, oxidized carbon or the nanotube itself. However, better control of the storage medium was necessary to achieve further improvement of the reliability and performance of the memory elements. Correspondingly, in the following years a significant effort was and still is being applied to modify CNTFET memory cells. For instance, Sakurai and coworkers suggested to replace the dielectric insulator by a thin ferroelectric film [532]. The current in the nanotube was controlled by spontaneous polarization inherent to the film. Another type of device uses a CNT or nanowire functionalized with redox-active molecules, which are charged by means of the applied bias or gate voltage as shown in Fig. 6.8. These charged redox molecules gate the system to a logic "on" state with higher channel conductance or "off" state which is characterized by the lower conductance [533]. Similar devices using SWNTs as channels, where the conductance is switched "on" and "off" guided by chemical reactivity of the nanotubes in sulfur acid, were described in later works [534, 535].

The most of CNT-based nanoelectromechanical systems are exploiting the CNTs propensity to efficiently respond to capacitive forces. These forces can develop between the nanotube and the gate electrode when the voltage is applied to the gate. The gate voltage may induce the rearrangement of the electric charge on the nanotube surface which results in the specific electric field appearing between the nanotube and the gate. The effect of this field could be described introducing the capacitive force. Due to the high flexibility inherent to carbon nanotubes, the capacitive force may bend it towards the gate electrode. The ultimate shape and position of the nanotube is determined by the combined effect of the capacitive, elastic, van der Waals, and short-range forces. The delicate balance of all these forces is necessary to make possible the functioning of nanotube-based nanoelectromechanical devices, and this is not easy to achieve. Nevertheless, quasi-one-dimensional structure and high aspect ratio of CNTs combined with exceptional electronic and mechanical properties make them promising building blocks for nanoelectromechanical memory elements and switches [536, 537].

One of the first concepts of a nanoelectromechanical memory cell using single-wall carbon nanotubes was proposed by Rueckes and coworkers a decade ago [538].The proposed device used CWNT crossbar array. The crossbar consisted of two subarrays of CNTs. One set of parallel nanotubes was placed on a substrate, and the nanotubes belonging to the second set were suspended on periodically arranged supports. The crossbar bistability resulted from the balance between the elastic energy whose minimum corresponded to the situation when the nanotubes were not bent, and the energy occurring due to attractive van der Waals interactions between the nanotubes belonging to the different subsets. This energy reached its minimum when the upper nanotubes were bent towards the lower ones. Two states of the system were distinguished by the presence/absence of the electric current flowing through the system when the bias voltage was applied. The system is found in "on" state if the upper nanotubes are bent, which provides the possibilities for the charge carriers to tunnel from them to the lower nanotubes. Correspondingly, the electron current flows through the crossbar between the source and drain electrodes. On the contrary, when the upper nanotubes remain straight, they are separated from the lower ones, and the electron tunneling is hindered. This arrangement is associated with the "off" state when no current flows through the system. The device may be switched between the "on" and "off" states by transiently charging the nanotubes to produce attractive or repulsive electrostatic forces. The suggested device architecture was shown to be tolerant to structure variations. Also, the difference in the resistances of the system being in the "on" and "off" states is very significant for the probability of electron tunneling exponentially decreases as the separation between the relevant nanotubes grows.

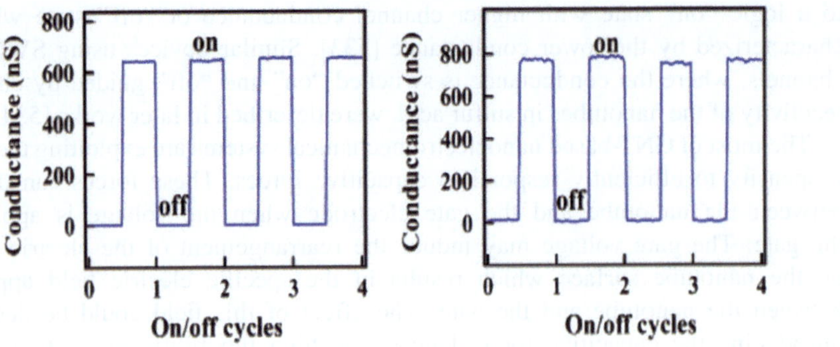

Fig. 6.9 Reversible "on" and "off" switching of the device shown in the previous figure [533]. Reprinted with the permission of the ACS

Another example of a nanoelectromechanical CNT-based memory unit was suggested in [539]. The key element of the considered device was a carbon nanotube suspended in between two self-aligned electrodes which could be mechanically switched by the electrostatic forces. Numerous variations of CNT-based nanoelectromechanical memory cells were and still are suggested and discussed (see Fig. 6.9). Theoretical and experimental investigations have been carried out to optimize the characteristics of these devices to make them better suitable for practical applications. As was mentioned above, the properties of CNT-based nanoelectromechanical memory elements are controlled by the relation of elastic, van der Waals, short-range (adhesive), and electrostatic(capacitive) forces. It was shown that van der Waals forces reveal a tendency to reduce the pull-in voltage, that is, the voltage required to bend the nanotube. The importance of van der Waals forces reduces as the tube length and/or diameter increases. However, the transition between bent and straight nanotube positions is very sensitive to the effect off short-range adhesive forces. These forces cause the adhesion which hinders the breaking of the established contact between the nanotube and some other element of the system. Due to the adhesion a device may be impossible to use, as it remains stuck in one position. The sticking is a general issue encountered in various nanotube-based devices. The adhesion inhibits usefulness of single-wall CNTs for they are too flexible. Multiwall CNTs or single-wall CNT bundles appear to be better candidates for designing nanoswitches and memory elements working as electromechanical devices [540]. Multiwall CNTs possess another important feature which makes them attractive for use as elements of nanoscale memory cells. It was shown that the internal friction forces between the shells are very weak, so the inner core of an open CNT can move nearly freely along the tube axis. This property opens up new possibilities for design of nanodevices. For instance, Deshpande and coworkers [541] proposed a switching nanodevice including two open-ended MWCNT segments separated by a nanometer-scale gap. The switching is provided by electrostatically induced sliding of inner nanotubes shells. The shells move towards each other to close the gap. As the gap is closed, the current starts to flow through the system indicating its "on" state (Fig. 6.10). The reversed

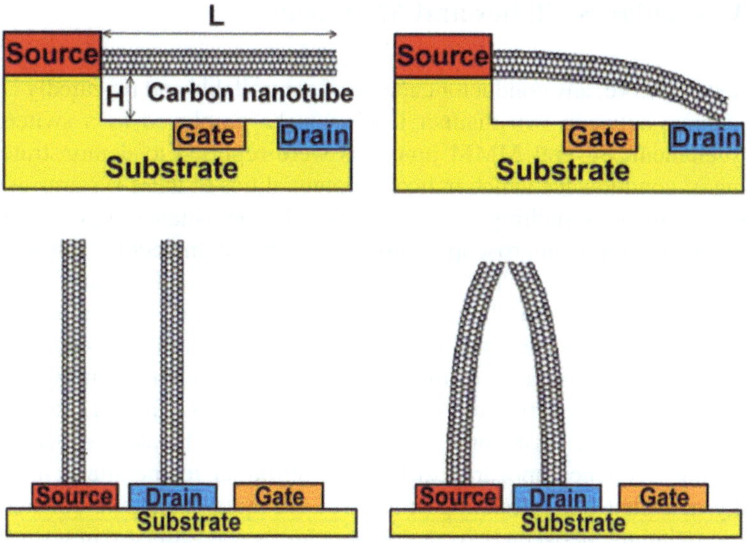

Fig. 6.10 Schematics of nanoelectromechanical CNT-based memory cells. Nonconducting ("off") and conducting ("on") states of the devices are shown in the *left* and *right panels*, respectively. (Courtesy of Dr. Yu. Lozovik)

motion of the MWCNT inner shells opens the gap and the system passes into the "off" state with zero conductance. Some other structures for switches and memory cells exploiting the exceptionally low friction between MWCNT inner shells were proposed [542, 543].

Electromigration CNT-based memory elements were first suggested more than decade ago [544]. The key element of this and later proposed electromigration-based memory devices is an ionized fullerene C_{60} enclosed in a CNT or in a larger fullerene. It was experimentally proven that the ends of the nanotube could correspond to minima in the energy landscape for the endohedral fullerene encapsulated in the tube. Therefore, the fullerene is striving to place itself near one of the tube ends. Its switching between the preferred positions may be provided for by the voltage applied across the nanotube. These positions are associated with the "on" and "off" states of the system. While the described concept provides a relatively simple physical mechanism for storing information, practical measurement of thc intrinsic "on" and "off" states of the system meets some serious difficulties which hinder practical use of the memory elements based on the electromigration mechanism. Nevertheless, the concept which relies on the electromigration of an encapsulated architecture from one local energy minimum to another, its positions indicating the "on" and "off" states, looks very promising. Such devices should have very short switching times, and they should be highly robust in regard to the thermal and electrical stability. There are grounds to expect that future explorations will facilitate fabrication of electromigration-based nanoswitches and memory elements using carbon nanotubes as building blocks.

6.4 Molecular Switches and Memories

As discussed before, any conductor capable of controllably and repeatedly changing its impedance between two distinct levels can be employed as a switch and/or memory element. Several MMM junctions were reported to demonstrate binary switching of conductance which indicates potentialities of these systems as switching devices. Binary switching of single-molecule conductance was first observed in scanning tunneling microscopy studies of isolated molecules adsorbed on a surface [545, 546]. In the experiments reported in [546], alkanedithiol molecules were embedded in alkanethiol monolayer on a gold surface. These molecules were imaged as bright spots, and the blinking of these bright spots was revealed in studies of consecutive images of the same area. This phenomenon was explained as an evidence of two different conductance states ("on" an "off" states) of the alkanedithiol molecules associated with the formation and breaking of $Au–S$ bonds. The blinking frequency appeared to increase at higher temperatures indicating a thermally activated bond-breaking process at the metal–molecule contacts. Also, stochastic switching between two conductance levels was observed in several MMM junctions in break-junction experiments [10, 547]. The switching was attributed to thermal fluctuations of electrode-molecule bonding sites and corresponding changes in metal–molecule coupling strengths. These experiments gave ground to believe that binary switches could be manufactured by using the geometrical dependence of molecular conductance through developing a way to control the molecule conformations as well as configurations of metal–molecule contacts [548]. One more STM experiment revealing binary switching of a single-molecule conductance was reported in [549]. In this case, considered MMM junctions included perylene tetracarboxylic acid bismudes molecules modified with pyridine and methylthiol groups attached to gold leads.

As was shown, switching between the states characterized by different molecular conductances could be achieved following several ways. One of these ways implies the employment of mechanical forces. Switching by mechanical movement of an atom (or a few atoms) in the molecule is being discussed for a long time. A molecular electromechanical device, where the displacement of a certain atom could bring a crucial change in the conductance of an atomic wire where the atom belongs, was first proposed two decades ago [550]. Theoretical estimates suggested that for a carbon or silicon atomic chain, the shift of an atom by a single atomic diameter would change the conductance through the chain by several orders of magnitude. An early practical version of such device was demonstrated using an STM-type molecular junction with the fullerene C_{60} as a linker. Minute compressions of the fullerene by the STM tip resulted in very significant changes in the current flowing through the junction [551]. Later, a molecular switching device based on the rotation of the di-butyl-phenyl leg in a Cu-tetra-3.5-di-tertiary-butyl-phenyl-porphyrin molecule was proposed [552]. The leg motion was mechanically controlled by the tip apex of an atomic force microscope. Similar effect could be manifested in π-conjugated oligomers if one of monomers is twisted away from a planar configuration of a

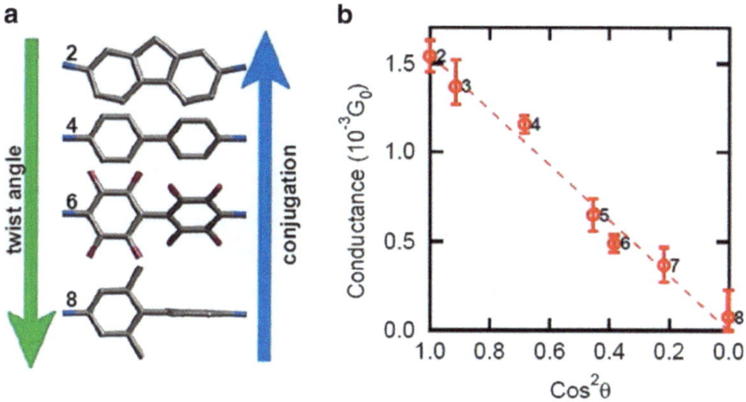

Fig. 6.11 Bi-phenyl-based molecules with different torsion angles between the two phenyl rings. (**a**) Conductance measured with the STM break junction versus the torsion angle (**b**) (after [14]). Reprinted with the permission of the NPG

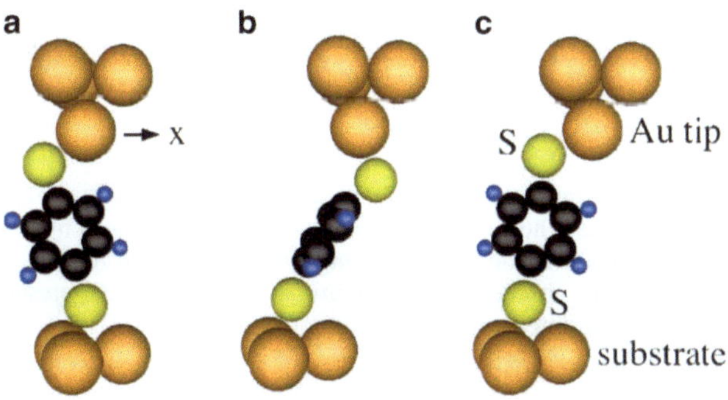

Fig. 6.12 Changes in the molecular bridge geometry provoked by sinusoidal motion of the STM tip along "x" direction. (**a**) Ground-state conformation for STM tip at $x = 0$ Å. (**b**) Ground-state conformation for $x = 3$ Å. (**c**) Metastable edge conformation for $x = 3$ Å (after [548]). Reprinted with the permission of the APS

molecule. Twisting the monomer breaks the conjugation along the backbone, thus significantly reducing the conductance through the molecule. It was shown that the conductance in such molecules decreases proportionally to $\cos^2\theta$, θ being the torsion (twist) angle between the tilted away monomer and the plane of the molecule [14, 193], as presented in Fig. 6.11.

Another early example of a mechanically controlled molecular switch is the one discussed by Emberly and Kirchenow [548]. The device is shown in Fig. 6.12. In this case, binary conductance switching occurs due to the sinusoidal modulation of the tip motion which repeatedly deforms metal–molecule contact structure between two stable configurations. In later works, several other mechanically controlled single-molecule switches were suggested. Recently, switching in MMM junctions including 4,4′-bipyridine molecules bridging gold electrodes was reported [553].

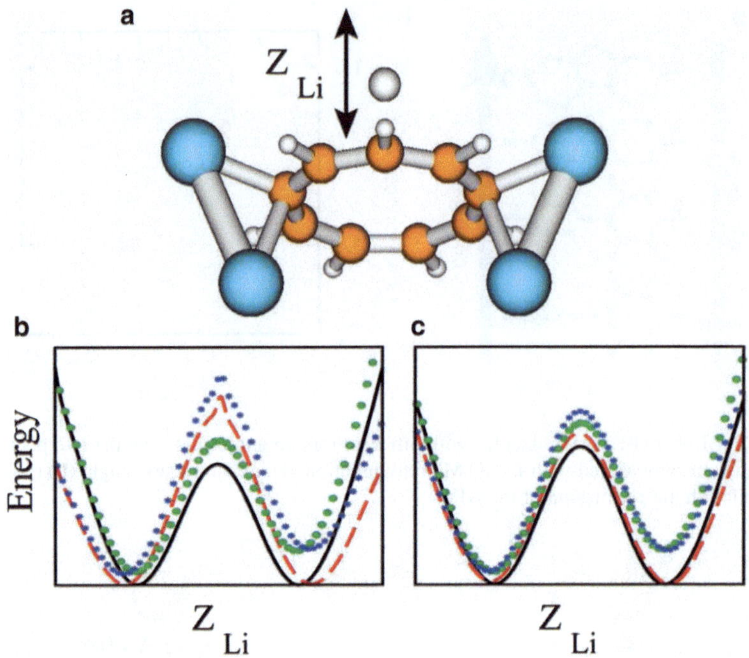

Fig. 6.13 The system consisting of a lithium atom, the carbon ring, and two aluminum nanoclusters, each including two atoms. The direction of the lithium shift is indicated by Z axis (**a**). Potential energy profiles along Z axis with the electrodes including one (**b**) and four (**c**) aluminum atoms. The profiles are plotted in the absence (*solid* and *dashed lines*) and in the presence (*dotted lines*) of the bias voltage [556]. Reprinted with the permission of the ACS

It was observed that these junctions could switch back and forth between two states with different conductances as one of the electrodes performs a several-angstrom sinusoidal motion. Elastic deformation of the bridge was recognized as the mechanism controlling the compression-induced transitions of bipyridine molecules between two different binding sites on the surface of the gold tip. Also, mechanically controlled conductance switching was recently reported in Au/hexanedithiol/Au systems [554]. In this case, the conductance variations were inspired by a sinusoidal displacement of the tip electrode position with 3 Å amplitude. Again, the change in the conduction was caused by the change of coupling between the molecule and the tip.

Several molecular switches basing on the use of reversible conformational changes in molecules occurring due to the application of the electric field were suggested and demonstrated (see, e.g., [13, 197, 555–557]). These are bistable molecular devices characterized by "on" and "off" states which have their analogs in semiconductor-based electronics. In general, such devices are based on molecules consisting of two parts where one part is allowed to move around and/or along the other one. A simple example of such bistable system was recently described by

Jorn and Seideman [556] (Fig. 6.13). The system consists of a single lithium atom bound to a cyclononatetraene molecule which is connected to aluminum clusters including a few atoms. These clusters take part of nanoleads of the system. The lithium is held to the cyclononatetraene ring by the ionic interaction in a double potential well. The wells positions correspond to the two locations of the lithium atom. Both locations are out of the ring plane. Application of the gate voltage breaks the symmetry of the double well and increases the probability for the lithium to remain in the corresponding position above or below the ring depending on the voltage polarity. When the polarity is reversed, the lithium should shift to another equilibrium position which is located on the other side of the ring.

Catenane and rotaxane also belong to this kind of molecules. These and similar molecules may adopt one of the two possible configurations depending on their redox states. Applying a moderate external voltage across the molecule ($\sim$1 V), one may trigger the conformational change in the molecule, which is manifested as the displacement of its mobile part leading to the minimization of the total energy. Different configurations of the molecule are characterized with different values of the molecule conduction. Therefore, these configurations may be associated with the "on" and "off" states of the bistable system. Using catenane and rotaxane molecules, Chen and coworkers had demonstrated a 64 bit nonvolatile memory crossbar with an integration density of 6.4 Gbit/cm^2 (about ten times greater than the corresponding characteristics of presently existing silicon chips) [558]. The fabrication yield of these molecular memory devices was about 85 %, the data retention time was close to 24 h, and they had been able to endure from 50 to 100 write-erase cycles before the collapse of the on–off ratio to unity. Later, a 160 kBit memory elements using the same molecules was demonstrated by Green and coworkers [559]. In these memory elements, 25 % of tested memory points have revealed the on–off ratio of about 1.5 with an average retention time close to 1 h.

One more example of a molecular switch driven by the applied electric field is shown in Fig. 6.14. Here, a C_{60} fullerene molecule is placed between two gold electrodes. The coupling between the fullerene and the leads is supposed to be weak. So, a transient attachment of the electron to the fullerene leading to attraction between the charge localized in the molecule and its image in the nearest electrode. Due to the weakness of the original coupling of the molecule to the leads, this attraction causes a shift of the fullerene center of masses $Z_{C_{60}}$ from its original position towards the corresponding electrode. As the molecule moves closer to the lead, the probability of the extra electron tunneling away from it increases. Eventually, the electron leaves from the molecule to the electrode, and the fullerene becomes neutral again. The potential energy associated with the interaction between the molecule and the electrode reaches its minimum at different distances between the molecule center of masses and the electrode for the charged and neutral states of the molecule, which results in difference in molecular conductances. Therefore, in this case, the changes in the conductance through the junction do not appear as a result of conformational changes in the molecule. They originate from a motion of the whole fullerene between the leads.

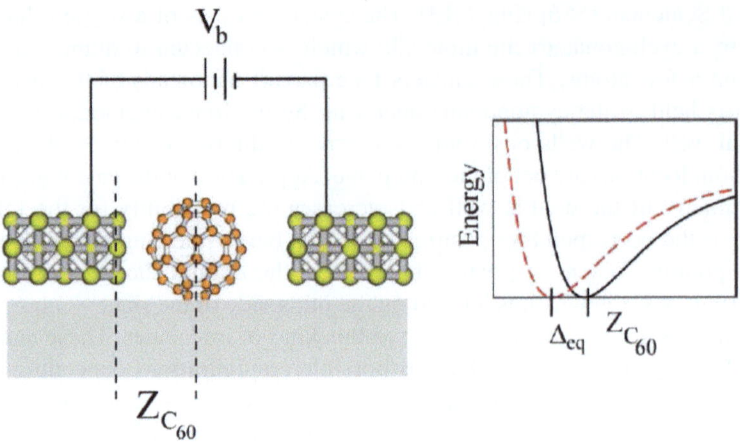

Fig. 6.14 Schematic illustration of a C_{60} fullerene junction, where the molecule is weakly connected to the leads (*left*) and potential energy profiles for the interaction between the fullerene and the nearest electrode in the neutral (*solid line*) and charged (*dashed line*) states of the molecule (*right*; after [556]). Reprinted with the permission of the ACS

It is worth to mention that electrically triggered switching behavior had been observed by using simple alkyl chains instead of bistable molecules [560, 561]. However, it is likely that in this case and other similar cases, the switching behavior occurs due to the formation and breaking of metallic microfilaments introduced in the molecular layer during the device preparation [481], and the switching effect may not be attributed to molecules.

Another efficient way to trigger configuration changes in molecules is by light. Recently, reversible photoswitching devices employing diarylethene and azobenzene derivatives were demonstrated [255, 563, 564], which show potential applications of metal–molecule–metal junctions in designing optically controlled switches. A light-driven molecular switch based on a conformational change in azobenzene molecules is shown in Fig. 6.15. These switches could work due to the ability of some molecules to make transitions between different configurations upon absorption of photons. For example, azobenzene molecules use to convert from "*trans*" to "*cis*" configuration while exposed to ultraviolet radiation with wavelength close to 360 nm. Also, these molecules show reversible isomerization under blue light ($\lambda \sim 480$ nm). The azobenzene structural sensitivity to electromagnetic radiation makes these molecules promising building blocks for light-controlled molecular switches and memories. However, there exist some difficulties hindering their practical applications. As known, isomerizable molecules need to be electronically decoupled from the metal electrode surface to perform switching between the relevant isomers. In several STM experiments the photoisomerizable azobenzene SAMs were attached to the substrate by various spacers including short alkyl chains [563], phenyl or thiophene moieties [562,564], or ethylene bonds [565]. The choice

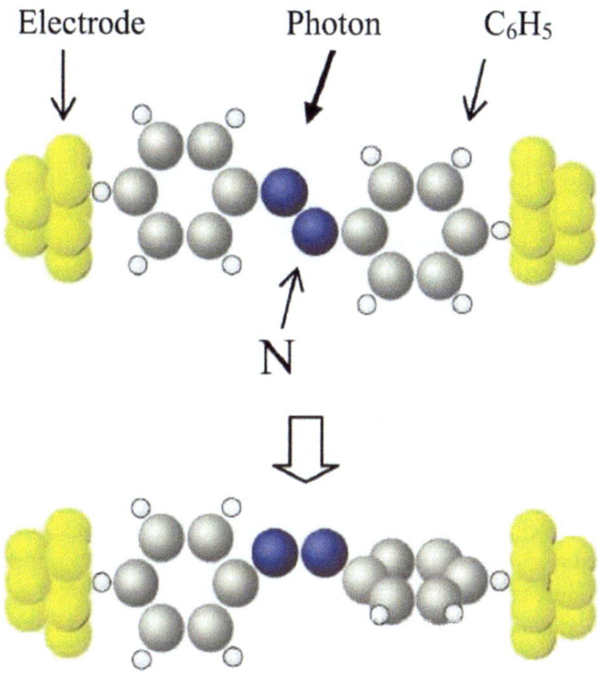

Fig. 6.15 Schematics of a light-driven molecular switch based on the conformational change of an azobenzene molecule in contact with two metallic electrodes (after [255]). Reprinted with the permission of the APS

of this unavoidable linker appeared to be very important. In some cases the very manifestation of the switching behavior depended on the nature of the spacer used to build the junction [566].

Another important factor is the coupling of the molecule to the electrodes. For instance, Nozaki and Cuniberti recently showed basing on first-principles electron structure calculations that the conductance for "*trans*" configuration of azobenzene molecules placed between CNT electrodes should exceed the conductance through these molecules being in "*cis*" configuration. At the same time, the relation between the conductances was reversed for a junction with silicon electrodes [567]. Also, the spacer length could affect the properties of a MMM junction including a photoisomerizable molecule. A short spacer should favor the electron transport through the junction increasing its conductance, whereas a longer spacer could allow more room for the switching. So, there exist several factors, which could affect the performance of a light-controlled molecular switch, and one must take all of them into account to fabricate a good quality device.

Unlike azobenzene that only changes its length by *trans–cis* isomerization, diarylethene molecules can undergo reversible ring-opening and ring-closing processes when exposed to visible or ultraviolet light. More importantly, electrical characteristics of these molecules drastically change upon their transformation from the nonconjugated open form to the conjugated close form [568]. In these molecules, the ratio of conductances between open and closed isomers could reach the values of the order of 10–100 [569]. Furthermore, photochromic reactions of diarylethenes were observed to proceed within picoseconds in a solution, which suggests the switching speed of the order of a few THz [570]. These unique characteristics of diarylethenes indicate their potentialities for ultrafast single-molecule photo-switch applications.

Single-molecule conductance of diarylethene derivatives had been measured using break-junction techniques. Dulic and coworkers [571] examined light-induced conductance switching of dithienylethenes including thiophene rings connected at both sides and a dithiol anchoring group. Single-molecule junctions were lithographically formed at room temperatures. Dithienylethene bridges were initially in the closed form with the electron density of states delocalized over the whole molecules. When these systems were exposed to the visible light, three orders of magnitude drop of the conductance were observed. This conductance drop was attributed to the ring-opening reaction in the dithienylethene units. However, despite the excellent on–off conductance ratio demonstrated in these experiments, the switching speed occurred to be slow. The on-to-off switching happened only about 10 s to a few minutes after the light irradiation, and the reverse (off-to-on) switching was not observed at all. The lack of reversibility was attributed to quenching of the photo-excited closing reaction caused by the overlap of the open isomer HOMO with the high density of states at the gold electrode which facilitated fast electron transfer. Reversible photochromic switching in MMM junctions with diarylethene bridges and gold electrodes was achieved by replacing thiophene spacer at both sides of dithienylethene by phenyl rings [572]. Although this replacement brought the decrease of the on–off ratio by several factors, the phenylene spacer enabled the molecule to switch back to the closed form by ultraviolet irradiation. Also, reversible photochromic switching was attained using SWCNTs as electrodes [573]. These results give grounds to hope that the diarylethene-based molecular switches with reasonably high switching speeds could be manufactured.

Recently, the synthesis and characterization of a new molecular switch using the azobenzene moiety linked to a bithiophene spacer and a short alkanethiol chain was reported [574]. This device seems a success for it combines the benefits of a rather long spacer providing a high on–off ratio (up to 7×10^3) between the "*cis*" and "*trans*" configurations and a high level of electron current. The analysis of the experimental results based on the first-principles calculations of the molecular orbitals proves that the high value of the photoinduced on–off ratio originates from a synergetic combination of SAM thickness and energy offset between the molecular LUMO and the chemical potentials of the electrodes. This gives grounds to expect the discussed device to be fitting for practical low-cost molecular switches. Redox-active molecules such as metallocene and porphyrin were shown to act as charge-storing entities [575–577]. They could be used as memory elements. In this case the information carrying medium is the electric charge whose amount on the molecule could be varied by driving the latter into different chemically reduced or oxidized states. These charge-based memory elements were demonstrated to offer possibilities of multibit storage, and their retention times were reasonably (about a minute) long [577].

So, these exists a scientifically provoking suggestion to employ the bistable charge-based molecular devices to design memory cells and logic elements. However, the properties of charge-based molecular memories and logic elements are rather unfavorable for their use in large computing systems. The critical point is that they operate in biased molecular junctions close to the threshold. Therefore, the addition/removal of a small input may cause large changes in the output. Being an element of a circuit among other elements, the charge-based molecular switches and memories inevitably become subjects to the ambient noise, and this may easily damage the signal. This puts certain limits on the possible applications of molecular junctions as building blocks for switching devices and memory cells, using the electric charge as the storage medium. At the same time, molecular junctions including magnetic molecules are presently considered as very promising candidates for nanoswitches and memories based on the electron spin as an information carrier. As was already mentioned in previous chapters, spintronics is a fast developing branch of nanoelectronics. Electron spin state possesses important advantages for technological applications. These include long relaxation times in the absence of external field (long storage times) and the possibility of easy spin manipulation by an external magnetic field and some other means (such as the changes in the electric charge on the magnetic molecule). Potentialities of magnetic molecules in designing of memory elements were recognized more than a decade ago [291–295]. It was shown that magnetic molecules may be used to build memory devices where the spin states of the molecule were employed to store information. Such devices whose working parts are based on Fe_8 or Mn_{12} magnetic molecules, and writing and reading of information is carried out by applying of the oscillating magnetic field for a fixed duration, are expected to be very dense and efficient. It was suggested that fast magnetic field pulses should allow to decode and read up to 10^5 numbers with access time as short as 10^{-10} s [156, 578]. Transport properties of magnetic molecules including those relevant for spintronic applications are intensively studied. Some recent results of these studies were already discussed in Sect. 3.9.

6.5 Chemoselective Sensors

The integration of nanotechnology with molecular recognition poses numerous possibilities for the design and creation of novel types of sensing devices. A chemosensor is commonly defined as a device, which, in response to the presence of some kind of molecules in the close environment, changes its readily monitored characteristics. These changes occur due to chemical reactions between the ambient molecules (analyte) and the device elements. Presently, several types of nanoelectronic chemosensoring devices are being explored. One of the promising

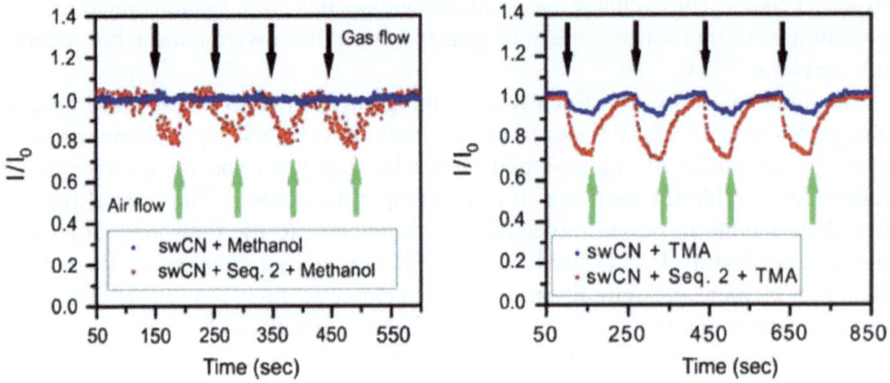

Fig. 6.16 The response of a sole SWNT and a DNA-decorated SWNT-based sensor to methanol (*left panel*) and trimethylamine (*right panel*) vapors [582]. Reprinted with the permission of the ACS

types of molecular recognizing sensors are those using sp^2-bonded carbon materials (fullerenes, carbon nanotubes, and graphene sheets). Electron transport through these materials is known to be very sensitive to absorbed molecules which make them good candidates for working elements in chemoselecting devices. In the last few years, a significant effort was applied to build nanosensors basing on changes in the conductance or capacitance of single-wall CNTs, which happen when some molecular species interact with the CNT sidewalls [579–581]. An efficient chemical sensor may be constructed using a single-wall CNT functionalized with biomolecular complexes [582]. In such device, the chemical recognition is performed by biomolecules (e.g., single-stranded DNA molecules), whereas a single-wall CNT-based field-effect transistor works as a read-out component (see Fig. 6.16). Nucleic acid biopolymers appear to be very useful for molecular targeting since they could be chosen and/or engineered for affinity to a wide variety of targets including small molecules and specific proteins [583]. The above-described devices have demonstrated a good sensitivity to the presence of several substances such as methanol propionic acid, trimethylamine, and dinitrotoluene.

So far, exact details of the interaction chemistry, which develops between ambient molecular species and CNT sidewalls, remain elusive. However, it is clear that molecular interactions with defect sites in the sidewalls can significantly contribute to the CNT electrical response. Therefore, one may enhance both sensitivity and specificity of a CNT-based chemoselective sensor by controlling the defects density. The optimal defects density is that which balances the gains in the device sensitivity against the losses in the conductivity due to defects. A graphene sheet decorated with oxygen groups is a material containing a diverse range of surface defect sites whose density could be easily controlled. Therefore, it was suggested that the graphene oxide could be a very promising material where the optimal defects density may be engineered [584]. Being in the as-oxidized state, the graphene oxide shows a poor conductivity hindering its potential usefulness

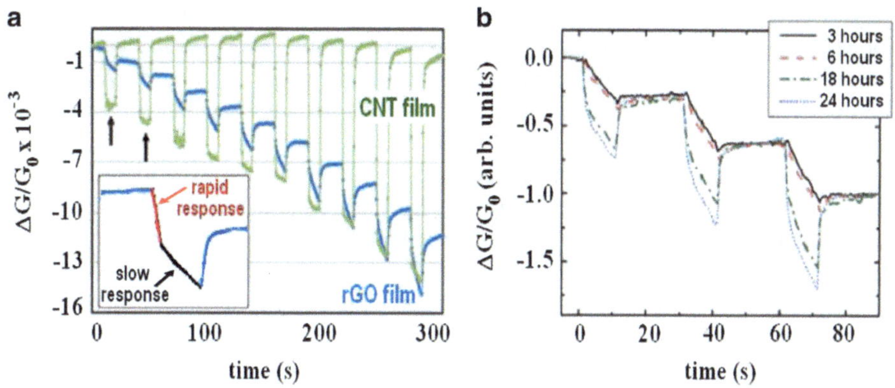

Fig. 6.17 The response of a SWNT-network-based sensor (*left panel*) and the reduced-graphene-oxide (rGO) network sensor (*right panel*) to 5 s acetone pulses of increasing concentration with arrows indicating the first two pulses. *Inset*: close-up of the fifth pulse highlighting the rapid and slow response of the rGO device (*left panel*). The plot of the normalized conductance response of rGO device of 5 s acetone pulses for the devices reduced with hydrazine for 3, 6, 18, and 24 h (*right panel*). As the reduction time increases, the fast response increases while the slow response decreases [584]. Reprinted with the permission of the ACS

for fabrication of conduction-based sensors. However, its conductivity could be significantly increased by removal of oxygen and recovery of aromatic double-bonded carbon [585]. The reduced graphene oxide (rGO) obtained in this manner is a reasonably conductive material, which still has chemically active defect sites. It was suggested that high-performance molecular sensor could be fabricated using networks of reduced graphene oxide platelets [584]. The corresponding device was demonstrated, and it showed good sensitivity of the conductance response to molecular adsorbates as shown in Fig. 6.17. Another advantage of the reduced-graphene-oxide-based sensor is the low level of low-frequency noise. The noise arises from fluctuations in the mobility and/or density of charge carriers, and it is imminent is solid-state devices. In carbon nanotubes these fluctuations and the accompanying noise are manifested as well, although in these cases, they are mildly pronounced compared to the solid-state devices. However, in graphene these fluctuations are strongly suppressed, as shown in the recent work [586]. This provides the low-frequency noise level orders of magnitude lower than that typical for single-wall carbon-nanotube-based sensors.

The above-described sensors work as field-effect transistors where the conductivity of the channel varies when the device is exposed to certain ambient molecular species. There exists another kind of miniaturized chemosensors using carbon nanotubes. These devices feature the electrical breakdown in the presence of some gases and gas mixtures near the CNT tips. An ionization sensor of this kind was suggested by Modi and coworkers [587]. The proposed device included a set of vertically aligned carbon nanotubes which took a part of the anode and the cathode made out of Al sheet. The electrodes were separated by a thin layer of glass. Due to the nanometer-scale diameters of nanotube tips, a high electric field appears near the tips even at reasonably low voltage applied across the electrodes. In consequence, the conducting filament of highly ionized gas is being formed about the tips promoting an electron avalanche of plasma streamer which bridges the gap between the electrodes. For this device, the breakdown voltage magnitude had the order of 10^2 V, and its exact value indicated the nature of the gas to be identified. The recognition ability of the device originates from the established fact that every gas is characterized by a unique breakdown electric field provided that both temperature and pressure remain the same. This operating mechanism is employed in fabrication of several ionization sensors using carbon nanotubes. Recent research developments on these devices are reviewed in [588].

Metal/metal oxide, silicon, and polymer nanowires are also widely recognized as potentially and actually useful elements in fabrication of chemoselective sensing devices. One of the commonly employed platforms for sensing systems based on nanoscale materials is the use of conductivity changes in nanowires spanning two electrodes [473, 514, 589–599]. Usually, these sensors operate in a way similar to that typical for CNT-based sensors working as field-effect transistors with the variable conductance through the channel. Two electrodes linked by the nanowire can behave as the source and drain reservoirs of the FET-like device. Assuming that the constant resistances are sufficiently small, binding of charged molecules to the nanowires surfaces may induce a field effect, thus bringing changes in the nanowires conductivity. This approach was adopted to make a variety of nanowire-based sensors operating in liquids and used for chemical and/or biological detection, as well as for gas sensors operating in air.

Nanowires transport properties and their subsequent sensing potentialities could be modified by doping. For instance, undoped tin dioxide (SnO_2) nanowires are proven to be useful elements of gas sensors detecting O_2 and CO_2 [594]. At the same time, Sb doped SnO_2 wires were demonstrated to be responsible to ethanol [600], and Ru-doped ones appeared to be sensitive to the presence of NO_2 molecules [601]. Also, doping-dependent sensing behavior was reported for In_2O_3 nanowires. Zhang and coworkers suggested that the response of these wires towards ammonia vapors was strongly correlated with their oxygen vacancy doping concentrations. In heavily doped wires, electrons were transferred from the wire to the adsorbed NH_3 species which resulted in the reduction of the charge carrier concentration accompanied by the suppression of the nanowire conductance. On the contrary, at low doping concentration the carriers were transferred from the adsorbed molecules into the wire leading to the increase in the conductivity [602].

Due to the large surface-to-volume ratio typical for nanowires, their charge carrier densities could be controlled by chemical processes developing at their surfaces. In turn, these processes could be modified or even entirely halted by applying a suitable gate voltage. Following this way one may manipulate the nanowire sensitivity towards certain molecular species. This was demonstrated for the case of SnO_2 wires response towards CO [603].

In some cases, the changes in the polycrystalline nano- and mesowire conductivity were shown to be unrelated to the surface adsorption. Instead, the conductivity variations were caused by structural alterations in the wire. Such conductivity controlling mechanism was suggested by Penner and coworkers for hydrogen sensors based on electrodeposited palladium mesowire arrays [604]. They showed that the exposure to the hydrogen leaded to Pd grains swelling in the wire interior. When the hydrogen was absent, the grains did return to their original dimensions, and nanoscale gaps separating the grains were opening up. Therefore, the wires demonstrated better conductance in response to the ambient hydrogen than in the absence of the latter. In general, it should be noted that the grain sizes in polycrystalline nanowires may noticeably affect sensibility and reversibility of nanowire-based chemoselective sensors. This happens because the reduction of the grains dimensions increases the surface-to-volume ratio and creates better conditions for the molecular adsorption [599].

An interesting design for a chemoselective device combining a carbon nanofiber with some kind of molecular junction was reported in the recent work [598]. This sensor was intended for a digital signal generation. Digital signals are known to provide better immunities against the noise, so detection systems intrinsically generating such signals may sometimes have advantages over the analog sensors some of which were described above. In the suggested device, the nanowire linking the electrodes was attached to them by certain molecular assemblies. There exist various molecules which could be used to link the wire to the electrodes. However, suitable molecules must include chemoresponsive parts which cause conformational changes or cleavage events in response to a particular analytical target. Therefore, in the presence of the target molecules, some molecular links between the nanowire and the substrate become broken. At sufficient concentration of the target molecules, the nanowire becomes cleaved from the substrate, and this causes a sudden fall of the current through the system. In the reported experiments, biotin–avidin molecular linkages were used to build the links between carbon nanowires and gold contacts. The molecules included disulfide bonds which provided chemical selectivity by cleaving in response to 2-mercaptoethanol molecular targets [598].

Currently, there exists a strong appeal for using molecules themselves as sensoring entities. This appeal derives from the ability to employ chemical synthesis for preparation of molecules that are designed to specifically interact with certain molecular species. Molecule–analyte interactions can be engineered to produce changes in readily monitored properties. These advantages give grounds to expect that sensor technology could be miniaturized down to the scale, where a single-molecule or MMM junction could act as an individual chemoselective sensor. A major challenge in building up molecule-based sensors originates from difficulties arising in the process of assembling and interconnecting molecules to create nanoscale devices in general and molecule-based sensors in particular. Also, there are problems with measuring responses at molecular level. Nevertheless, some molecule-based chemoselective sensing devices have been already demonstrated. For instance, exploring transport properties of a three-dimensional molecular network consisting of gold nanoparticles linked by oligophenylenevinulene (OPV) molecules, it was shown that the network conductance responded towards trinitro-toluene vapors [42].

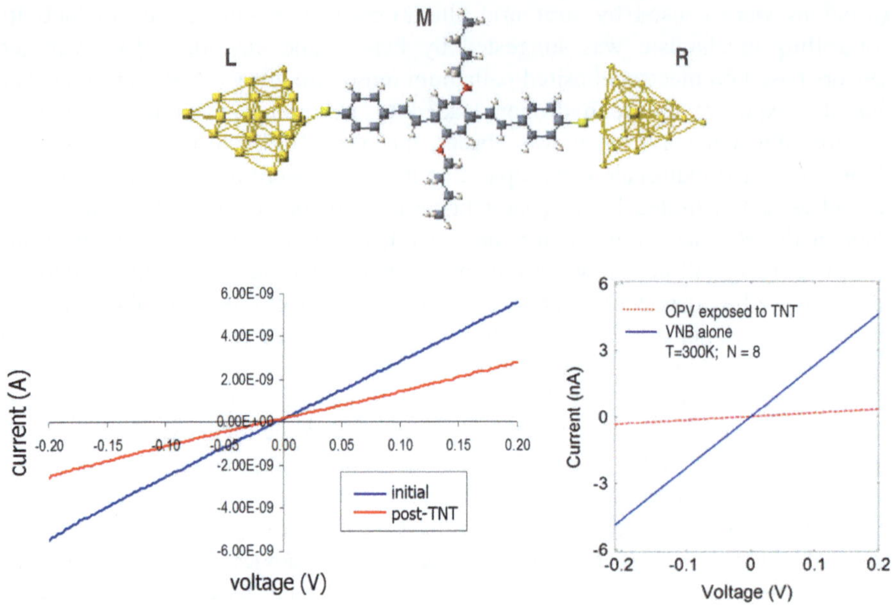

Fig. 6.18 *Left panel*: the nanoparticles network element and current–voltage characteristics for the network alone (*upper blue line*) and the network exposed to TNT vapors (*lower red line*) obtained in the experiments. *Right panel*: current–voltage characteristics calculated for a chain of nine gold nanoparticles linked by OPVs alone (*solid line*) and by OPV-TNT complexes (*dashed line*). The number of linkers N = 8 [42]. Reprinted with the permission of the AIP

The molecular network was built using a cowpea mosaic virus (CPMV) as a scaffold. The CPMV is a particle of an icosahedral geometry made out of sixty copies of a protein subunit whose average diameter is about 60 nm. To use the CPMV as a scaffold for the molecular network, it was genetically engineered in order to provide cysteine residues at selected positions. Gold nanoparticles with diameters ∼5 nm were attached to these residues. The CPMVs decorated with gold nanoparticles were exposed to OPV molecules which linked the gold particles producing a conductive network on the viruses. To measure the conductance of a network built on a single virus, the latter was assembled between proximal probe electrodes. Strong changes in the network conduction were observed when the latter was exposed to trinitrotoluene (TNT) vapors. The exposure resulted in a significant decrease in the conduction at rather small values of the bias voltage applied across the network (−0.2 to 0.2 V). The molecular recognition was carried out by the OPV linkers, whose conductance appeared to be significantly reduced when they had been loaded with TNT molecules (see Fig. 6.18). Very recently a similarly arranged nanosensor was suggested and demonstrated [605]. The device was based on the above-described molecular network mounted on a CPMV. However, additional recognition elements (biotin molecules) were attached to the virus in close proximity to the molecular linkers. Exposure to charged target molecules (avidin, NeutrAvidin, and streptavidin) resulted in binding these targets to the recognition elements triggering significant changes in the network conductivity. The conductivity increased in response to the presence of avidin molecular targets and decreased in the case of NeutrAvidin and/or streptavidin. It was suggested that these conductivity changes occurred due to the alterations in the HOMO energy induced by the nearby changes in the protein target molecules attached to the biotin recognition elements. This suggestion agrees with the previous results of Venkataraman and coworkers [192] which show that molecular conductance in aromatic molecular systems varies as the ionization potential changes.

As mentioned above, molecule-based sensors must undergo changes in monitored properties in response to analyte-sensor binding. Variations in the conduction through the system represent one kind of such response which is very important but by no means unique. There exists a variety of the molecular properties amenable for monitoring which could be interpreted in terms of the sensor-analyte response. Among these variable properties, photoluminescence is particularly appealing [606]. Luminescent molecular-based sensors could be easier manufactured than those basing on the molecular conductance for the difficulties accompanying the process of preparation of contacts between the molecule and the miniaturized electrodes are avoided. Also, these devices could offer high signal and spatial resolution, as well as inherently large signal-to-noise ratios and high degree of sensitivity [607]. All these properties provide a solid basis for design of molecule-based chemoselective sensors which respond to analytes by changes in photoluminescence. To successfully design molecular sensors with the analyte-dependent luminescence, it is necessary to understand the methods for manipulating the emission properties of luminescent molecular units (luminophores). Some important factors are known to be able to affect emission properties of luminophores either disturbing the energetics of excited states or quenching the latter. The excited states could be altered by bonding interactions between the luminophores and other molecules and ions or by changes in the electrostatic properties of the close environment. Resulting alterations in the emission characteristics could range from slight shifts in emission energies to switching between emissive and nonemissive states. For example, in the case of azaaromatic molecules, the lowest excited state is usually a nonemissive one. However, the contact with cations such as H^+ or metal ions leads to the formation of Lewis acid-base adducts changing the arrangement of molecular orbitals in such a manner that the lowest excited state becomes emissive. This type of luminescence switching has been used to fabricate several luminescence molecule-based sensors for pH and metal cations [607].

Specific characteristics of the deactivation of luminophores excited states could be used for chemosensory applications, as well. For this purpose, the most suitable approach is that assuming deactivation of the excited states occurring due to the interaction of the luminophore with other moieties, which possess the ability to quench the excited state via energy and/or electron transfer. Energy transfer-related quenching happens when the excitation energy is passed from the luminophore to other species being in the ground state. There exist many luminescent gas sensors for oxygen based on the quenching of luminophores with triplet excited states by means of the energy transfer [608]. Another quenching process involves electron transfer reactions. In this case, the excited state deactivation is provided by removal/addition of an electron from/to a singly occupied frontier orbital associated with the excited state. The photoinduced electron transfer leaves the luminophore in an oxidized or reduced state that is no longer emissive. In molecular sensors based on photoinduced electron transfer reactions, the luminophore and the receptor parts are separated by a spacer allowing for efficient electron transfer process.

The analyte-binding events modify relative energies of the donor and acceptor levels, thus bringing variations to the luminescence. For instance, the complexation of metal cations with macrocyclic receptors such as crown ethers can alter the redox potential of the receptor so that the latter loses its ability to transfer an electron to the luminophore. As a result, the sensor switches to a highly luminescent form indicating the presence of the target. In other cases, the analyte acts as an electron transfer quencher switching off the luminescence. Luminescent chemosensors built using the electron-transfer reactions to control the light emission are used to detect various redox-active analytes ranging from metal ions to organic species [606, 607, 609–612]. In some devices, the photoinduced-electron-transfer-based switching of the luminescence appears due to the dependence of the transfer on the distance separating the luminophore and the quencher. Within this approach, the relevant distance is changed when the target molecule is bound to the receptor site. These distance modulations can bring the target-induced-luminescence switching [607]. Miniaturized chemoselective devices are becoming increasingly important for numerous environmental, biomedical, and industrial applications. This explains the unceasing interest of the research community to these devices. In particular, the integration of nanotechnology with molecular recognition offers numerous possibilities which presently are being intensively explored.

6.6 Thermoelectric Devices

Thermoelectricity is known to enable direct conversion of heat energy to electricity and vice versa without use of mechanical components such as compressors and turbines. At present, there is a growing interest in understanding of thermoelectric transport in nanosystems including metal–molecular junctions. MMM junctions are considered to be promising candidates for energy harvesting [613]. The efficiency of a thermoelectric device is characterized by a dimensionless quantity ZT which is known as the figure of merit. If the figure of merit takes on greater values, this indicates an improvement of the thermoelectric device efficiency. The figure of merit is directly proportional to the electric conductivity of the considered material and inversely proportional to its thermal conductivity [see Eq. (4.73)]. Therefore, to increase this quantity, one needs to simultaneously enhance the former and reduce the latter. In bulk materials, this goal is intrinsically difficult to achieve. In these materials, a rise in the electrical conductivity is always accompanied by the increase of the charge carrier contribution to the thermal conductivity, as follows from the Wiedemann Franz law. So, in bulk materials ZT usually remains smaller than one.

It was suggested that the thermoelectric figure of merit in low-dimensional structures may accept values surpassing those typical for bulk materials. Quantum well superlattices appear to be very promising systems. In these systems, the thermal conductivity may be independently controlled through the increase of phonon scattering rates at the interfaces [614, 615]. Using this effect, rather high figure of merit ($ZT > 2$) has been reported for Be_2Te_3/Sb_2Te_3 superlattices [616]. Another way to enhance ZT is to turn to another kind of systems such as nanowires, quantum dots, and molecules. Experiments conducted on various nanowires and carbon nanotubes [337–344] suggest that in these systems it is possible to either increase the thermopower (by increasing the electron density of states) or reduce the thermal conductance (by reducing the length of a nanowire/nanotube so that it becomes smaller than the phonon mean free path still remaining larger than the corresponding mean free path for the electrons). Both ways lead to the enhancement of ZT. In some cases the thermopower increase and the decrease of thermoconductivity may be simultaneously achieved. For instance, such combined effect was recently observed in Si nanowires [617]. In quantum dots and molecules, the electron DOS is distributed in series of sharp peaks associated with discrete energy levels, and this characteristic property could be used to obtain high thermopower [359]. At the same time, small size of these systems could be used to design systems where the effect of phonons is reduced and, correspondingly, the thermal conductivity is low.

Advances in theoretical modeling of electrical and thermal conductances and thermopower of a single-molecule junction suggest a way to improve the figure of merit by increasing the power factor gS^2 in the expression for ZT while keeping the thermoconductivity low [345, 613]. Thermoelectric devices are supposed to operate at zero bias voltage. So, ignoring for simplicity the effect of phonons on electron transport through the system, one may use a simple expression for the electron conductance at the Fermi level:

$$g = \frac{e^2}{\pi\hbar} T(E_F). \tag{6.4}$$

Also, one may assume that the electron transmission function $T(E)$ is determined by the frontier molecular orbital levels E_{HOMO} and E_{LUMO} and by the coupling of the molecule to the left and right lead. Then the transmission could be approximated by the expression

$$T(E) = \sum_i \frac{\Gamma_L \Gamma_R}{(E - E_i)^2 + \Gamma^2/4} \tag{6.5}$$

where $\Gamma = \Gamma_L + \Gamma_R$ and E_i is supposed to take on values equal to E_{HOMO} and E_{LUMO}. The transmission reaches its maxima when either HOMO or LUMO energy level aligns with the electrochemical potential of the considered junction E_F. Omitting electron–phonon interactions, the expression for thermopower S has the form (4.71). However, a simpler approximation for thermopower could be used under far-from-resonance conditions, namely [617],

$$S = -\frac{\pi^2 k_B^2 T_0^2}{3|e|T(E)} \frac{\partial T(E)}{\partial E}\bigg|_{E=E_F}. \tag{6.6}$$

Here, T_0 is the temperature of the junction and k_B is the Boltzmann's constant. Lorentzian-shaped transmission curves peaked at the HOMO and LUMO energy levels indicate that the derivative $\partial T(E)/\partial E$ should reveal strong resonance features near the resonance states. This simple consideration qualitatively predicts that both conductance and thermopower could be maximized by bringing HOMO–LUMO level of the molecular bridge close to the Fermi energy of the electrodes. This should bring significant increase in the thermoelectric figure of merit characterizing the system. More thorough analysis carried out in [361] leads to the conclusion that large values of thermopower, and ZT are reached in MMM junctions when Fano resonances are moved close to the Fermi level regardless of the physical mechanism controlling this shift. Recent experiments have demonstrated that in MMM junctions the off-resonance thermopower behaves in accordance with the theoretical predictions [361, 613], thus confirming that MMM junctions possess potential usefulness as thermoelectric nanodevices.

In principle, thermoelectric transport through a MMM junction could be measured in a relatively simple way [15]. One can build a molecular bridge by approaching a metal tip to a conductive substrate coated with a self-assembled monolayer of target molecules until the conductance through the tip-substrate system reaches a set value indicating that the MMM junctions is formed. The temperature gradient ΔT should be applied by heating the substrate while keeping the tip at an ambient temperature. Then the thermoelectric voltage ΔV across the junction appears, and the thermoelectric current starts to flow. Usually, the thermoelectric voltage is measured by applying the bias voltage of the opposite polarity and of such a magnitude that it is able to completely cancel the thermoelectric current. This procedure could be repeated many times resulting in a set of histograms of the thermovoltage, each corresponding to a certain value of the temperature difference ΔT. A few of the histograms are shown in Fig. 6.19. Basing on these histograms, one may determine the most probable values of the thermovoltage (which is indicated by the position of the peaks) and then explore the dependence of ΔV of the temperature gradient ΔT. The results obtained in this way by Reddy and coworkers [15] are presented in Fig. 6.19 along with those depicting the effect of the molecule's length of the thermopower. The value of S could be extracted from the experimental results by plotting the most probable voltage against the temperature and by fitting the resulting curve with a linear fit. The slope of the obtained straight line yields the thermopower S.

Molecular length dependence of the thermopower was studied in several other works (see, e.g., [619]). It is well established that an increase in the molecule length L leads to exponential decay of the tunneling conductance g. In contrast,

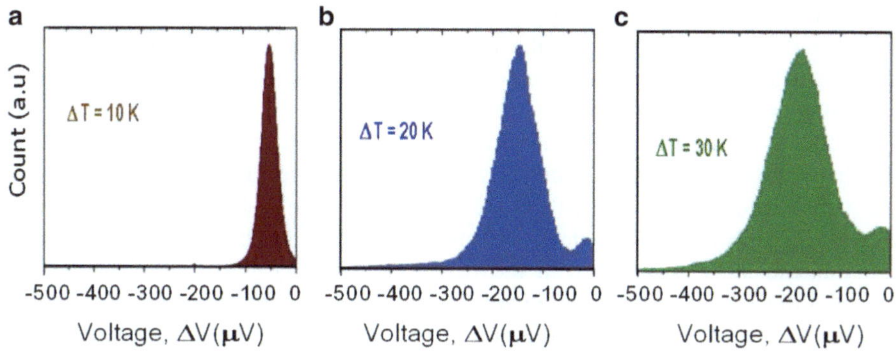

Fig. 6.19 Histograms obtained by analyzing about 1,000 consecutive thermoelectric voltage curves obtained in measurements of Au-benzenedithiol-Au junctions. The difference in the temperature between the tip and the substrate was 10 K (**a**), 20 K (**b**), and 30 K (**c**), respectively (after [16]). Reprinted with the permission of the American Association for the Advancement of Science (AAAS)

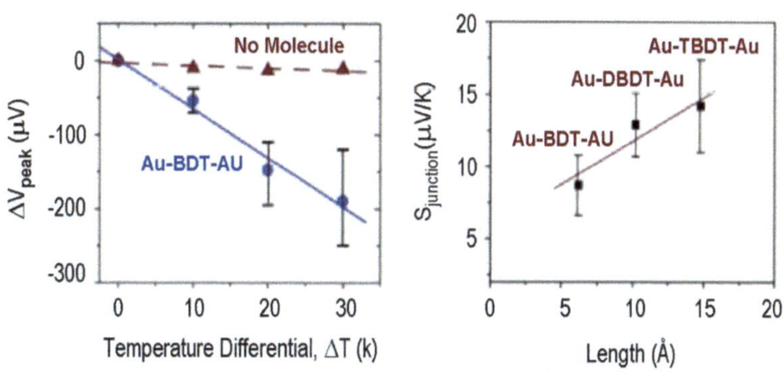

Fig. 6.20 Plot of the thermoelectric voltage ΔV versus temperature difference between the tip and the substrate ΔT for Au-benzenedithiol-Au junctions (*left*) and the dependence of thermopower of the length of a molecular bridge for diverse molecules linking gold leads (*right*; after [16]). Reprinted with the permission of the AAAS

the thermopower enhances as the length L linearly increases. This linear S–L dependence could be explained as originating from the shrinking of HOMO–LUMO gap with the increase of the molecule length that reduced the energy gap between the frontier molecule's orbital levels and the Fermi energy of the system [619]. Although, the thermopower increases as the molecular bridge lengthens, it is accompanied by the exponential decrease in the conductance. Therefore, the overall effect is the worsening rather than improvement of the power factor gS^2 and, consequently, of the figure of merit. So, the MMM junctions with long molecular linkers may not be good building blocks for manufacturing of molecular thermoelectric devices with high ZT (Fig. 6.20). As was demonstrated in the experiments, the thermoelectric voltage is not sensitive to the number of molecules bridging the electrodes [16]. This is in sharp contrast with the electrical conductance of molecular junctions which is known to enhance as the number of molecular linkers increases. As the voltage ΔV is measured and the temperature gradient ΔT is known, the thermopower may be computed in accordance with its definition (4.71).

Within a simple Landauer-based approach represented by Eqs. (6.5) and (6.6), the thermopower appears to be closely related to the electronic properties of the considered system. Specifically, in the case of metal–molecular junctions, S strongly depends of the electrochemical potentials of the leads (for unbiased junction the latter coincide with the Fermi energy E_{F}) with respect to the gap between the HOMO and LUMO energy levels. Basing on the expression (6.6) or another similar expression for the thermopower, it was suggested [338,618] that one can probe the positions of these levels by measuring and analyzing S. Combining experimental results with DFT-based electron structure computations and using the Landauer approach, it was concluded that the position of HOMO–LUMO gap can be thus determined. However, the applicability of the linear response theory in this analysis is questionable due to the fact that temperature differences required to reliably measure the thermovoltage can reach tens of Kelvins and the corresponding energy may appear to be not the smallest energy scale in the experiment. In particular, this energy may be of the same order as the coupling energy between the molecules and the substrate [15]. Also, nonequilibrium effects which are not taken into account within the linear response theory may come into play. Possible significance of the effects is indicated by extra peaks visible in the thermovoltage histograms at $\Delta V = 0$. These features are treated as manifestations of nonlinear effects. Specifically, they are attributed to variations in contact geometry, orbital hybridization, and intermolecular interactions in the considered systems [342].

Despite its shortcomings, the simple linear theory could be successfully applied to determine the charge-carrier type (electrons/holes) in a given MMM junctions. As follows from Eq. (6.6), the thermopower takes on nearly zero values when the E_{F} is situated in the middle of the HOMO–LUMO gap. When the Fermi energy is shifted towards the HOMO level, $\partial T/\partial E$ becomes negative which makes S positive. On the contrary, when the Fermi energy is shifted towards the LUMO, the thermopower takes on a negative value. Conversely, positive S indicates that E_{F} is closer to HOMO, so the charge carriers involved in the transport process should be holes, and negative S points at electrons as the charge carriers [618]. In the experiments on metal–molecule junctions including 1,4-benzenedithiol bridges connecting gold electrodes [15], the thermopower occurred to be positive, thus suggesting HOMO contributions to the conductance. This was in agreement with another results obtained by the gate modulation of the molecular orbitals. Recently, both positive and negative thermopower for MMM junctions with Au–NH_2 and Au-pyridine contacts were reported suggesting transport via HOMO and LUMO, respectively [620].

As follows from experimental and theoretical studies, a suitable way to bring the enhancement of the power factor is to adjust the Fermi level alignment with the transport channel (i.e., with HOMO–LUMO level on the molecular bridge). This may be achieved either by gate modulation of the molecular orbitals or by chemical tuning of the latter. The last method was employed by Baheti and coworkers [338].

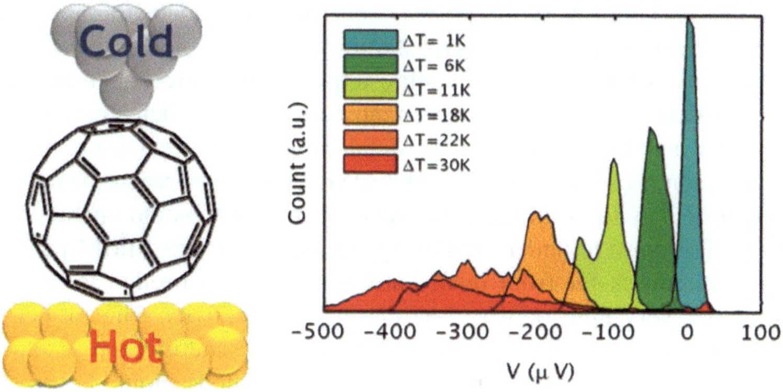

Fig. 6.21 Schematics of a junction including a C_{60} molecule as a linker between hot gold substrate and cold STM tip (*left*) and thermoelectric voltage histograms observed in this system for several temperature differences (*right*; after [621]). Reprinted with the permission of the ACS

They tuned the HOMO level of 1,4-benzenedithiols by dressing the molecules with various donor and/or acceptor substituents, thus varying the thermopower. In principle, this method could be used for simultaneous raising of thermopower and conductance of the MMM junctions. Recently, high thermopower was reported for n-type C_{60} fullerene molecules placed in between gold substrate and SMT tip, as illustrated in Fig. 6.21 [621]. Further improvement of the power factor of these systems by means of chemical doping strategy is expected. Also, DNA molecules show a large potential in thermoelectronics. Using a model Hamiltonian to describe an atomic chain which mimics a DNA molecule, it was shown that under certain conditions MMM junctions with DNA linkers may have quite high thermopower [622]. The latter may reach values of the order of hundreds $\mu\text{V}/\text{k}$ compared to a few $\mu\text{V}/\text{k}$ which is typical for the most of MMM junctions studied so far [623, 624].

Aiming at design of thermoelectric devices characterized by high values of the figure of merit, one must take care of heat transport through the considered system keeping thermal conductivity low. As mentioned before, there are grounds to expect κ to be reasonably low in metal–molecule junctions. Indirect estimations of single-molecule's thermal conductance of several kinds of molecules reported so far confirm these expectations. For example, thermal conductance of alkanethiols was estimated as being of about 50 pW/k [625] and for benzenedithiol it was reported to take on value close to 40 pW/k [626]. Available data on the thermoconductivity of molecules are still insufficient because experiments aiming at measuring and studying the heat flow at nanoscale are very difficult to carry on. There is no way to directly measure the energy flow. Only indirect estimations of thermal conductivity are possible. This increases the likelihood of erroneous estimations. Also, the scale of achievable thermal conductivities is generally much smaller than that of electric conductivities, which further complicates the matter. Furthermost investigations should be carried on to bring deeper understanding of heat transfer process in nanosystems (including MMM junctions) to properly evaluate their potentialities in designing nanoscale thermoelectric devices.

In the last decade, several proposals concerning thermoelectric applications of nanoscale systems were put forward. Along with heat-voltage convertors which were discussed before, other interesting applications were suggested including thermal transistors, thermal rectifiers, nanoscale radiation detectors, and thermal memory cells and logic gates [613]. Metal–molecule junctions may be used to manufacture some of these devices. However, in spite of recent advances, there still is the long way to go to reach the state of art which would enable to actually build them. The quest to understand energy transport through MMM junctions and other nanosystems is far from over. Further research is needed to find systems showing either very high or very low thermal conductivity. The former may be useful for nanoelectronic applications and the latter for thermoelectric applications at nanoscale. As for thermoelectricity, theoretical methods should be substantially advanced to properly elucidate the statistical nature of the relevant experiments and to account for electron–electron and electron–phonon interactions on equal footing. Also, to correctly describe the heat transfer in nanosystems, it is necessary to deal with the systems out of equilibrium, so dynamic effects should be included in consideration. Presently, however, both science and technology are rapidly developing fields, and there is no doubt that their advances will help to design and manufacture thermoelectric nanodevices including those based on metal–molecule junctions.

6.7 Molecules on Silicon Surfaces

An interesting application of molecules is their employment as passivates used to tie up surface states on silicon. This is significant because decreasing sizes of devices make them increasingly susceptible to scattering processes at their surfaces due to the effect of dangling bonds. Properly chosen organic molecules assembled at the oxide-channel interfaces of silicon-based electronic devices can create a soft boundary for electron scattering. This happens because the molecules covalently bond to the silicon surface remove the surface π–π^* states through bonding–antibonding splitting. Also, these molecules could control the threshold voltage of an underlying FET by their dipolar electrostatics.

Molecular attachment on the top of a low-doped silicon transistor channel with a backgate may bring significant alterations in the threshold voltage of this channel. More importantly, a correlation occurs between the shift of the threshold voltage and the electronegativity of the attached molecular monolayers headgroups [458]. This opens up a possibility for organic molecules to control electronic properties of the underlying silicon FET. Conversely, modulations of the FET characteristics could be used to detect the attached organic species. The overall threshold voltage shift is determined by two different mechanisms. The first of these is the charge-transfer doping driven by the difference in the work functions of the silicon and the molecular backbone. The second mechanism is the electrostatic gating driven by the dipolar

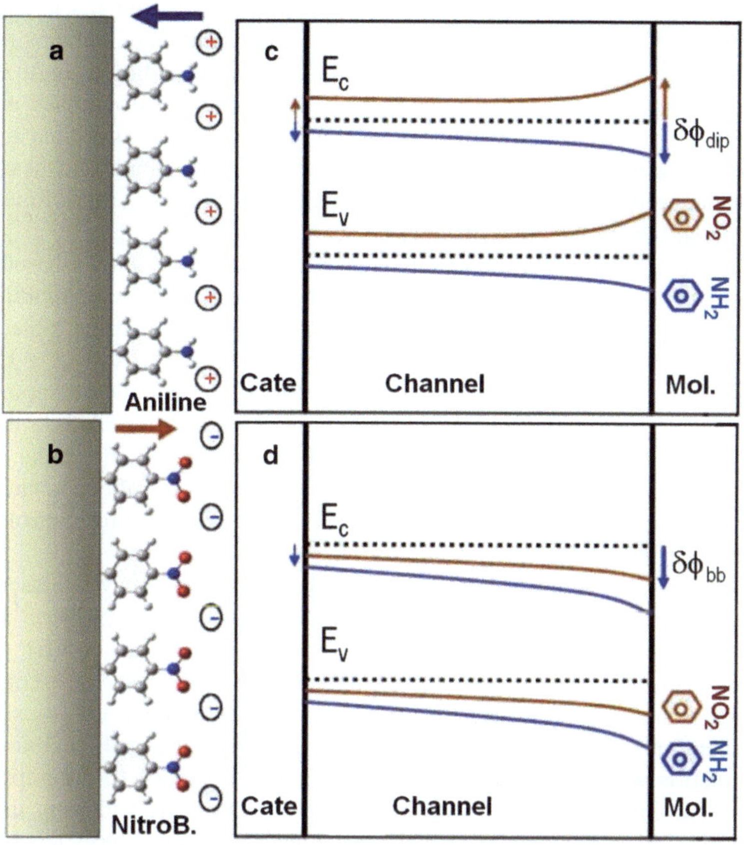

Fig. 6.22 Band bending contributions driven by surface adsorbed molecules. Dipolar headgroups from amine and nitro moieties give rise to opposite bendings of the band, while charge transfer between the molecules and the silicon surface causes band bending in the same direction. The contribution from the charge transfer predominates and creates an overall shift relative to the control in addition to the monotonic dependence of the dipolar headgroups [458]. Reprinted with the permission of the AIP

headgroups of molecules. This is illustrated in Fig. 6.22. The corresponding shifts in potentials at the silicon surface are described by the following expressions [458]:

$$\delta\Phi_{\rm ct} = \frac{E_{\rm F} - E_N}{1 + 1/U_0 D_0}, \tag{6.7}$$

$$\delta\Phi_{\rm dip} = \frac{n_{\rm mol}\mu_d \cos\theta}{\epsilon}. \tag{6.8}$$

Here, $n_{\rm mol}$ is the areal density of the molecules on the silicon surface, μ_d is the dipole moment per molecule, θ is the angle between the molecular dipole axis and the silicon surface, ϵ is the molecule dielectric constant, and E_N is the charge neutrality level of the molecule. The denominator in Eq. (6.7) includes the term describing the Coulomb cost of charging the molecule by adding a single electron at the neutrality level (U_0) and the average density of electron states near this level over the energy window $\delta\Phi_{\text{ct}}$. It follows from this expression that for small charging energies ($U_0 D_0 \ll 1$), the charge transfer caused shift vanishes. Within the opposite limit, $\delta\Phi_{\text{ct}}$ becomes close to the difference between work functions of the silicon and the molecule. The latter is identified with the E_N.

So, one way for a molecule to influence the channel of a silicon-based device is the electrostatic gating. Another and more efficient way is to modulate the conductivity of the channel. This may be achieved provided that the molecule has a sufficiently strong wave-function overlap with the channel, so that a series of molecule-specific traps appear in the bandgap. Stochastic population/depopulation of these levels may give rise to characteristic noise known as random telegraph signals (RTS). Analysis of the RTS may lead to detection of molecular species. In a scattering configuration, molecules can impose their specific "signatures" on the current of an underlying gate channel. This could be useful for metrology and electronic sensing.

Sequence of events leading to molecular characterization may be described as follows. Occurrence of a molecular trap in the bandgap of the silicon FET (to which the molecule is bonded) brings an RTS component into the FET current–voltage curves. The noise appears at certain values of the bias which correspond to the trap states. The recorded RTS could be Fourier transformed to extract the power spectrum. In turn, this power spectrum can yield information concerning a shape function which is unique for given number and position of the traps. Shape functions significantly differ for different kinds of molecules. Plotting the shape function versus gate voltage, one can directly read off the trap energy levels.

In many cases, RTS noise may be sufficiently explained as a result of multiple trapping events happening randomly and independently of each other. In these cases, the trapping events could be successfully described by Poissonian statistics. However, in some cases noticeable deviations from this statistics occur, which indicates that trapping events may be correlated. It was suggested that the deviations from the Poissonian distribution may be used to indirectly characterize correlation effects between the traps [627]. Besides these relatively weak correlations, the traps may strongly and directly interact. A direct effect occurs when one trap actually neutralizes the other through Coulomb blockade mechanism. Such "passivation" events are more likely to happen in low-dimensional systems where the wave-function overlaps are stronger and rapidly varying field lines may cause a certain trap to energetically overtake another one and neutralize its impact.

This effect was observed in long CNTs with a backgate. The current dependences of the gate voltage showed a region of zero current indicating a complete blockade of the channel. This blockade was attributed to an RTS sequence [628]. However, a slight change in the gate voltage was reported to bring a complete reversal of the blockade, a backward RTS sequence being characterized with entirely different time constants. These results may be explained by assuming the presence of two separate and distinct traps which can strongly interact with each other. While one trap blocks the channel, the other may neutralize it, thus restoring conduction. This hypothesis was confirmed by self-consistent calculations of a CNT density of states. Critical point of this theoretical analysis (carried out in [628]) appeared to be accounting for the self-imposed corrections (SIC). A trap does not experience a potential created by itself, so self-interactions should be excluded from mutual Coulomb costs. Carrying on the corresponding computations, good agreement between the theory and experimental results was demonstrated.

Typical current–voltage characteristics of molecules put in contact with semiconductors show rectification due to Schottky barriers appearing at the regions of contact. The barriers height and width are determined by voltage distributions over the molecules. So, under these conditions, molecules usually behave as tunnel barriers. However, when certain molecule's energy levels are situated near the Fermi level or band edge of the semiconductor, the molecule can actively participate in transport processes with these levels serving as transport channels. Application of an external voltage of an appropriate polarity may shift these levels into the bandgap of the semiconductor, thus cutting off the current. A prominent NDR kink should appear in the current–voltage characteristic indicating this effect. Under the described conditions, the molecule behaves as an analog of a resonance tunnel diode, except that the rectification occurs due to the slippage of molecular levels past the band edge of a passivated silicon or some other semiconductor. This gives grounds to conjecture that a tunable molecule-driven and integrated on silicon resonance tunnel diode may be built [5].

There are three critical conditions to be satisfied to actually build such a device. First, a molecule should have energy level proximal with either band edge of silicon. Secondly, a bias voltage should be highly localized to be able to substantially shift the molecular level, so that the NDR appears. Finally, the silicon should be heavily doped to allow sufficient electric charge transfer into the molecule. A significant characteristic of the molecular tunnel diode is that in such a device the NDR should switch polarity depending on the type of dopant (n vs. p) in the semiconductor. Also, one may expect the NDR to be more distinct in systems including a single molecule than in those including a SAM. The expected reduction of NDR intensity on SAMs originates from unavoidable variations of the NDR voltage associated with individual molecules.

Experimental discovering of exact locations of molecular orbitals of an adsorbed molecule is a complicated task, although regular and two-photon photoemission and transport spectra give some relevant information. The thing is that this information is indirect, and the indirectness has to do with an accurate interpretation of the measured levels. It is not easy to find out either these levels are associated with quasiparticle states or with excitations. The matter becomes even more complicated if one takes into account various effects of external factors such as image charges from contacts and vibrational side bands. Also, it is known that the NDR may appear due to several reasons; therefore, the very fact of its occurrence may not indicate the molecular tunnel diode action. Nevertheless, during the last decade some experimental results were obtained which could be interpreted as signatures of molecular tunnel diode action. For instance, Guisenger and coworkers observed NDR through individual organic molecules on silicon characterized by the dependence of the polarity of the dopant type [629].

Also, the NDR dilution on a molecular monolayer was observed [172] as well as some other effects predicted by theory. Surprisingly, in these experiments the NDR occurred at very low values of the bias voltage. This feature could be possibly explained by a drastic shrinking of the bandgap following the molecule attachment. A significant reduction of the bandgap which appear to be sensitive to the presence of adsorbed molecules was often observed in experiments [630]. It is known that threshold voltage of regular MOSFETs vary by several volts in the presence of a few mobile ions. The available experimental data may have different explanations, but they do not disagree with the concept of molecular-based tunnel diode. Further experimental and theoretical studies could show either such devices could be actually built.

Chapter 7
Conclusion

At present, the electron transport through molecule-scale systems is being intensively studied both theoretically and experimentally. Largely, unceasing efforts of the research community to further advance these studies are motivated by important application potentialities of single molecules as active elements of various nanodevices intended to compliment and/or to replace the silicon-based electronics. Elucidation of physics underlying electron transport through molecules is necessary for efficient designing and operating of molecule-based devices. The first measurement of practical molecular transport junctions are less than two decades old. In this short time the methods of preparation and characterization of metal–molecular junctions have demonstrated a very rapid development. The constant improvement of experimental capabilities brings new and better opportunities to measure and study transport properties of MMM junctions and other similar systems. Already good quality measurements of transport characteristics of metal–molecular junctions in the Kondo and Coulomb blockade regimes did appear, and various manifestations of electron–vibron interactions were observed. Also, interesting experimental results concerning electron transport through magnetic molecules have been reported. These experiments reveal interesting transport characteristics of molecular magnets which could be used in applications combining molecular electronics and spintronics.

Functionality and reliability of molecular devices crucially depend on their thermal properties. At present, there exists a significant interest in studies of phenomena related to the heat transfer through quantum dots and molecules. Besides its importance for practical applications, the exploration of thermal properties of molecule-scale systems has a notable fundamental interest. The quantization of the energy levels of the molecule/quantum dot bridging the electrodes in the junction and Coulomb interactions between the electrons on the bridge give rise to specific effects which cannot appear in bulk materials. Some of these effects such as the thermopower and thermal conductance oscillations accompanying variations of the gate voltage were observed in heat transfer through quantum dots, and there are solid grounds to expect similar oscillations to occur in the heat transfer through MMM junctions.

N.A. Zimbovskaya, *Transport Properties of Molecular Junctions*, Springer Tracts in Modern Physics 254, DOI 10.1007/978-1-4614-8011-2_7,

As the experimental techniques develop, transport experiments become more sophisticated and sometimes challenging. However, while good measurements continue to appear, the theoretical analysis of the reported experimental results remains incomplete so far. Mostly, difficulties in the development of the theory of electron transport through molecular junctions originate from the specific features of the appropriate general formalism (NEGF) which is extensively introduced into the studies of electron transport through molecules. Being applied to practical molecules, this approach leads to extreme computational difficulties, especially in the case when one needs to simultaneously take into account the effects of electron–electron and electron–phonon interactions.

Another commonly used approach based on the rate ("master") equations for the electron and/or phonon occupation probabilities for various states of the system of electrons and phonons associated with the electron transport through the junction provides a significantly less time-consuming computational method. However, it is justified only while transport properties of weakly coupled junctions are being considered. In the present book, we briefly describe both NEGF and "master" equation-based formalisms, and we discuss some applications of these approaches to specific problems. Also, we describe and discuss a phenomenological approach, where the transport through the system is treated as a multichannel scattering problem. This approach based on the Buttiker's model for the incoherent/inelastic quantum transport has an advantage over more sophisticated methods. It allows to comparatively easily analyze electron transport through MMM junctions including multilevel bridges. Besides, the phenomenological parameters inserted into the Buttiker's model could be expressed in terms of the relevant energies. This provides means to qualitatively analyze some important effects originating from electron–vibron interactions such as vibration-induced features in the inelastic electron tunneling spectra. These and other characteristics of the electron transport through multilevel molecular bridges could be theoretically analyzed within the NEGF formalism only with great difficulties.

In the present book, we focused on a general description of major problems currently existing in the research field covering the transport properties of molecules. Besides the overview of the commonly used theoretical approaches, we concentrated on the physics of the mechanisms controlling electron transport through molecules and the relevant effects manifested in the transport characteristics with brief remarks on their experimental observation. Due to the vastness of the field, we had to be very schematic in some areas restricting ourselves with rough outlines of the corresponding theory and modeling approaches. For example, we only briefly discussed such interesting phenomena as the Franck–Condon blockade, which could be manifested in the electron transport through molecules provided that there exists strong coupling between the electron and vibrational degrees of freedom. Also, we only slightly touched upon a vast research field of spin transport through molecules.

Furthermore, along with the electron–phonon and electron–electron interactions, the electron transport through molecules may be affected by the electron–photon interactions. Photo-assisted transport through molecular junctions was demonstrated in experiments and theoretically addressed using several techniques. It is known that optical pumping could give rise to the charge flow in an unbiased MMM junction and that light emission in biased current-carrying junctions could occur. At present, the photo-assisted transport through molecules is intensely studied. It is directly related to the practical problem of designing of quantum light-harvesting devices intended for the solar energy conversion. Here, we touch upon some aspects of the photo-assisted transport through molecules, but the major part of this research area is beyond the scope of the present book. However, we remark that the similar statistics of phonons and photons implies that the main equations and approaches used to describe the effects of electron–phonon interactions could be employed to analyze problems of photoexcitation and photoemission in molecular junctions.

First principles electronic structure calculations present an indispensable basis for the theory of electron transport through molecules. When the MMM junctions including relatively simple molecular linkers are considered, the necessary information could be obtained by using elaborate quantum chemistry computational approaches. However, in the cases of large molecules with complex geometries which now are often used to bridge the electrodes, these approaches appear to be too time consuming to be useful. In these cases, the density-functional theory which is characterized by a relatively low computational cost combined with a reasonable accuracy becomes a top choice approach to electron structure calculations. Accordingly, we briefly describe the main equations and a few applications of the density-functional theory.

Finally, practical molecular junctions are complex systems, and significant effort is necessary to bring electron transport calculations to the results which could be successfully compared with the experimental data. For this purpose, one needs to compute molecular states and voltage distribution over the considered system, to get sufficient information on the vibrational spectrum of the molecule and on the electron–phonon and electron–electron coupling strengths. Also, one needs a good quantitative computational scheme for transport calculations where all these data are accounted for. Currently, there remain some challenges which have not been properly addressed by the theory. Therefore, comparison between theoretical and experimental results on the molecular conductance sometimes does not bring satisfactory results. Nevertheless, there are solid grounds to believe that further effort applied by the research community will result in detailed understanding of all important aspects of the molecular conductance. Such understanding is paramount for furthermost development of molecular electronics.

in experiments and theoretically addressed using several techniques. It is known that optical pumping could give rise to the charge flow in an unbiased MMM junction, and that light emission in biased current-carrying junctions could occur. At present, the photo-assisted transport through molecules is intensely studied. It is directly relevant to the theoretical prediction of designing of quantum light-harvesting devices intended for the solar energy conversion. Here, we touch upon some aspects of the photo-assisted transport through molecules, but the major part of this research area is beyond the scope of the present book. However, we remark that the similar statistics of phonons and photons implies that the main equations and approaches used to describe the effects of electron–phonon interactions could be employed to analyze the effects of photoexcitation and photoemission in molecular junctions.

[illegible]

[illegible]

Appendix A
MATLAB Codes Used to Generate Text Figures

Figure 1.2 (left and middle panels:)

```
lower_limit_integral= −10;
upper_limit_integral=10;
function_name='trann1';
lower_parameter=0;
upper_parameter=2;
number_subdivisions=101;
v=linspace(lower_parameter,upper_parameter,number_subdivisions);
len3=zeros(1,number_subdivisions);
for k=1:number_subdivisions
len3(k)=quad(function_name,
    lower_limit_integral,upper_limit_integral,[],[],v(k));
end subplot(1,2,1)
h=plot(v,len3);
set(h,'LineWidth',1.2,'LineStyle','-')
set(h,'Color','b')
xlabel('Voltage (V)','FontSize',14)
ylabel('Current (μA)','FontSize',14)
axis([0 2 0 1.24])
set(gca,'xtick',[0 1 2],'FontSize',13)
set(gca,'ytick',[0 0.3 .6 .9 1.2 ],'FontSize',13)
axis square
lower_limit_integral= −10;
upper_limit_integral=10;
function_name='rat2a';
lower_parameter=0;
upper_parameter=2;
number_subdivisions=101;
v=linspace(lower_parameter,upper_parameter,number_subdivisions);
len3=zeros(1,number_subdivisions);
```

N.A. Zimbovskaya, *Transport Properties of Molecular Junctions*, Springer Tracts in Modern Physics 254, DOI 10.1007/978-1-4614-8011-2,

```
for k=1:number_subdivisions
len3(k)=quad(function_name,lower_limit_integral,
   upper_limit_integral,[],[],v(k));
end
subplot(1,2,2)
h=plot(v,len3);
set(h,'LineWidth',1.2,'LineStyle','-')
set(h,'Color','b')
xlabel('Voltage (V)','FontSize',14)
ylabel('Current (μ A)','FontSize',14)
axis([0 2 0 1.24])
set(gca,'xtick',[0 1 2],'FontSize',13)
set(gca,'ytick',[0 0.3 .6 .9 1.2 ],'FontSize',13)
axis square
```

Supplementary files:

trann1

```
function f=trann1(t,v);
kT=0.0005;
fEl=1./(1+exp((t-0.5*v)./kT));
fEr=1./(1+exp((t+0.5*v)./kT));
y1=0.0004./((t+0.4).^2+0.0004);
y2=0.026*(0.026+0.7*sqrt((t+0.4).^2+0.185))./(1.3*sqrt((t+0.4).^2
   +0.185)+0.14).^2;
y3=0.034*(0.034+0.3*sqrt((t+0.4).^2+0.0006))./(1.7*sqrt((t+0.4).^2
   +0.0006)+0.06).^2;
y4=0.038*(0.038+0.1*sqrt((t+0.5).^2+0.0004))./(1.9*sqrt((t+0.5).^2
   +0.0004)+0.02).^2;
f=25.6*(fEl-fEr).*y4;
```

```
y1=sqrt(t.^2-4*(b.^2));
y2=(t+y1).^(N-1).*(t+y1+i*g).^2-(t-y1).^(N-1).*(t-y1+i*g).^2;
G=4*(2*b).^(N-1).*(y1)./(y2);
S=(g).^2.*G.*conj(G);
y3=(S).^(0.5);
T=(2-e).*y3.*(1+y3-(1-e).*(1-y3))./((1+y3+(1-e).*(1-y3)).^2);
fEl=1./(1+exp((t-0.5*v)./kT));
fEr=1./(1+exp((t+0.5*v)./kT));
f=7.8*(fEl-fEr).*T;
```

(right panel)

max2

```
lower_limit_integral= -10;
upper_limit_integral=10;
```

```
function_name='max2';
lower_parameter=0;
upper_parameter=10;
number_subdivisions=101;
v=linspace(lower_parameter,upper_parameter,number_subdivisions);
len3=zeros(1,number_subdivisions);
for k=1:number_subdivisions
len3(k)=quad(function_name,lower_limit_integral,upper_limit_integral,
   [],[],v(k));
end
h=plot(v,len3);
set(h,'LineWidth',2,'LineStyle','-')
set(h,'Color','b')
axis([0 10 0 2.5])
set(gca,'xtick',[0 2.5 5 7.5 10],'FontSize',14)
set(gca,'ytick',[0.5 1 1.5 2 2.5],'FontSize',14)
xlabel('Voltage (V)','FontSize',16)
ylabel('Current (μ A)','FontSize',16)
axis square
```

Supplementary file:

rat2a

```
function f=max2(E,v)
zplus=i*0.01;
kT=0.0026;
fEl=1./(1+exp((E-0.5*v)./kT));
fEr=1./(1+exp((E+0.5*v)./kT));
s3l=1.74*1.74./(E+4.34+zplus);
s4l=0.51*0.51./(E+4.34+zplus);
s3r=0.66*0.66./(E+3.73+zplus);
s4r=2.44*2.44./(E+3.73+zplus);
g3=1./(E+4.71-s3l-s3r);
g4=1./(E+4.48-s4l-s4r);
T=imag(s3l).*imag(s3r).*(g3).*conj(g3)+imag(s4l).*imag(s4r) .*(g4).
   *conj(g4);
f=256*(fEl-fEr).*T;
```

max2

Figure 1.5

```
t= 0 : 0.001 : 1;
gl=0.02
```

```
;gr=0.02;
U=0.5;
e0= −0.2;
a=gr./(gl+gr);
kT=0.0066;
fl=1./(1+exp((e0−a.*t)./kT));
fll=1./(1+exp((e0+U−a.*t)./kT));
fr=1./(1+exp((e0+(1−a).*t)./kT));
frr=1./(1+exp((e0+U+(1−a).*t)./kT));
y1=gl*((1+fl−fll).*(gl.*(1−fll)+gr.*(1−frr))−(1−fll). *(gl.*(1+fl−fll)+gr.
    *(1+fr−frr)));
y2=gl.*(1+fl−fll)+gr.*(1+fr−frr);
y3=450.*y1./y2;
fl1=1./(1+exp((e0−a.*t)./kT));
fll1=1./(1+exp((e0−a.*t)./kT));
fr1=1./(1+exp((e0+(1−a).*t)./kT));
frr1=1./(1+exp((e0+(1−a).*t)./kT));
y4=gl*((1+fl1−fll1).*(gl.*(1−fll1)+gr.*(1−frr1))−(1−fll1).*(gl.*(1+fl1−fll1)
    +gr.*(1+fr1−frr1)));
y5=gl.*(1+fl1−fll1)+gr.*(1+fr1−frr1);
y6=450.*y4./y5;
subplot (1,2,1)
h=plot(t,y3,t,y6)
axis([0.2 0.8 0 6])
set(h,'LineWidth',1.5,'LineStyle','-';'–')
set(h,'Color','b';'r')
set(gca,'xtick',[ 0.2 0.4 0.6 0.8 ],'FontSize',14)
set(gca,'ytick',[0 2 4 6 ],'FontSize',14)
xlabel('Voltage (V)','FontSize',16)
ylabel('Current (μA)','FontSize',16)
axis square
t=0:0.001:4;
gl=0.02;gr=0.01;grr=0.005;
U=0.5;e0= −0.2; a=gr./(gl+gr);a1=grr./(gl+grr)
kT=0.0066;
fl=1./(1+exp((e0−a.*t)./kT));
fll=1./(1+exp((e0+U−a.*t)./kT));
fr=1./(1+exp((e0+(1−a).*t)./kT));
frr=1./(1+exp((e0+U+(1−a).*t)./kT));
y1=gl*((1+fl−fll).*(gl.*(1−fll)+gr.*(1−frr))−(1−fll).*(gl.*(1+fl−fll)+gr.
    (1+fr−frr)));
y2=gl.*(1+fl−fll)+gr.*(1+fr−frr);
y3=450.*y1./y2;
fl1=1./(1+exp((e0−a1.*t)./kT));
fll1=1./(1+exp((e0+U−a1.*t)./kT));
```

```
fr1=1./(1+exp((e0+(1−a1).*t)./kT));
frr1=1./(1+exp((e0+U+(1−a1).*t)./kT));
y4=gl*((1+fl1−fll1).*(gl.*(1−fll1)+grr.*(1−frr1))−(1−fll1).
   (gl.*(1+fl1−fll1) +grr.*(1+fr1−frr1)));
y5=gl.*(1+fl1−fll1)+grr.*(1+fr1−frr1);
y3=450.*y1./y2;
y6=450.*y4./y5;
subplot (1,2,2)
h=plot(t,y3,t,y6)
axis([0 2 0 4])
set(h,'LineWidth',1.5,'LineStyle','-';'–')
set(h,'Color','b';'r')
set(gca,'xtick',[ 0 0.5 1 1.5 2 ],'FontSize',14)
set(gca,'ytick',[0 1 2 3 4 ],'FontSize',14)
xlabel('Voltage (V)','FontSize',16)
ylabel('Current (μA)','FontSize',16)
axis square
```

Figure 2.6

```
subplot(1,2,2)
t= −2:0.0001:2;
b=0.6;
g=0.2;
N=5;
e=0;
e1=0.5;
e2=1;
y1=sqrt(t.^2 − 4 * (b.^2));
y2=(t+y1).^(N−1).*(t+y1+i*g).^2−(t−y1).^(N−1) .*(t−y1+i*g).^2;
G=4*(2*b).^(N−1).*(y1)./(y2);
S=(g).^2.*G.*conj(G);
y3=(S).^(0.5);
T=(2−e).*y3.*(1+y3−(1−e).*(1−y3))./((1+y3+(1−e).*(1−y3)).^2);
T1=(2−e1).*y3.*(1+y3−(1−e1).*(1−y3))./((1+y3+(1−e1).*(1−y3)).^2);
T2=(2−e2).*y3.*(1+y3−(1−e2).*(1−y3))./((1+y3+(1−e2).*(1−y3)).^2);
h=plot(t,T,t,T1,t,T2);
set(h,'LineWidth',1.5,'LineStyle','-.';'–';'-')
set(h,'Color','r';'b';'k')
axis([−1.5 1.5 0 1.15])
set(gca,'xtick',[ −1.5 −1 −0.5 0 0.5 1 1.5],'FontSize',14)
set(gca,'ytick',[ 0 0.2 0.4 0.6 0.8 1 ],'FontSize',14)
xlabel('E(eV)','FontSize',16)
ylabel('Transmission','FontSize',16)
```

```
axis square
subplot(1,2,1)
t= −1:.01:0;
y1=0.0004./((t+0.4).^2+0.0004);
y2=0.026*(0.026+0.7*sqrt((t+0.4).^2+0.185))./(1.3*sqrt((t+0.4).^2+0.185)
   +0.14).^2;
y3=0.034*(0.034+0.3*sqrt((t+0.4).^2+0.0006))./(1.7*sqrt((t+0.4).^2+0.0006)
   +0.06).^2;
y4=0.038*(0.038+0.1*sqrt((t+0.4).^2+0.0004))./(1.9*sqrt((t+0.4).^2+0.0004)
   +0.02).^2;
h=plot(t,y1,t,y2,t,y3,t,y4);
set(h,'LineWidth',1.5,'LineStyle','-';'-';'-';'-')
set(h,'Color','r';'k';'m';'b')
axis([−.8,0,0,1.15])
set(gca,'xtick',[−1.0 −.8 −.6 −.4 −.2 0],'FontSize',14)
set(gca,'ytick',[0 .2 .4 .6 .8 1.0],'FontSize',14);
xlabel('E(eV)','FontSize',16)
ylabel('Transmission','FontSize',16)
axis square
```

Figure 3.8 (left panel)

```
t= −2:0.0001:2;
c=0.5;
b=0.51;
a=0;
y=c.^2.*(t-1).^2./(((t-1).*(t-a)-b.^2).^2+c.^2.*(t-1).^2);
z=0.1*(10).^y;
subplot(1,2,1)
h=plot(t,y);
axis([−1.2 1.8 0 1.2])
set(h,'LineWidth',2,'LineStyle','-')
set(h,'Color','b')
set(gca,'xtick',[−1 −0.5 0 0.5 1 1.5],'FontSize',14)
set(gca,'ytick',[0 0.3 0.6 0.9 1.2],'FontSize',14)
xlabel('E_F/E_s','FontSize',16)
ylabel('g(E_F)/G_0','FontSize',16)
axis square;
```

Figure 3.8 (right panel)

```
lower_limit_integral= −10;
upper_limit_integral=10;
```

```
function_name='fano2';
lower_parameter=0;
upper_parameter=4;
number_subdivisions=121; v=linspace(lower_parameter,upper_parameter,
    number_subdivisions);
len3=zeros(1,number_subdivisions);
for k=1:number_subdivisions len3(k)=quad(function_name,lower_limit_integral,
   upper_limit_integral,[],[],v(k));
end
subplot(1,2,2)
h=plot(v,len3);
set(h,'LineWidth',2,'LineStyle','-')
set(h,'Color','b')
axis([0 4 0 0.8])
set(gca,'xtick',[0 1 2 3 4],'FontSize',14)
set(gca,'ytick',[0 0.2 0.4 0.6 0.8],'FontSize',14)
xlabel('Voltage (V)','FontSize',16)
ylabel('Current (μA)','FontSize',16)
axis square
```

Supplementary files:

```
function f=fano2(t,v)
c=0.1;
b=0.5;
a=0;
kT=0.0026;
fEl=1./(1+exp((t-0.5*v)./kT));
fEr=1./(1+exp((t+0.5*v)./kT));
y=c.^2.*(t-1).^2./(((t-1).*(t-a)-b.^2).^2+c.^2.*(t-1).^2);
f=2.40*y.*(fEl-fEr);
```

Figure 3.10

```
lower_limit_integral= -50;
upper_limit_integral=50;
function_name='ferro1';
lower_parameter= -0.8;
upper_parameter=0.8;
number_subdivisions=201;
v=linspace(lower_parameter,upper_parameter,number_subdivisions);
len1=zeros(1,number_subdivisions);
for k=1:number_subdivisions len1(k)=quad(function_name,lower_limit_
    integral, upper_limit_integral,[],[],v(k));
end
```

```
lower_limit_integral= −50;
upper_limit_integral=50;
function_name='ferro3';
lower_parameter = −0.8;
upper_parameter=0.8;
numbersubdivisions=201; v=linspace(lowerparameter,upperparameter,
    number_subdivisions);
len3=zeros(1,number_subdivisions);
for k=1:number_subdivisions len3(k)=quad(function_name,
    lower_limit_integral,upper_limit_integral,[],[],v(k));
end
lower_limit_integral= −50;
upper_limit_integral=50;
function_name='ferro6';
lower_parameter= −0.8;
upper_parameter=0.8;
numbersubdivisions=201; v=linspace(lowerparameter,upperparameter,
    numbersubdivisions);
len2=zeros(1,number_subdivisions);
for k=1:number_subdivisions len2(k)=quad(function_name,lower
    _limit_integral,upper_limit_integral,[],[],v(k));
end
lower_limit_integral= −50;
upper_limit_integral=50;
function_name='ferro5';
lower_parameter= −0.8;
upper_parameter=0.8;
number_subdivisions=201; v=linspace(lower_parameter,upper_parameter,
    number_subdivisions);
len4=zeros(1,number_subdivisions);
for k=1:number_subdivisions len4(k)=quad(function_name,
    lower_limit_integral,upper_limit_integral,[],[],v(k));
end
subplot(1,2,1);
h=plot(v,len3,v,len1);
set(h,'LineWidth',2,'LineStyle','–';'-')
set(h,'Color','b';'r')
axis([−0.8 0.8 −5000000 1000000])
set(gca,'xtick',[−0.8 −0.4 0 0.4 0.8],'FontSize',16)
set(gca,'ytick',[−5000000 −3000000 −1000000 1000000],'FontSize',16)
xlabel('Voltage (V)','FontSize',16)
ylabel('Current (A/cm^2)','FontSize',16)
axis square
subplot(1,2,2);
h1=plot(v,len2,v,len4);
```

```
set(h1,'LineWidth',2,'LineStyle','–';'-')
set(h1,'Color','b';'r')
axis([−0.8 0.8 −1000000 5000000])
set(gca,'xtick',[−0.8 −0.4 0 0.4 0.8],'FontSize',18)
set(gca,'ytick',[−1000000 1000000 3000000 5000000],'FontSize',16)
xlabel('Voltage (V)','FontSize',16)
ylabel('Current (A/cm^2)','FontSize',16)
axis square
```

Supplementary files:

ferro1

```
function f=ferro1(E,v);
gl=0.75;
gr=0.25;
kT=0.0025;
U=0.8;
fEl=1./(1+exp((E−gr.*v./(gl+gr))./kT));
fEr=1./(1+exp((E+gl.*v./(gl+gr))./kT));
y1=exp(−7.524*sqrt(U−E));
y3=7.524*sqrt(U−E);
f=0.479*(10^8).*y1.*(y3+1).*(fEl−fEr);
```

ferro2

```
t= −0.000004:0.0000001:0.000004;
v=0.07+10^(−7)*t;
plot(v,t)
```

ferro3

```
function f=ferro3(E,v);
gl=0.75;
gr=0.25;
kT=0.0025;
U=0.6;
fEl=1./(1+exp((E−gr.*v./(gl+gr))./kT));
fEr=1./(1+exp((E+gl.*v./(gl+gr))./kT));
y1=exp(−7.524*sqrt(U−E));
y3=7.524*sqrt(U−E);
f=0.479*(10^8).*y1.*(y3+1).*(fEl−fEr);
```

ferro4

```
function f=ferro4(E,v);
gl=0.5;
gr=0.5;
kT=0.0025;
U=0.5;
```

```
fEl=1./(1+exp((E−gr.*v./(gl+gr))./kT));
fEr=1./(1+exp((E+gl.*v./(gl+gr))./kT));
y1=exp(−7.524*sqrt(U−E));
y3=7.524*sqrt(U−E);
f=1.5*1.28*(10^8).*y1.*(y3+1).*(fEl-fEr);
```

ferro5

```
function f=ferro5(E,v);
gl=0.25;
gr=0.75;
kT=0.0025;
U=0.8;
fEl=1./(1+exp((E−gr.*v./(gl+gr))./kT));
fEr=1./(1+exp((E+gl.*v./(gl+gr))./kT));
y1=exp(−7.524*sqrt(U−E));
y3=7.524*sqrt(U−E);
f=0.479*(10^8).*y1.*(y3+1).*(fEl−fEr);
```

ferro6

```
function f=ferro5(E,v);
gl=0.25;
gr=0.75;
kT=0.0025;
U=0.8;
fEl=1./(1+exp((E−gr.*v./(gl+gr))./kT));
fEr=1./(1+exp((E+gl.*v./(gl+gr))./kT));
y1=exp(−7.524*sqrt(U−E));
y3=7.524*sqrt(U−E);
f=0.479*(10^8).*y1.*(y3+1).*(fEl-fEr);
```

Figure 3.15

```
zplus=i*0.05;
kT=0.00026;
tau=0.04;
U=0.5;
v=0;
a=0;
ep0= −0.2;
ep1= −0.075;
ep2=0.05;
ep3=0.175;
ep4=0.3;
```

```
s0= −i*0.005;
s1= −i*0.005;
NE=5001;
E=linspace(−1,1,NE);
dE=E(2)−E(1);
fE=1./(1+exp((E−a)./kT));
dfe=exp((E−a)./(kT))./((kT)*(1+exp((E−a)./(kT))).^2);
s2= −0.005*i*fE;
s3= −0.005*i*(1−1./(1+exp((E−a−0.1)./kT)));
g1=(E−a−ep0−U−s0−s0−s1+0.455*U)./((E−a−ep0−s0).
    (E−a−ep0−U−s0−s0−s1) +U*(s2+s3));
g2=(E−a−ep1−U−s0−s0−s1+0.455*U)./((E−a−ep1−s0).
    (E−a−ep1−U−s0−s0−s1) +U*(s2+s3));
g3=(E−a−ep2−U−s0−s0−s1+0.455*U)./((E−a−ep2−s0).
    (E−a−ep2−U−s0−s0−s1) +U*(s2+s3));
g4=(E−a−ep3−U−s0−s0−s1+0.455*U)./((E−a−ep3−s0).
    (E−a−ep3−U−s0−s0−s1) +U*(s2+s3));
g5=(E−a−ep4−U−s0−s0−s1+0.455*U)./((E−a−ep4−s0).
    (E−a−ep4−U−s0−s0−s1) +U*(s2+s3));
A1=i*(g1−conj(g1))./(2*pi);
A2=i*(g1+g2+g3+g4+g5−conj(g1+g2+g3+g4+g5))./(2*pi);
f1=0.01*A1;
f2=0.01*A2;
subplot(1,2,1)
h=plot(E,f1)
set(h,'LineWidth',1.5,'LineStyle','-')
set(h,'Color','b')
axis([−0.4 0.6 0 0.3])
set(gca,'xtick',[−0.4 −.2 0 .2 0.4 0.6],'FontSize',14)
set(gca,'ytick',[0 0.1 0.2 0.3 0.4],'FontSize',14)
ylabel('Conductance (e^2/h)','FontSize',16);
xlabel('Chemical potential (eV)','Fontsize',16)
axis square;
subplot(1,2,2)
h1=plot(E,f2)
set(h1,'LineWidth',1.5,'LineStyle','-')
set(h1,'Color','b')
axis([−0.4 0.6 0 0.4])
set(gca,'xtick',[−0.4 −.2 0 .2 0.4 0.6],'FontSize',14)
set(gca,'ytick',[0 0.1 0.2 0.3 0.4],'FontSize',14)
ylabel('Conductance (e^2/h)','FontSize',16);
xlabel('Chemical potential (eV)','Fontsize',16)
axis square;
```

Figure 4.2

```
lower_limit_integral= −2;
upper_limit_integral=2;
function_name='phontran6';
lower_parameter=0;
upper_parameter=3.5;
number_subdivisions=711;
t= −1:0.005:0.5;
e=(0.3/0.44).^2; b=0.02;c=0.0002;d=0.22;
x=2*c*(exp(−e)./((t+0.5).^2+b^2)+e*exp(−e)./((t+0.5−d).^(2)+b^2)+0.5*e*e*
   exp(−e)./((t+0.5−2*d).^2+b^2)+e*e*e*exp(−e)./(6*((t+0.5−3*d).^2+b^2))
   +e.^4exp(−e)./(24*((t+0.5−4*d).^2+b^2))+e.^5.*exp(−e)./
   (120*((t+0.5−5*d).^2+b^2)));
y=2*c./((t+0.3).^2+b^2);
subplot(1,2,1)
h=plot(t,x,t,y);
set(h,'LineWidth',2,'LineStyle','-';'–')
set(h,'Color','b';'r')
axis([−0.8 0.4 0 1.05])
set(gca,'xtick',[−0.8 −0.4 0 0.4],'FontSize',14)
set(gca,'ytick',[0 0.2 0.4 0.6 0.8 1],'FontSize',14)
xlabel('E (eV)','FontSize',16)
ylabel('Transmission','FontSize',16)
axis square
subplot(1,2,2);
a=linspace(lower_parameter,upper_parameter,number_subdivisions);
len=zeros(1,number_subdivisions);
for k=3:number_subdivisions
len(k)=(quad(function_name,lower_limit_integral,upper_limit_integral,[],[],
   a(k))−quad(function_name,lower_limit_integral,upper_limit_integral,
   [],[],a(k−1)))./(a(k)−a(k−1));
end
lower_limit_integral= −2;
upper_limit_integral=2;
function_name='phontran5';
lower_parameter=0;
upper_parameter=3.5;
number_subdivisions=711;
a=linspace(lower_parameter,upper_parameter,number_subdivisions);
len1=zeros(1,number_subdivisions);
for k=3:number_subdivisions
len1(k)=(quad(function_name,lower_limit_integral,upper_limit_integral,[],[],
   a(k)) −quad(function_name,lower_limit_integral,upper_limit_integral,[],[],
   a(k−1)))./(a(k)−a(k−1));
```

```
end
h=plot(a,len,a,len1);
set(h,'LineWidth',2,'LineStyle','-';'–')
set(h,'Color','b';'r')
axis square
axis([0 2 0 6])
set(gca,'xtick',[0 1 2],'FontSize',14)
set(gca,'ytick',[0 2 4 6 8],'FontSize',14)
xlabel('Voltage (V)','FontSize',16)
ylabel('dI/dV (μ A/V)','FontSize',16)
```

Supplementary files:

```
function f=phontran5(E,v);
kT=0.025;
fEl=1./(1+exp((E−0.5*v)./kT));
fEr=1./(1+exp((E+0.5*v)./kT));
a=(0.3/0.44).^2; b=0.02; c=0.0002; d=0.22;
x=2*c*(exp(−a)./((E−0.3).^2+b^2)+a*exp(−a)./((E−0.3−d).^(2) +b^2)+0.5*a*a*
   exp(−a)./((E−0.3−2*d).^2+b^2)+a*a*a* exp(−a)./(6*((E−0.3−3*d).
   ^2+b^2))+a.^4.*exp(−a)./(24*((E−0.3−4*d).^2+b^2))+a.^5.*
   exp(−a)./(120*((E−0.3−5*d).^2+b^2)));
y=2*c./((E−0.5).^2+b^2);
f=25.6*(fEl−fEr).*y;
```

```
function f=phontran6(E,v);
kT=0.00075;
fEl=1./(1+exp((E−0.5*v)./kT));
fEr=1./(1+exp((E+0.5*v)./kT));
a=(0.3/0.44).^2; b=0.01; c=0.00002; d=0.22;
x=2*c*(exp(−a)./((E−0.3).^2 +b^2)+a*exp(−a)./((E−0.3−d).^(2) + b^2)
   +0.5*a*a*exp(−a)./((E−0.3−2*d).^2+b^2)+a*a*a*exp(−a)./
   (6*((E−0.3−3*d).^2+b^2))+a.^4.* exp(−a)./(24*((E−0.3−4*d).^2+b^2))+a.^5.
   *exp(−a)./(120*((E−0.3−5*d).^2+b^2)));
y=2*c./((E−0.5).^2+b^2);
f=25.6*(fEl−fEr).*x;
```

Figure 4.3

```
lower_limit_integral= −2;
upper_limit_integral=2;
function_name='phontran6';
lower_parameter=0;
upper_parameter=2.5;
number_subdivisions=151;
a=linspace(lower_parameter,upper_parameter,number_subdivisions);
```

```
len=zeros(1,number_subdivisions);
for k=3:number_subdivisions
len(k)=((quad(function_name,lower_limit_integral,upper_limit_integral,[],[],
   a(k))−quad(function_name,lower_limit_integral,upper_limit_integral,[],[],
   a(k−1)))./(a(k)−a(k−1))−(quad(function_name,lower_limit_integral,
   upper_limit_integral,[],[], a(k−1))−quad(function_name,
   lower_limit_integral,upper_limit_integral,[],[],
   a(k−2)))./(a(k−1)−a(k−2)))./(a(k)−a(k−1));
end
subplot(1,2,1)
h=plot(a,len);
set(h,'LineWidth',1.5,'LineStyle','-')
set(h,'Color','b')
axis square
axis([0 2 −100 100])
set(gca,'xtick',[0 .5 1 1.5 2],'FontSize',14)
set(gca,'ytick',[−100 −50 0 50 100],'FontSize',14)
ylabel('d^2I/dV^2 (μA/V^2)','FontSize',16)
xlabel('Voltage(V)','FontSize',16)
```

Supplementary file:

```
function f=phontran6(E,v);
kT=0.00075;
fEl=1./(1+exp((E−0.5*v)./kT));
fEr=1./(1+exp((E+0.5*v)./kT));
a=(0.3/0.44).^2; b=0.01; c=0.00002; d=0.22;
x=2*c*(exp(−a)./((E−0.3).^2 +b^2)+a*exp(−a)./((E−0.3−d).^(2) + b^2)
   +0.5*a*a* exp(−a)./((E−0.3−2*d).^2+b^2)+a*a*a*exp(−a)./
   (6*((E−0.3−3*d).^2+b^2))+a.^4.* exp(−a)./(24*((E−0.3−4*d).^2+b^2))+a.^5*.
   exp(−a)./(120*((E−0.3−5*d).^2+b^2)));
y=2*c./((E−0.5).^2+b^2);
f=25.6*(fEl−fEr).*x;
```

Figure 4.14 (right panel)

```
lower_limit_integral= −1;
upper_limit_integral=1;
function_name='rat11a';
lower_parameter=0.05;
upper_parameter=1.5;
number_subdivisions=8611;
a=linspace(lower_parameter,upper_parameter,number_subdivisions);
len=zeros(1,number_subdivisions);
for k=3:number_subdivisions
```

```
len(k)=((quad(function_name,lower_limit_integral,upper_limit_integral,[],[],
   a(k))−quad(function_name,lower_limit_integral,upper_limit_integral,[],[],
   a(k−1)))./(a(k)−a(k−1))−(quad(function_name,lower_limit_integral,
   upper_limit_integral,[],[],a(k−1))−quad(function_name,lower_limit_integral,
   upper_limit_integral,[],[],a(k−2)))./(a(k−1)−a(k−2)))./(a(k)−a(k−1));
end
lower_limit_integral= −1;
upper_limit_integral=1;
function_name='rat11';
lower_parameter=0.05;
upper_parameter=1.5;
number_subdivisions=8611;
a=linspace(lower_parameter,upper_parameter,number_subdivisions);
len1=zeros(1,number_subdivisions);
for k=3:number_subdivisions
len1(k)=((quad(function_name,lower_limit_integral,upper_limit_integral,[],[],
   a(k))−quad(function_name,lower_limit_integral,upper_limit_integral,[],[],
   a(k−1)))./(a(k)−a(k−1))−(quad(function_name,lower_limit_integral,
   upper_limit_integral,[],[],a(k−1))−quad(function_name,lower_limit_integral,
   upper_limit_integral,[],[],a(k−2)))./(a(k−1)−a(k−2)))./(a(k)−a(k−1));
end
subplot(1,2,2)
h=plot(a,len,a,len1);
set(h,'LineWidth',1.5,'LineStyle','-';'–')
set(h,'Color','b';'r')
axis square
axis([0.21 0.25 −1.5 1.5])
set(gca,'xtick',[0.21 0.23 0.25],'FontSize',14)
set(gca,'ytick',[−1 0 1],'FontSize',14)
ylabel('d²I/dV² (μ A/V²)','FontSize',16)
xlabel('Voltage (V)','FontSize',16)
```

Supplementary files:

```
function f=rat11(t,v)
b=0.4;
g=0.2;
N=1;
ω =0.22;
zplus=0.003;
M=0.3;
kT=0.00075;
e=0;
t0=0;
m=1;
```

```
fEl=1./(1+exp((t-0.5*v)./kT));
fEr=1./(1+exp((t+0.5*v)./kT));
y1=sqrt((t-t0).^2-4*(b.^2));
y2=(t-t0+y1).^(N-1).*(t-t0+y1+i*g).^2-(t-t0-y1).^(N-1).*(t-t0-y1+i*g).^2;
G=4*(2*b).^(N-1).*(y1)./(y2);
y4=sqrt((t-t0+omega).^2-4*(b.^2));
y5=sqrt((t-t0-omega).^2-4*(b.^2));
y8=(t-t0+y1).^(m-1).*(t-t0+y1+i*g)-(t-t0-y1).^(m-1).*(t-t0-y1+i*g);
y9=(t-t0+y1).^(m-1).*(t-t0+y1+i*g)+(t-t0-y1).^(m-1).*(t-t0-y1+i*g);
y6=(t-t0+omega+y4).^(m-1).*(t-t0+omega+y4+i*g)
   -(t-t0+omega-y4).^(m-1).*(t-t0+omega-y4+i*g);
y7=(t-t0-omega+y5).^(m-1).*(t-t0-omega+y5+i*g)
   -(t-t0-omega-y5).^(m-1).*(t-t0-omega-y5+i*g);
y10=(t-t0+omega+y4).^(m-1).*(t-t0+omega+y4+i*g)
   +(t-t0+omega-y4).^(m-1).*(t-t0+omega-y4+i*g);
y11=(t-t0-omega+y5).^(m-1).*(t-t0-omega+y5+i*g)
   +(t-t0-omega-y5).^(m-1).*(t-t0-omega-y5+i*g);
D= -imag((y1).^(-1).*(y8)./(y9))./pi;
D1= -imag((y4).^(-1).*(y6)./(y10))./pi;
D2= -imag((y5).^(-1).*(y7)./(y11))./pi;
x1=1./(exp((t-t0+omega-0.55*v)./(zplus))+1);
x2=1./(exp((omega-0.45*v-t+t0)./(zplus))+1);
Gph=M*M*sqrt((D1.*x1+D2.*x2).^2);
ep=Gph./((2*b)+Gph);
S=0.55*0.45*G.*conj(G);
y3=(S).^(0.5);
T1=(2-ep).*y3.*(1+y3-(1-ep).*(1-y3))./((1+y3+(1-ep).*(1-y3)).^2);
f=0.25*77.*(10.^(-6)).*(fEl-fEr).*(T1);

function f=rat11a(t,v)
b=0.4;
g=0.2;
N=1;
ω =0.22;
zplus=0.003;
M=0.3;
kT=0.00075;
e=0;
t0=0;
m=1;
fEl=1./(1+exp((t-0.5*v)./kT));
fEr=1./(1+exp((t+0.5*v)./kT));
y1=sqrt((t-t0).^2-4*(b.^2));
y2=(t-t0+y1).^(N-1).*(t-t0+y1+i*g).^2-(t-t0-y1).^(N-1).*(t-t0-y1+i*g).^2;
G=4*(2*b).^(N-1).*(y1)./(y2);
```

```
y4=sqrt((t-t0+omega).^2-4*(b.^2));
y5=sqrt((t-t0-omega).^2-4*(b.^2));
y8=(t-t0+y1).^(m-1).*(t-t0+y1+i*g)-(t-t0-y1).^(m-1).*(t-t0-y1+i*g);
y9=(t-t0+y1).^(m-1).*(t-t0+y1+i*g)+(t-t0-y1).^(m-1).*(t-t0-y1+i*g);
y6=(t-t0+omega+y4).^(m-1).*(t-t0+omega+y4+i*g)
   -(t-t0+omega-y4).^(m-1).*(t-t0+omega-y4+i*g);
y7=(t-t0-omega+y5).^(m-1).*(t-t0-omega+y5+i*g)
   -(t-t0-omega-y5).^(m-1).*(t-t0-omega-y5+i*g);
y10=(t-t0+omega+y4).^(m-1).*(t-t0+omega+y4+i*g)
   +(t-t0+omega-y4).^(m-1).*(t-t0+omega-y4+i*g);
y11=(t-t0-omega+y5).^(m-1).*(t-t0-omega+y5+i*g)
   +(t-t0-omega-y5).^(m-1).*(t-t0-omega-y5+i*g);
D= -imag((y1).^(-1).*(y8)./(y9))./pi;
D1= -imag((y4).^(-1).*(y6)./(y10))./pi;
D2= -imag((y5).^(-1).*(y7)./(y11))./pi;
x1=1./(exp((t-t0+omega-0.55*v)./(zplus))+1);
x2=1./(exp((omega-0.45*v-t+t0)./(zplus))+1);
Gph=M*M*sqrt((D1.*x1+D2.*x2).^2);
ep=Gph./((2*b)+Gph);
S=0.55*0.45*G.*conj(G);
y3=(S).^(0.5);
T1=(2-ep).*y3.*(1+y3-(1-ep).*(1-y3))./((1+y3+(1-ep).*(1-y3)).^2);
f= -0.125*77.*(10.^(-6)).*(fEl-fEr).*(T1);
```

Figure 4.14

```
t=0:0.001:1.5;
y1=1.9425*t./((1+sqrt(1+1.9425*t)).^2);
y2=3.885*t./((1+sqrt(1+3.885*t)).^2);
y3=7.77*t./((1+sqrt(1+7.77*t)).^2);
y4=11.655*t./((1+sqrt(1+11.655*t)).^2);
y5=1.9425*t./(1+sqrt(1+1.9425*t));
y6=3.885*t./(1+sqrt(1+3.885*t));
y7=7.77*t./(1+sqrt(1+7.77*t));
y8=11.655*t./(1+sqrt(1+11.655*t));
y9=2./(2+y5);
y10=2./(2+y6);
y11=2./(2+y7);
y12=2./(2+y8);
z1=y9.*(2+y1).*(y9.*(2+y1)+y1)./((y9.*y1+2+y1).^2);
z2=y10.*(2+y2).*(y10.*(2+y2)+y2)./((y10.*y2+2+y2).^2);
z3=y11.*(2+y3).*(y11.*(2+y3)+y3)./((y11.*y3+2+y3).^2);
z4=y12.*(2+y4).*(y12.*(2+y4)+y4)./((y12.*y4+2+y4).^2);
subplot(1,2,1)
```

```
h=plot(t,y1,t,y2,t,y3,t,y4);
set(h,'LineWidth',1.49,'LineStyle',':';'-.';'-';'-')
set(h,'Color','r';'g';'b';'k')
axis([0 1.5 0 0.75])
axis square;
set(gca,'xtick',[0 0.5 1 1.5],'FontSize',14)
set(gca,'ytick',[0 0.25 0.5 0.75],'FontSize',14);
xlabel('T/T_0','FontSize',16)
ylabel('\epsilon(T/T_0)','FontSize',16)
subplot(1,2,2);
h=plot(t,z1,t,z2,t,z3,t,z4);
set(h,'LineWidth',1.2,'LineStyle',':';'-.';'-';'-')
set(h,'Color','r';'g';'b';'k')
axis([0 1.5 0 1])
axis square;
set(gca,'xtick',[0 0.5 1 1.5],'FontSize',14)
set(gca,'ytick',[0.0 0.25 0.5 .75 1],'FontSize',14);
xlabel('T/T_0','FontSize',16)
ylabel('Transmission','FontSize',16)
```

Figure 4.15, (left panel)

```
lower_limit_integral= -20;
upper_limit_integral=20;
function_name='poli16';
lower_parameter=0;
upper_parameter=3;
number_subdivisions=301;
v=linspace(lower_parameter,upper_parameter,number_subdivisions);
len1=zeros(1,number_subdivisions);
for k=1:number_subdivisions
len1(k)=quad(function_name,lower_limit_integral,upper_limit_integral
   ,[],[],v(k));
end
lower_limit_integral= -20;
upper_limit_integral=20;
function_name='poli17'; lower_parameter=0;
upper_parameter=3;
number_subdivisions=301;
v=linspace(lower_parameter,upper_parameter,number_subdivisions);
len2=zeros(1,number_subdivisions);
for k=1:number_subdivisions
len2(k)=quad(function_name,lower_limit_integral,upper_limit_integral,
   [],[],v(k));
```

```
end lower_limit_integral= -20;
upper_limit_integral=20;
function_name='poli18';
lower_parameter=0;
upper_parameter=3;
number_subdivisions=301;
v=linspace(lower_parameter,upper_parameter,number_subdivisions);
len3=zeros(1,number_subdivisions);
for k=1:number_subdivisions
len3(k)=quad(function_name,lower_limit_integral,upper_limit_integral,
   [],[],v(k));
end lower_limit_integral= -20;
upper_limit_integral=20;
function_name='poli19';
lower_parameter=0;
upper_parameter=3;
number_subdivisions=301;
v=linspace(lower_parameter,upper_parameter,number_subdivisions);
len4=zeros(1,number_subdivisions);
for k=1:number_subdivisions
len4(k)=quad(function_name,lower_limit_integral,upper_limit_integral,
   [],[],v(k));
end
subplot(1,2,1)
h=plot(v,len1,v,len2,v,len3,v,len4);
set(h,'LineWidth',1.2,'LineStyle',':';'-.';'-';'-')
set(h,'Color','r';'m';'b';'k')
axis([0 3 0 1.4])
set(gca,'xtick',[0 1 2 3],'FontSize',13)
set(gca,'ytick',[0 0.4 .8 1.2],'FontSize',13)
xlabel('Voltage (V)','FontSize',14)
ylabel('Current (μA)','FontSize',14)
axis square
```

Supplementary files:

poli16

```
function f=poli16(t,v);
U=0.004;
kT=0.00431;
x1=1./(exp((U*t-0.025*v)./kT)+1);
x2=1./(exp((U*t+0.025*v)./kT)+1);
y1=4.96./((t+10).^2+14.31);
y3=9.91./((t+10).^2+9.36);
y2=2./sqrt((t+10).^2+(2+y3).^2);
```

```
z1=y2.*(2+y1).*(y2.*(2+y1)+y1)./((y2.*y1+2+y1).^2);
f=0.326*z1.*(x1-x2);
```

poli17

```
function f=poli17(t,v);
U=0.004;
kT=0.00431;
x1=1./(exp((U*t-0.025*v)./kT)+1);
x2=1./(exp((U*t+0.025*v)./kT)+1);
y1=20.77./((t+10).^2+39);
y3=41.54./((t+10).^2+18.27);
y2=2./sqrt((t+10).^2+(2+y3).^2);
z1=y2.*(2+y1).*(y2.*(2+y1)+y1)./((y2.*y1+2+y1).^2);
f=0.326*z1.*(x1-x2);
```

poli18

```
function f=poli18(t,v);
U=0.004;
kT=0.00431;
x1=1./(exp((U*t-0.025*v)./kT)+1);
x2=1./(exp((U*t+0.025*v)./kT)+1);
y1=41.69./((t+10).^2+68.19);
y3=83.38./((t+10).^2+26.49);
y2=2./sqrt((t+10).^2+(2+y3).^2);
z1=y2.*(2+y1).*(y2.*(2+y1)+y1)./((y2.*y1+2+y1).^2);
f=0.326*z1.*(x1-x2);
```

poli19

```
function f=poli19(t,v);
U=0.004;
kT=0.00431;
x1=1./(exp((U*t-0.025*v)./kT)+1);
x2=1./(exp((U*t+0.025*v)./kT)+1);
y1=66.52./((t+10).^2+100.93);
y3=133./((t+10).^2+34.34);
y2=2./sqrt((t+10).^2+(2+y3).^2);
z1=y2.*(2+y1).*(y2.*(2+y1)+y1)./((y2.*y1+2+y1).^2);
f=0.326*z1.*(x1-x2);
```

Figure 4.15, (right panel)

```
lower_limit_integral= -20;
upper_limit_integral=20;
function_name='poli2';
lower_parameter=0;
upper_parameter=3;
```

```
number_subdivisions=201;
v=linspace(lower_parameter,upper_parameter,number_subdivisions);
len1=zeros(1,number_subdivisions);
for k=1:number_subdivisions
len1(k)=quad(function_name,lower_limit_integral,upper_limit_integral,
   [],[],v(k));
end
lower_limit_integral= -20;
upper_limit_integral=20;
function_name='poli3';
lower_parameter=0;
upper_parameter=3;
number_subdivisions=201;
v=linspace(lower_parameter,upper_parameter,number_subdivisions);
len2=zeros(1,number_subdivisions);
for k=1:number_subdivisions len2(k)=quad(function_name,
   lower_limit_integral,upper_limit_integral,[],[],v(k));
end
lower_limit_integral= -20;
upper_limit_integral=20;
function_name='poli4';
lower_parameter=0;
upper_parameter=3;
number_subdivisions=201;
v=linspace(lower_parameter,upper_parameter,number_subdivisions);
len3=zeros(1,number_subdivisions);
for k=1:number_subdivisions
len3(k)=quad(function_name,lower_limit_integral,upper_limit_integral,
   [],[],v(k));
end
lower_limit_integral= -20;
upper_limit_integral=20;
function_name='poli5';
lower_parameter=0;
upper_parameter=3;
number_subdivisions=201;
v=linspace(lower_parameter,upper_parameter,number_subdivisions);
len4=zeros(1,number_subdivisions);
for k=1:number_subdivisions
len4(k)=quad(function_name,lower_limit_integral,upper_limit_integral,
   [],[],v(k));
end
subplot(1,2,2)
h=plot(v,len1,v,len2,v,len3,v,len4);
set(h,'LineWidth',1.2,'LineStyle',':';'-.';'-';'-')
```

```
set(h,'Color','r';'m';'b';'k')
axis([0 3 0 1.4])
set(gca,'xtick',[0 1 2 3],'FontSize',13)
set(gca,'ytick',[0 0.4 .8 1.2],'FontSize',13)
xlabel('Voltage (V)','FontSize',14)
ylabel('dI/dV (μA/V)','FontSize',14)
axis square
```

<u>Supplementary files:</u>

<u>poli2</u>

```
function f=poli2(t,v);
U=0.004;
kT=0.00431;
x1=1./(exp((U*t-0.025*v)./kT)+1);
x2=1./(exp((U*t+0.025*v)./kT)+1);
y1=4.96./((t+10).^2+14.31);
y3=9.91./((t+10).^2+9.36);
y2=2./sqrt((t+10).^2+(2+y3).^2);
z1=y2.*(2+y1).*(y2.*(2+y1)+y1)./((y2.*y1+2+y1).^2);
x3=z1./(exp((U*t-0.025*v)./(2*kT))+exp((0.025*v-U*t)./(2*kT))).^2
x4=z1./(exp((U*t+0.025*v)./(2*kT))+exp(-(0.025*v+U*t)./(2*kT))).^2
f=0.025*0.4*(x3+x4)./kT;
```

<u>poli3</u>

```
function f=poli3(t,v);
U=0.004;
kT=0.00431;
x1=1./(exp((U*t-0.025*v)./kT)+1);
x2=1./(exp((U*t+0.025*v)./kT)+1);
y1=20.77./((t+10).^2+39);
y3=41.54./((t+10).^2+18.27);
y2=2./sqrt((t+10).^2+(2+y3).^2);
z1=y2.*(2+y1).*(y2.*(2+y1)+y1)./((y2.*y1+2+y1).^2);
x3=z1./(exp((U*t-0.025*v)./(2*kT))+exp((0.025*v-U*t)./(2*kT))).^2
x4=z1./(exp((U*t+0.025*v)./(2*kT))+exp(-(0.025*v+U*t)./(2*kT))).^2
f=0.025*0.4*(x3+x4)./kT;
```

<u>poli4</u>

```
function f=poli4(t,v);
U=0.004;
kT=0.00431;
x1=1./(exp((U*t-0.025*v)./kT)+1);
x2=1./(exp((U*t+0.025*v)./kT)+1);
y1=41.69./((t+10).^2+68.19);
```

```
y3=83.38./((t+10).^2+26.49);
y2=2./sqrt((t+10).^2+(2+y3).^2);
z1=y2.*(2+y1).*(y2.*(2+y1)+y1)./((y2.*y1+2+y1).^2);
x3=z1./(exp((U*t-0.025*v)./(2*kT))+exp((0.025*v-U*t)./(2*kT))).^2
x4=z1./(exp((U*t+0.025*v)./(2*kT))+exp(-(0.025*v+U*t)./(2*kT))).^2
f=0.025*0.4*(x3+x4)./kT;
```

poli5

```
function f=poli5(t,v);
U=0.004;
kT=0.00431;
x1=1./(exp((U*t-0.025*v)./kT)+1);
x2=1./(exp((U*t+0.025*v)./kT)+1);
y1=66.52./((t+10).^2+100.93);
y3=133./((t+10).^2+34.34);
y2=2./sqrt((t+10).^2+(2+y3).^2);
z1=y2.*(2+y1).*(y2.*(2+y1)+y1)./((y2.*y1+2+y1).^2);
x3=z1./(exp((U*t-0.025*v)./(2*kT))+exp((0.025*v-U*t)./(2*kT))).^2
x4=z1./(exp((U*t+0.025*v)./(2*kT))+exp(-(0.025*v+U*t)./(2*kT))).^2
f=0.025*0.4*(x3+x4)./kT;
```

Figure 4.16

```
lower_limit_integral= -20;
upper_limit_integral=20;
function_name='curr1a';
lower_parameter=0;
upper_parameter=4;
number_subdivisions=201;
y=linspace(lower_parameter,upper_parameter,number_subdivisions);
len1=zeros(1,number_subdivisions);
for k=1:number_subdivisions len1(k)=quad(function_name,lower_limit_integral,
   upper_limit_integral,[],[],y(k));
end
lower_limit_integral= -20
upper_limit_integral=20;
function_name='curr2a';
lower_parameter=0;
upper_parameter=4;
number_subdivisions=201;
y=linspace(lower_parameter,upper_parameter,number_subdivisions);
len2=zeros(1,number_subdivisions);
for k=1:number_subdivisions len2(k)=quad(function_name,lower_limit_integral,
   upper_limit_integral,[],[],y(k));
end
```

```
lower_limit_integral= −20
upper_limit_integral=20;
function_name='curr3a';
lower_parameter=0;
upper_parameter=4;
number_subdivisions=201;
y=linspace(lower_parameter,upper_parameter,number_subdivisions);
len3=zeros(1,number_subdivisions);
for k=1:number_subdivisions len3(k)=quad(function_name,lower_limit_integral,
    upper_limit_integral,[],[],y(k));
end
lower_limit_integral= −20
upper_limit_integral=20
function_name='curr4a';
lower_parameter=0;
upper_parameter=4;
number_subdivisions=201;
y=linspace(lower_parameter,upper_parameter,number_subdivisions);
len4=zeros(1,number_subdivisions);
for k=1:number_subdivisions len4(k)=quad(function_name,lower_limit_integral,
    upper_limit_integral,[],[],y(k));
end subplot(1,2,1)
h=plot(y,len1,y,len2,y,len3,y,len4)
set(h,'LineWidth',1.5,'LineStyle','-';'–';'−.';':')
set(h,'Color','k';'b';'g';'r')
axis([0 4 0 30])
set(gca,'xtick',[−4 −2 0 2 4],'FontSize',14)
set(gca,'ytick',[−30 −15 0 15 30],'FontSize',14)
xlabel('Voltage (V)','FontSize',14)
ylabel('Current (nA)','FontSize',14)
axis square
lower_limit_integral= −20;
upper_limit_integral=20;
function_name='con1a';
lower_parameter= −2;
upper_parameter=4;
number_subdivisions=201;
y1=linspace(lower_parameter,upper_parameter,number_subdivisions);
len5=zeros(1,number_subdivisions);
for k=1:number_subdivisions
len5(k)=quad(function_name,lower_limit_integral,upper_limit_integral,
   [],[],y1(k));
end
lower_limit_integral= −20
upper_limit_integral=20;
```

```
function_name='con2a';
lower_parameter= -2;
upper_parameter=4;
number_subdivisions=201;
y1=linspace(lower_parameter,upper_parameter,number_subdivisions);
len6=zeros(1,number_subdivisions);
for k=1:number_subdivisions len6(k)=quad(function_name,lower_limit_integral,
    upper_limit_integral,[],[],y1(k));
end
lower_limit_integral= -20
upper_limit_integral=20;
function_name='con3a';
lower_parameter= -2;
upper_parameter=4;
number_subdivisions=201;
y1=linspace(lower_parameter,upper_parameter,number_subdivisions);
len7=zeros(1,number_subdivisions);
for k=1:number_subdivisions len7(k)=quad(function_name,lower_limit_integral,
    upper_limit_integral,[],[],y1(k));
end lower_limit_integral= -20
upper_limit_integral=20;
function_name='con4a';
lower_parameter= -2;
upper_parameter=4;
number_subdivisions=201;
y1=linspace(lower_parameter,upper_parameter,number_subdivisions);
len8=zeros(1,number_subdivisions);
for k=1:number_subdivisions len8(k)=quad(function_name,lower_limit_integral,
    upper_limit_integral,[],[],y1(k));
end subplot(1,2,2)
h1=plot(y1,len5,y1,len6,y1,len7,y1,len8)
set(h1,'LineWidth',1.5,'LineStyle','-';'–';'–.';':')
set(h1,'Color','k';'b';'g';'r')
axis([-2 4 0 .04])
set(gca,'xtick',[ -2 0 2 4],'FontSize',14)
set(gca,'ytick',[.01 .02 .03 .04],'FontSize',14)
xlabel('Voltage (V)','FontSize',14)
ylabel('g/G0','FontSize',14)
axis square
```

Supplementary files:

curr1a

```
function f=curr1a(E,y);
x1=0.5*18.765*(1+sqrt(1+18.765))./((0.25*E).^2+0.5*(1+sqrt(1+18.765)).^3);
```

```
x2=4*18.765*(1+sqrt(1+18.765))./((0.25*E).^2+(1+sqrt(1+18.765)).^2);
x3=8./(sqrt((0.25*E).^2+(2+0.25.*x2).^2));
x4=x3.*(2-x1).*(x3.*(2-x1)+x1)./((x3.*x1+2-x1).^2);
x5=1./(exp((E-4-5*y)./(4.17))+1);
x6=1./(exp((E-4+5*y)./(4.17))+1);
f=2*0.136*(x5-x6).*x4;
```

curr2a

```
function f=curr2a(E,y);
x1=0.5*12.51*(1+sqrt(1+12.51))./((0.25*E).^2+0.5*(1+sqrt(1+12.51)).^3);
x2=4*12.51*(1+sqrt(1+12.51))./((0.25*E).^2+(1+sqrt(1+12.51)).^2);
x3=8./(sqrt((0.25*E).^2+(2+0.25.*x2).^2));
x4=x3.*(2-x1).*(x3.*(2-x1)+x1)./((x3.*x1+2-x1).^2);
x5=1./(exp((E-4-5*y)./(4.17))+1);
x6=1./(exp((E-4+5*y)./(4.17))+1);
f=2*0.136*(x5-x6).*x4;
```

curr3a

```
function f=curr3a(E,y);
x1=0.5*6.255*(1+sqrt(1+6.255))./((0.25*E).^2+0.5*(1+sqrt(1+6.255)).^3);
x2=4*6.255*(1+sqrt(1+6.255))./((0.25*E).^2+(1+sqrt(1+6.255)).^2);
x3=8./(sqrt((0.25*E).^2+(2+0.25.*x2).^2));
x4=x3.*(2-x1).*(x3.*(2-x1)+x1)./((x3.*x1+2-x1).^2);
x5=1./(exp((E-4-5*y)./(4.17))+1);
x6=1./(exp((E-4+5*y)./(4.17))+1);
f=2*0.136*(x5-x6).*x4;
```

curr4a

```
function f=curr4a(E,y);
x1=0.5*3.1275*(1+sqrt(1+3.1275))./((0.25*E).^2+0.5*(1+sqrt(1+3.1275)).^3);
x2=4*3.1275*(1+sqrt(1+3.1275))./((0.25*E).^2+(1+sqrt(1+3.1275)).^2);
x3=8./(sqrt((0.25*E).^2+(2+0.25.*x2).^2));
x4=x3.*(2-x1).*(x3.*(2-x1)+x1)./((x3.*x1+2-x1).^2);
x5=1./(exp((E-4-5*y)./(4.17))+1);
x6=1./(exp((E-4+5*y)./(4.17))+1);
f=2*0.136*(x5-x6).*x4;
```

con1a

```
function f=con1a(E,y1);
x1=0.5*18.765*(1+sqrt(1+18.765))./((0.25*E).^2+0.5*(1+sqrt(1+18.765)).^3);
x2=4*18.765*(1+sqrt(1+18.765))./((0.25*E).^2+(1+sqrt(1+18.765)).^2);
x3=8./(sqrt((0.25*E).^2+(2+0.25.*x2).^2));
x4=x3.*(2-x1).*(x3.*(2-x1)+x1)./((x3.*x1+2-x1).^2);
x5=1./(cosh((E-4+5*y1)./(2*4.17))).^2;
x6=1./(cosh((E-4+5*y1)./(2*4.17))).^2;
f=0.0003*(x5+x6).*x4;
```

con2a

```
function f=con2a(E,y1);
x1=0.5*12.51*(1+sqrt(1+12.51))./((0.25*E).^2+0.5*(1+sqrt(1+12.51)).^3);
x2=4*12.51*(1+sqrt(1+12.51))./((0.25*E).^2+(1+sqrt(1+12.51)).^2);
x3=8./(sqrt((0.25*E).^2+(2+0.25.*x2).^2));
x4=x3.*(2-x1).*(x3.*(2-x1)+x1)./((x3.*x1+2-x1).^2);
x5=1./(cosh((E-4-5*y1)./(2*4.17))).^2;
x6=1./(cosh((E-4+5*y1)./(2*4.17))).^2;
f=0.0003*(x5+x6).*x4;
```

con3a function f=con3a(E,y1);

```
x1=0.5*6.255*(1+sqrt(1+6.255))./((0.25*E).^2+0.5*(1+sqrt(1+6.255)).^3);
x2=6.255*4*(1+sqrt(1+6.255))./((0.25*E).^2+(1+sqrt(1+6.255)).^2);
x3=8./(sqrt((0.25*E).^2+(2+0.25.*x2).^2));
x4=x3.*(2-x1).*(x3.*(2-x1)+x1)./((x3.*x1+2-x1).^2);
x5=1./(cosh((E-4-5*y1)./(2*4.17))).^2;
x6=1./(cosh((E-4+5*y1)./(2*4.17))).^2;
f=0.0003*(x5+x6).*x4;
```

con4a

```
function f=con4a(E,y1);
x1=0.5*3.1275*(1+sqrt(1+3.1275))./((0.25*E).^2+0.5*(1+sqrt(1+3.1275)).^3);
x2=3.1275*4*(1+sqrt(1+3.1275))./((0.25*E).^2+(1+sqrt(1+3.1275)).^2);
x3=8./(sqrt((0.25*E).^2+(2+0.25.*x2).^2));
x4=x3.*(2-x1).*(x3.*(2-x1)+x1)./((x3.*x1+2-x1).^2);
x5=1./(cosh((E-4-5*y1)./(2*4.17))).^2;
x6=1./(cosh((E-4+5*y1)./(2*4.17))).^2;
f=0.0003*(x5+x6).*x4;
```

Figure 4.17

```
lower_limit_integral= -20;
upper_limit_integral=20;
function_name='poli2';
lower_parameter=0;
upper_parameter=1;
number_subdivisions=401;
v=linspace(lower_parameter,upper_parameter,number_subdivisions);
len1=zeros(1,number_subdivisions);
for k=1:number_subdivisions
len1(k)=quad(function_name,lower_limit_integral,upper_limit_integral,
   [],[],v(k));
end
```

```
lower_limit_integral= −20;
upper_limit_integral=20;
function_name='poli5';
lower_parameter=0;
upper_parameter=1;
number_subdivisions=401;
v=linspace(lower_parameter,upper_parameter,number_subdivisions);
len4=zeros(1,number_subdivisions);
for k=1:number_subdivisions;
len4(k)=quad(function_name,lower_limit_integral,upper_limit_integral,
   [],[],v(k));
end
subplot(1,2,1)
h=plot(v,len1,v,len4)
set(h,'LineWidth',1.5,'LineStyle','–';'-')
set(h,'Color','r';'b')
axis([0 1 0.17 0.63])
set(gca,'xtick',[0 0.5 1 ],'FontSize',14)
set(gca,'ytick',[0.2 0.4 0.6],'FontSize',14)
xlabel('Voltage (V)','FontSize',16)
ylabel('dI/dV (μ A/V)','FontSize',16)
axis square
t= −5:0.005:5;
t1=1:0.05:3;
y1=1./((t.^2).*(1−9.8*exp(−4.9*t.^2)).^2+15.4*exp(−9.8*t.^2));
y3=1./((t.^2).*(1−29.4*exp(−14.7*t.^2)).^2+46.18*exp(−29.4*t.^2));
subplot(1,2,2)
h=plot(t,y1,t,y3);
set(h,'LineWidth',1.5,'LineStyle','-';'–')
set(h,'Color','b';'r')
axis([−2.2 2.2 −2 25])
set(gca,'xtick',[−2 −1 0 1 2 ],'FontSize',14)
set(gca,'ytick',[0 10 20 ],'FontSize',14)
ylabel('T(E)w^2/4Δ^2','FontSize',16);
xlabel('E/w','FontSize',16)
axis square
```

Supplementary files:

```
function f=poli2(t,v);
U=0.004;
kT=0.00431;
x1=1./(exp((U*t−0.025*v)./kT)+1);
x2=1./(exp((U*t+0.025*v)./kT)+1);
```

```
y1=4.96./((t+10).^2+14.31);
y3=9.91./((t+10).^2+9.36);
y2=2./sqrt((t+10).^2+(2+y3).^2);
z1=y2.*(2+y1).*(y2.*(2+y1)+y1)./((y2.*y1+2+y1).^2);
x3=z1./(exp((U*t-0.025*v)./(2*kT))+exp((0.025*v-U*t)./(2*kT))).^2;
x4=z1./(exp((U*t+0.025*v)./(2*kT))+exp(-(0.025*v+U*t)./(2*kT))).^2;
f=0.025*0.4*(x3+x4)./kT;

function f=poli5(t,v);
U=0.004;
kT=0.00431;
x1=1./(exp((U*t-0.025*v)./kT)+1);
x2=1./(exp((U*t+0.025*v)./kT)+1);
y1=66.52./((t+10).^2+100.93);
y3=133./((t+10).^2+34.34);
y2=2./sqrt((t+10).^2+(2+y3).^2);
z1=y2.*(2+y1).*(y2.*(2+y1)+y1)./((y2.*y1+2+y1).^2);
x3=z1./(exp((U*t-0.025*v)./(2*kT))+exp((0.025*v-U*t)./(2*kT))).^2;
x4=z1./(exp((U*t+0.025*v)./(2*kT))+exp(-(0.025*v+U*t)./(2*kT))).^2;
f=0.025*0.4*(x3+x4)./kT;
```

Figure 4.17

```
t= -5:0.005:5;
t1=1:0.05:40;
y1=1./((t.^2).*(1-9.8*exp(-4.9*t.^2)).^2+15.4*exp(-9.8*t.^2));
y2=1./((t.^2).*(1-3.92*exp(-1.96*t.^2)).^2+6.16*exp(-3.92*t.^2));
y3=1./((t.^2).*(1-29.4*exp(-14.7*t.^2)).^2+46.18*exp(-29.4*t.^2));
y4=1./((t.^2).*(1-11.76*exp(-5.88*(t.^2))).^2+18.47*exp(-5.88*(t.^2)));
y5=0.0036./((t.^2).*(1-153.66*exp(-1900*t.^2)).^2+0.0036*(1+1294*
   exp(-1900*t.^2)).^2);
y6=0.0036./((t.^2).*(1-768.3*exp(-380*t.^2)).^2+0.0036*(1+578.7*
   exp(-380*t.^2)).^2);
y7 = 0.0036./((t.^2) + 0.0036);
z1 = 9.8*t.*exp(-4.9*t.^2);
z2 = -3.92*exp(-4.9*t.^2);
z3 = 29.4*t.*exp(-14.7*t.^2);
z4 = -6.8*exp(-14.7*t.^2);
z5 = - log((1 - 5 * (t1).^2.*exp(-0.00225*(t1).^2)).^2 + (1 + 0.14*(t1).^2.*
   exp(-0.00225*(t1).^2)).^2);
t1=1:0.05:3;
x1=1./((t.^2).*(1-9.8*exp(-4.9*t.^2)).^2+15.4 * exp(-9.8*t.^2));
x2=1./((t.^2). * (1-3.92 * exp(-1.96 * t.^2)).^2+6.16 * exp(-3.92 * t.^2));
x3=1./((t.^2). * (1-29.4 * exp(-14.7 * t.^2)).^2+46.18 * exp(-29.4 * t.^2));
x4=1./((t.^2). * (1-11.76 * exp(-5.88 * t.^2)).^2+18.47 * exp(-11.76 * t.^2));
```

```
x5=0.0036./((t.^2).*(1-153.66*exp(-1900*t.^2)).^2+0.0036*(1+1294
   *exp(-1900*t.^2)).^2);
x6=0.0036./((t.^2).*(1-768.3*exp(-380*t.^2)).^2+0.0036*(1+578.7
   *exp(-380*t.^2)).^2);
x7=1./((t.^2)+0.0036);
subplot(2,2,1);
h1=plot(t,z1,t,z3);
set(h1,'LineWidth',1.5,'LineStyle','-';'--')
set(h1,'Color','b';'r')
axis([-2 2 -4.2 4.2])
set(gca,'xtick',[-2 -1 0 1 2],'FontSize',14)
set(gca,'ytick',[-4 -2 0 2 4],'FontSize',14)
ylabel('w ReP(E)','FontSize',16)
xlabel('E/w','FontSize',16)
subplot(2,2,3);
h2=plot(t,z4,t,z2);
set(h2,'LineWidth',1.5,'LineStyle','--';'-')
set(h2,'Color','r';'b')
axis([-2 2 -8.4 0.96])
xlabel('E/w','FontSize',16)
ylabel('wImP(E)','FontSize',16)
set(gca,'xtick',[-2 -1 0 1 2],'FontSize',14)
set(gca,'ytick',[-8 -6 -4 -2 0 ],'FontSize',14)
subplot(2,2,2)
h=plot(t,y1,t,y3);
set(h,'LineWidth',1.5,'LineStyle','-';'--')
set(h,'Color','b';'r')
axis([-2 2 -1.3 25])
set(gca,'xtick',[-2 -1 0 1 2 ],'FontSize',14)
set(gca,'ytick',[0 5 10 15 20 25 ],'FontSize',14)
ylabel ('T(E)w^2/4Γ^2','FontSize',16);
xlabel('E/w','FontSize',16)
subplot(2,2,4);
h3=plot(t,y2,t,y4);
set(h3,'LineWidth',1.5,'LineStyle','-';'--')
set(h3,'Color','b';'r')
axis([-2 2 -1.3 25])
ylabel('T(E)w^2/4Γ^2','FontSize',16)
xlabel('E/w','FontSize',16)
set(gca,'xtick',[-2 -1 0 1 2],'FontSize',14)
set(gca,'ytick',[0 5 10 15 20 25],'FontSize',14)
```

References

1. A. Aviram, M.A. Ratner, Molecular rectifiers. Chem. Phys. Lett. **29**, 277–283 (1974)
2. A. Nitzan, Electron transmission through molecules and molecular interfaces. Annu. Rev. Phys. Chem. **52**, 681–750 (2001)
3. N. Agrait, A.L. Yeyati, J.M. van Ruitenbeek, Quantum properties of atomic-sized conductors. Phys. Rep. **377**, 81–129 (2003)
4. S. Datta, *Quantum Transport: Atom to Transistor* (Cambridge University Press, Cambridge, 2005)
5. A.W. Ghosh, Electronics with molecules, in *Comprehensive Semiconductor Science and Technology*, vol. 5, ed. by P. Brattacharya, R. Fornari, H. Kamimura (Elsevier, Amsterdam, 2011), pp. 383–478
6. J.C. Cuevas, E. Scheer, *Molecular Electronics: An Introduction to the Theory and Experiment* (World Scientific Publishing, New Jersey, 2010)
7. J. Reichert, R. Ochs, D. Beckmann, H.B. Weber, M. Mayor, H.v. Lohneysen, Driving current through a single organic molecule. Phys. Rev. Lett. **88**, 176804 (2002)
8. L.H. Yu, Z.K. Keane, J.W. Ciszek, L. Cheng, M.P. Stewart, J.M. Tour, D. Natelson, Inelastic electron tunneling via molecular vibrations in single-molecule transistors. Phys. Rev. Lett. **93**, 266802 (2004)
9. M. Poot, E. Osorio, K. O'Neil, J.M. Thijssen, D. Vanmaekelbergh, Temperature dependence of three-terminal molecular junctions with sulfur end-functionalized tercyclohexylidenes. Nano Lett. **6**, 1031–1035 (2006)
10. E. Lortscher, H.B. Weber, H. Riel, Statistical approach to investigating transport through single molecules. Phys. Rev. Lett. **98**, 176807 (2007)
11. J. Reichert, H.B. Weber, M. Mayor, H.v. Lohneysen, Low-temperature conductance measurements on single-molecules. Appl. Phys. Lett. **82**, 4137 (2003)
12. H.B. Weber, J. Reichert, F. Weigend, R. Ochs, D. Beckmann, M. Mayor, R. Ahlrichs, H.v. Lohneysen, Electronic transport through single conjugated molecules. Chem. Phys. **281**, 113 (2002)
13. P. Liljeroth, J. Repp, G. Meyer, Current-induced hydrogen tautomerization and conductance switching of naphthalocyanine molecules. Science **317**, 1203–1206 (2007)
14. L.V. Venkataraman, J.E. Klare, M.S. Hybertsen, C. Nuckolls, M.L. Steigerwald, Dependence of single-molecule junction conductance on molecular conformation. Nature **442**, 904–907 (2006)
15. L. Venkataraman, J.E. Klare, I.W. Tam, C. Nuckolls, M.S. Hybertsen, C. Nuckolls, M.L. Steigerwald, Single-molecule circuits with well-defined molecular conductance. Nano Lett. **6**, 458–462 (2006)

N.A. Zimbovskaya, *Transport Properties of Molecular Junctions*, Springer Tracts in Modern Physics 254, DOI 10.1007/978-1-4614-8011-2,

16. B. Reddy, S.-Y. Yang, R.A. Segalman, A. Majumdar, Thermoelectricity in molecular junctions. Science **315**, 1568–1571 (2007)
17. S.H. Choi, B. Kim, C.D. Frisbie, Electrical resistance of long conjugated molecular wires. Science **320**, 1482–1486 (2008)
18. G. Kurczenow, in *The Oxford Handbook of Nanoscience and Nanothechnology* (Oxford University Press, Oxford, 2009)
19. M. Galperin, M.A. Ratner, A. Nitzan, Hysteresis, switching, and negative differential resistance in molecular junctions: a polaron model. Nano Lett. **5**, 125–130 (2005)
20. R. Gutierrez, S. Mandal, G. Cuniberti, Dissipative effects in the electronic transport through DNA molecular wires. Phys. Rev. B **71**, 235116 (2005)
21. D. Segal, A. Nitzan, P. Hanggi, Thermal conductance through molecular wires. J. Chem. Phys. **119**, 6840–6855 (2003)
22. A. Mitra, I. Aleiner, A.J. Millis, Phonon effects in molecular transistors: quantal and classical treatment. Phys. Rev. B **69**, 245302 (2004)
23. T. Komeda, Y. Kim, M. Kawai, B.N.J. Persson, H. Ueba, Lateral hopping of molecules induced by excitation of internal vibration mode. Science **295**, 2055–2058 (2002)
24. T. Mii, S.G. Tikhodeev, H. Ueba, Spectral features of inelastic electron transport via a localized state. Phys. Rev. B **68**, 205406 (2003)
25. M. Galperin, M.A. Ratner, A. Nitzan, Inelastic electron tunneling spectroscopy in molecular junctions: peaks and dips. J. Chem. Phys. **121**, 11965–11979 (2004)
26. M. Galperin, M.A. Ratner, A. Nitzan, Molecular transport junctions: vibrational effects. J. Phys.: Condens. Matter **19**, 103201 (2007)
27. J. Park, A.N. Pasupathy, J.I. Goldsmith, C. Chang, Y. Yaish, J.R. Petta, M. Rinkovski, J.P. Sithna, H.D. Abruna, P.L. McEuen, D.C. Ralph, Coulomb blockade and the Kondo effect in single-atom transistors. Nature **417**, 722–725 (2002)
28. W.J. Liang, M.P. Shores, M.P. Bockrath, J.R. Long, H. Park, Kondo resonance in a single-molecule transistor. Nature **417**, 725–729 (2002)
29. N.B. Zhitenev, H. Meng, Z. Bao, Conductance of small molecular junctions. Phys. Rev. Lett. **88**, 226801 (2002)
30. B.Q. Xu, P.M. Zhang, X.L. Li, N.J. Tao, Direct conductance measurement of single DNA molecules in aqueous solution. Nano Lett. **4**, 1105–1108 (2004)
31. L.H. Yu, D. Natelson, Kondo physics in C60 single-molecule transistors. Nano Lett. **4**, 79–83 (2004)
32. B.J. LeRoy, S.G. Lemay, J. Kong, C. Dekker, Electrical generation and absorption of phonons in carbon nanotubes. Nature **432**, 371–374 (2004)
33. K. Grove-Rasmussen, H.I. Jorgensena, P.E. Lindelof, Fabry Perot interference, Kondo effect and Coulomb blockade in carbon nanotubes. Physica E **40**, 92–98 (2007)
34. M. Galperin, A. Nitzan, M.A. Ratner, Inelastic effects in molecular junctions in the Coulomb and Kondo regimes: nonequilibrium equation-of-motion approach. Phys. Rev. B **76**, 035301 (2007)
35. N.A. Zimbovskaya, Electron transport through a quantum dot in the Coulomb blockade regime: nonequilibrium Green's function based model. Phys. Rev. B **78**, 035331 (2008)
36. A. Rosch, J. Kroha, P. Wolfle, Kondo effect in quantum dots at high voltage: universality and scaling. Phys. Rev. Lett. **87**, 156802 (2001)
37. A. Rosch, J. Paaske, J. Kroha, P. Wolfle, The Kondo effect in non-equilibrium quantum dots: perturbative renormalization group. J. Phys. Soc. Japan **74**, 118 (2005)
38. H.B. Heersche, Z. de Groot, J.A. Folk, S.H.J. van der Zant, C. Romeike, M.R. Wegewijs, L. Zobby, D. Barreca, E. Tondello, A. Cornia, Electron transport through single Mn12 molecular magnets. Phys. Rev. Lett. **96**, 206801 (2006)
39. C. Romeike, M.R. Wegewijs, M. Ruben, W. Wenzel, H. Schoeller, Charge-switchable molecular magnet and spin blockade of tunneling. Phys. Rev. B **75**, 064404 (2007)
40. A.S. Blum, J. Yang, R. Shashidhar, B.R. Ratna, Comparing the conductivity of molecular wires with the scanning tunneling microscope. Appl. Phys. Lett. **82**, 3322–3324 (2003)

41. A.S. Blum, C.M. Soto, C.D. Wilson, T.L. Brower, S.K. Pollack, T.L. Schull, A. Chatterji, T. Lin, J.E. Johnson, C. Amsinck, P. Franzon, R. Shashidhar, B.R. Ratna, An engineered virus as a scaffold for three-dimensional self-assembly on the nanoscale. Small **1**, 702–706 (2005)
42. N.A. Zimbovskaya, M.R. Pederson, A.S. Blum, B.R. Ratna, R. Allen, Nanoparticle networks as chemoselective sensing devices. J. Chem. Phys. **130**, 094702 (2009)
43. A.M. Kuznetsov, I. Ulstrup, *Electron Transfer in Physics and Biology. An Introduction to the Theory* (Wiley, Chichester, 1999)
44. F. Chen, N.J. Tao, Electron transport in single molecules: from benzene to graphene. Acc. Chem. Res. **42**, 429–438 (2009)
45. V. Mujica, M. Kemp, M.A. Ratner, Electron conduction in molecular wires. I. A scattering formalism. J. Chem. Phys. **101**, 6849–6855 (1994)
46. V. Mujica, M. Kemp, M.A. Ratner, Electron conduction in molecular wires. II. Application to scanning tunneling microscopy. J. Chem. Phys. **101**, 6856–6864 (1994)
47. K.V. Mikkelsen, M.A. Ratner, Electron tunneling in solid-state electron-transfer reactions. Chem. Rev. **87**, 113–153 (1987)
48. A. Nitzan, A relationship between electron-transfer rates and molecular conduction. J. Phys. Chem. A **105**, 2677–2679 (2001)
49. S. Yeganeh, M.A. Ratner, V. Mujica, Dynamics of charge transfer: rate processes formulated with nonequilibrium Green's functions. J. Chem. Phys. **126**, 161103 (2007)
50. Yu. Dahnovsky, Modulating electron dynamics: modified spin-boson approach. Phys. Rev. B **73**, 144303 (2006)
51. N.A. Zimbovskaya, Low temperature electronic transport and electron transfer through organic macromolecules. J. Chem. Phys. **118**, 4–7 (2003)
52. N.A. Zimbovskaya, Low-temperature electronic transport through macromolecules and characteristics of intramolecular electron transfer. J. Chem. Phys. **123**, 114708 (2005)
53. N.D. Lang, Ph. Avouris, Carbon-atom wires: charge-transfer doping, voltage drop, and the effect of distortions. Phys. Rev. Lett. **84**, 358–361 (2000)
54. N.D. Lang, Ph. Avouris, Electrical conductance of parallel atomic wires. Phys. Rev. B **62**, 7325–7329 (2000)
55. M. Di Ventra, S.T. Pantelides, N.D. Lang, First-principles calculation of transport properties of a molecular device. Phys. Rev. Lett. **84**, 979–982 (2000)
56. P.S. Damle, A.W. Ghosh, S. Datta, Unified description of molecular conduction: from molecules to metallic wires. Phys. Rev. B **64**, 201403(R) (2001)
57. V. Mujica, A.E. Roitberg, M.A. Ratner, Molecular wire conductance: electrostatic potential spatial profile. J. Chem. Phys. **112**, 6834–6839 (2000)
58. Y. Xue, S. Datta, M.A. Ratner, Charge transfer and "band lineup" in molecular electronic devices: a chemical and numerical interpretation. J. Chem. Phys. **115**, 4292–4299 (2001)
59. Y. Xue, M.A. Ratner, Microscopic study of electrical transport through individual molecules with metallic contacts. I. Band lineup, voltage drop, and high-field transport. Phys. Rev. B **68**, 115406 (2003)
60. Y. Xue, M.A. Ratner, Microscopic study of electrical transport through individual molecules with metallic contacts. II. Effect of the interface structure. Phys. Rev. B **68**, 115407 (2003)
61. Y. Xue, M.A. Ratner, End group effect on electrical transport through individual molecules: a microscopic study. Phys. Rev. B **69**, 085403 (2004)
62. S.H. Ke, H.U. Baranger, W. Yang, Contact atomic structure and electron transport through molecules. J. Chem. Phys. **122**, 074704 (2005)
63. M. Galperin, A. Nitzan, M.A. Ratner, Molecular transport junctions: current from electronic excitations in the leads. Phys. Rev. Lett. **96**, 166803 (2006)
64. S. Datta, Electrical resistance: an atomistic view. Nanotechnology **15**, S433–S465 (2004)
65. R. Landauer, Electrical resistance of disordered one-dimensional lattices. Phil. Mag. **21**, 863–867 (1970)
66. N.S. Wingreen, A.P. Jauho, Y. Meir, Time-dependent transport through a mesoscopic structure. Phys. Rev. B **48**, 8487–8490 (1993)

67. J.L. D'Amato, G.M. Pastawski, Conductance of a disordered linear chain including inelastic scattering events. Phys. Rev. B **41**, 7411–7420 (1990)
68. N.J. Tao, Electron transport in molecular junctions. Nat. Nanotechnol. **1**, 173–181 (2006)
69. C. Goldman, Long-range electron transfer in proteins: a renormalized-perturbation-expansion approach. Phys. Rev. A **43**, 4500–4511 (1991)
70. X.-Q. Li, Y.J. Yan, Scattering matrix approach to electronic dephasing in long-range electron transfer. J. Chem. Phys. **115**, 4169–4174 (2001)
71. N.A. Zimbovskaya, G. Gumbs, Long-range electron transfer and electronic transport through macromolecules. Appl. Phys. Lett. **81**, 1518–1520 (2002)
72. M. Buttiker, Role of quantum coherence in series resistors. Phys. Rev. B **33**, 3020–3026 (1986)
73. D.N. Beratan, J.N. Onuchic, J.R. Winkler, H.B. Gray, Electron tunneling pathways in proteins. Science **258**, 1740–1741 (1992)
74. C.W.J. Beenakker, Theory of Coulomb-blockade oscillations in the conductance of a quantum dot. Phys. Rev. B **44**, 1646–1656 (1991)
75. D.V. Averin, A.N. Korotkov, K.K. Likharev, Theory of single-electron charging of quantum wells and dots. Phys. Rev. B **44**, 6199–6211 (1991)
76. N.B. Kopnin, Y.M. Galperin, V.M. Vinokur, Charge transport through weakly open one-dimensional quantum wires. Phys. Rev. B **79**, 035319 (2009)
77. P.W. Anderson, Localized magnetic states in metals. Phys. Rev. **124**, 41–53 (1961)
78. J.J. Palacios, Coulomb blockade in electron transport through a C60 molecule from first principles. Phys. Rev. B **72**, 125424 (2005)
79. C. Toher, A. Filippetti, S. Sanvito, K. Burke, Self-interaction errors in density-functional calculations of electronic transport. Phys. Rev. Lett. **95**, 146402 (2005)
80. N. Sai, M. Zwolak, G. Vignale, M. Di Ventra, Dynamical corrections to the DFT-LDA electron conductance in nanoscale systems. Phys. Rev. Lett. **94**, 186810 (2005)
81. N. Simonian, J.J. Li, K.K. Likharev, Negative differential resistance at sequential single-electron tunnelling through atoms and molecules. Nanotechnology **18**, 424006 (2007)
82. S. Datta, W. Tian, S. Hong, R. Reifenberger, J.I. Henderson, C.P. Kubiak, Current–voltage characteristics of self-assembled monolayers by scanning tunneling microscopy. Phys. Rev. Lett. **79**, 2530–2533 (1997)
83. B. Muralidharan, A.W. Ghosh, S. Datta, Probing electronic excitations in molecular conduction. Phys. Rev. B **73**, 155410 (2006)
84. F. Zahid, A.W. Ghosh, M. Paulsson, E. Polizzi, S. Datta, Charging-induced asymmetry in molecular conductors. Phys. Rev. B **70**, 245317 (2004)
85. S. Braig, P.W. Brouwer, Rate equations for Coulomb blockade with ferromagnetic leads. Phys. Rev. B **71**, 195324 (2005)
86. Y. Meir, N.S. Wingreen, P.A. Lee, Transport through a strongly interacting electron system: theory of periodic conductance oscillations. Phys. Rev. Lett. **66**, 3048–3051 (1991)
87. Y. Meir, N.S. Wingreen, P.A. Lee, Low-temperature transport through a quantum dot: the Anderson model out of equilibrium. Phys. Rev. Lett. **70**, 2601–2604 (1993)
88. R. Swirkowicz, J. Barnas, M. Wilczynski, Electron tunnelling in a double ferromagnetic junction with a magnetic dot as a spacer. J. Phys.: Condens. Matter **14**, 2011–2023 (2002)
89. R. Swirkowicz, J. Barnas, M. Wilczynski, Nonequilibrium Kondo effect in quantum dots. Phys. Rev. B **68**, 195318 (2003)
90. V. Kashcheyev, A. Aharony, O. Entin-Wohlman, Applicability of the equations-of-motion technique for quantum dots. Phys. Rev. B **73**, 125338 (2006)
91. A. Goker, Kondo resonance in an AC driven quantum dot subjected to finite bias. Solid State Commun. **148**, 230–233 (2008)
92. M. Krawiec, Compensation of the Kondo effect in quantum dots coupled to ferromagnetic leads within the equation of motion approach. J. Phys.: Condens. Matter **19**, 346234 (2007)
93. G.D. Mahan, *Many-Particle Physics* (Plenum, New York, 2000)
94. M. Buttiker, R. Landauer, Traversal time for tunneling. Phys. Scripta **32**, 429–434 (1985)

95. A.B. Kaiser, Electronic transport properties of conducting polymers and carbon nanotubes. Rep. Prog. Phys. **64**, 1–49 (2001)
96. A.N. Aleshin, H.J. Lee, Y.W. Park, K. Akagi, Coulomb-blockade transport in quasi-one dimensional polymer nanofibers. Phys. Rev. Lett. **93**, 196601 (2004)
97. A. Bachtold, M. de Jonge, K. Grove-Rasmussen, P.L. McEuen, M. Buitelaar, C. Schonenberger, Suppression of tunneling into multiwall carbon nanotubes. Phys. Rev. Lett. **87**, 166801 (2001)
98. P.S. Cornaglia, D.R. Grempel, H. Ness, Quantum transport through a deformable molecular transistor. Phys. Rev. B **71**, 075320 (2005)
99. S. Monturet, N. Lorente, Inelastic effects in electron transport studied with wave packet propagation. Phys. Rev. B **78**, 035445 (2008)
100. A. Donarini, M. Grifoni, K. Richter, Dynamical symmetry breaking in transport through molecules. Phys. Rev. Lett. **97**, 166801 (2006)
101. R. Egger, A.O. Gogolin, Vibration-induced correction to the current through a single molecule. Phys. Rev. B **77**, 113405 (2008)
102. D.A. Ryndyk, G. Cuniberti, Nonequilibrium resonant spectroscopy of molecular vibrons. Phys. Rev. B **76**, 155430 (2007)
103. L. Siddiqui, A.W. Ghosh, S. Datta, Phonon runaway in carbon nanotube quantum dots. Phys. Rev. B **76**, 085433 (2007)
104. A. Troisi, M.A. Ratner, Molecular transport junctions: propensity rules for inelastic electron tunneling spectra. Nano Lett. **6**, 1784–1788 (2006)
105. J. Zimmermann, P. Pavone, G. Cuniberti, Vibrational modes and low-temperature thermal properties of graphene and carbon nanotubes: minimal force-constant model. Phys. Rev. B **78**, 045410 (2008)
106. J.G. Kushmerick, J. Lazorcik, C.H. Patterson, R. Shashidhar, D.S. Seferos, G.C. Bazan, Vibronic contributions to charge transport across molecular junctions. Nano Lett. **4**, 639–642 (2004)
107. M. Cizek, M. Thoss, W. Domcke, Theory of vibrationally inelastic electron transport through molecular bridges. Phys. Rev. B **70**, 125406 (2004)
108. J. Mravlje, A. Ramsak, Kondo effect and channel mixing in oscillating molecules. Phys. Rev. B **78**, 235416 (2008)
109. N.A. Zimbovskaya, M.M. Kuklja, Vibration-induced inelastic effects in the electron transport through multisite molecular bridges. J. Chem. Phys. **131**, 114703 (2009)
110. D. Djukic, K.S. Thygesen, C. Untiedt, R.H.M. Smit, K.W. Jacobsen, J.M. van Ruitenbeek, Stretching dependence of the vibration modes of a single-molecule $Pt–H_2–Pt$ bridge. Phys. Rev. B **71**, 161402(R) (2005)
111. X.H. Qiu, G.V. Nazin, W. Ho, Vibronic states in single molecule electron transport. Phys. Rev. Lett. **92**, 206102 (2004)
112. J. Repp, G. Meyer, S. Paavilainen, F.E. Olsson, M. Persson, Scanning tunneling spectroscopy of Cl vacancies in NaCl films: strong electron-phonon coupling in double-barrier tunneling junctions. Phys. Rev. Lett. **95**, 225503 (2005)
113. R.H.M. Smit, C. Untiedt, J.M. van Ruitenbeek, The high bias stability of monoatomic chains. Nanotechnology **15**, S472–S478 (2004)
114. W. Wang, T. Lee, I. Kretzschmar, M.A. Reed, Inelastic electron tunneling spectroscopy of an alkanedithiol self-assembled monolayer. Nano Lett. **4**, 643–646 (2004)
115. S.W. Wu, G.V. Nazin, X. Chen, X.H. Qiu, W. Ho, Control of relative tunneling rates in single molecule bipolar electron transport. Phys. Rev. Lett. **93**, 236802 (2004)
116. M. Tsutsui, S. Kurokawa, A. Sakai, Bias-induced local heating in Au atom-sized contacts. Nanotechnology **17**, 5334–5338 (2006)
117. Z. Huang, F. Chen, R. D'Agosta, P.A. Bennett, M. Di Ventra, N. Tao, Local ionic and electron heating in single-molecule junctions. Nat. Nanotechnol. **2**, 698–703 (2007)
118. A. Mishchenko, D. Vonlanthen, V. Meded, M. Burkle, C. Li, I.V. Pobelov, A. Bagrets, J.K. Viljas, F. Pauly, F. Evers, M. Mayor, T. Wandlowski, Influence of conformation on conductance of biphenyl-dithiol single-molecule contacts. Nano. Lett. **10**, 156 (2010)

119. S. Sanvito, A.R. Rocha, Molecular spintronics: the art of driving spin through molecules. J. Comput. Theor. Nanosci. **3**, 624–668 (2006)
120. A.R. Rocha, V.M. Garcia-Suarez, S.W. Bailey, C.J. Lambert, J. Ferrer, S. Sanvito, Towards molecular spintronics. Nat. Mater. **4**, 335–340 (2005)
121. L. Bogani, W.L. Wernsdorfer, Molecular spintronics using single-molecule magnets. Nat. Mater. **7**, 179–186 (2008)
122. V.V. Maslyuk, A. Bagrets, V. Meded, A. Arnold, F. Evers, M. Brandbyge, T. Bredow, I. Mertig, Organometallic benzene-vanadium wire: a one-dimensional half-metallic ferromagnet. Phys. Rev. Lett. **97**, 097201 (2006)
123. A.B. Vorontsov, M.G. Vavilov, Spin relaxation in quantum dots due to electron exchange with leads. Phys. Rev. Lett. **101**, 226805 (2008)
124. M.R. Wasielewski, Energy, charge, and spin transport in molecules and self-assembled nanostructures inspired by photosynthesis. J. Org. Chem. **71**, 5051–5066 (2006)
125. L. Zhou, S.-W. Yang, M.-F. Ng, M.B. Sullivan, V.B.C. Tan, L. Shen, One-dimensional iron-cyclopentadienyl sandwich molecular wire with half metallic, negative differential resistance and high-spin filter efficiency properties. J. Am. Chem. Soc. **130**, 4023–4027 (2008)
126. L. Wang, Z. Cai, J. Wang, J. Lu, G. Luo, L. Lai, J. Zhou, R. Qin, Z. Gao, D. Yu, G. Li, W.N. Mei, S. Sanvito, Novel one-dimensional organometallic half metals: vanadium-cyclopentadienyl, vanadium-cyclopentadienyl-benzene, and vanadium-anthracene wires, Nano Lett. **8**, 3640–3644 (2008)
127. J.-C. Wu, X.-F. Wang, L. Zhou, H.-X. Da, K.H. Lim, S.-W. Yang, Z.-Y. Li, Manipulating spin transport via vanadium-iron cyclopentadienyl multidecker sandwich molecules. J. Phys. Chem. C **113**, 7913–7916 (2009)
128. J. Splettstoesser, M. Governale, J. Konig, M. Buttiker, Charge and spin dynamics in interacting quantum dots. Phys. Rev. B **81**, 165318 (2010)
129. K. Tsukagoshi, B.W. Alphenaar, H. Ago, Coherent transport of electron spin in a ferromagnetically contacted carbon nanotube. Nature **401**, 572–574 (1999)
130. J.M. Kikkawa, D.D. Awschalom, Resonant spin amplification in n-type GaAs. Phys. Rev. Lett. **80**, 4313–4316 (1998)
131. J.M. Kikkawa, D.D. Awschalom, Lateral drag of spin coherence in gallium arsenide. Nature **397**, 139–141 (1999)
132. D.P. DiVincenzo, Quantum computation. Science **270**, 255–257 (1995)
133. S. Datta, B. Das, Electronic analog of the electro-optic modulator. Appl. Phys. Lett. **56**, 665–667 (1990)
134. Y.K. Kato, R.C. Myers, A.C. Gossard, D.D. Awschalom, Observation of the spin hall effect in semiconductors. Sci. Exp. **306**, 1910–1913 (2004)
135. S. Murakami, N. Nagaosa, S.-C. Zhang, Dissipationless quantum spin current at room temperature. Science **301**, 1348–1351 (2003)
136. J. Sinova, D. Culcer, Q. Niu, N.A. Sinitsyn, T. Jungwirth, A.H. MacDonald, Universal intrinsic spin hall effect. Phys. Rev. Lett. **92**, 126603 (2004)
137. J. Wunderlich, B. Kaestner, J. Sinova, T. Jungwirth, Experimental observation of the spin-hall effect in a two-dimensional spin-orbit coupled semiconductor system. Phys. Rev. Lett. **94**, 047204 (2005)
138. G. Salis, D.T. Fuchs, J.M. Kikkawa, D.D. Awschalom, Y. Ohno, H. Ohno, Optical manipulation of nuclear spin by a two-dimensional electron gas. Phys. Rev. Lett. **86**, 2677–2680 (2001)
139. M.W. Wu, J.H. Jiang, M.Q. Weng, Spin dynamics in semiconductors. Phys. Rep.**493**, 61–236 (2010)
140. R. Pati, L. Senapati, P.M. Ajayan, S.K. Nayak, First-principles calculations of spin-polarized electron transport in a molecular wire: molecular spin valve. Phys. Rev. B **68**, 100407(R) (2003)
141. R. Pati, M. Mailman, L. Senapati, P.M. Ajayan, S.D. Mahanti, S.K. Nayak, Oscillatory spin-polarized conductance in carbon atom wires. Phys. Rev. B **68**, 014412 (2003)

142. S. Schmaus, A. Bagrets, Y. Nahas, T.K. Yamada, A. Bork, M. Bowen, E. Beaurepaire, F. Evers, W. Wulfhekel, Giant magnetoresistance through a single molecule. Nat. Nanotechnol. **6**, 185 (2011)
143. Q.-F. Sun, J. Wang, H. Guo, Quantum transport theory for nanostructures with Rashba spin-orbital interaction. Phys. Rev. B **71**, 165310 (2005)
144. D. Gatteschi, R. Sessoli, Quantum tunneling of magnetization and related phenomena in molecular materials. Angew. Chem. Int. Ed. **42**, 268–297 (2003)
145. J. Cirera, E. Ruiz, S. Alvarez, F. Neese, J. Kortus, How to build molecules with large magnetic anisotropy. Chem. Eur. J. **15**, 4078–4087 (2009)
146. C. Romeike, M.R. Wegewijs, W. Hofstetter, H. Schoeller, Quantum-tunneling-induced Kondo effect in single molecular magnets. Phys. Rev. Lett. **96**, 196601 (2006)
147. F. Elste, C. Timm, Transport through anisotropic magnetic molecules with partially ferromagnetic leads: spin-charge conversion and negative differential conductance. Phys. Rev. B **73**, 235305 (2006)
148. C. Timm, F. Elste, Spin amplification, reading, and writing in transport through anisotropic magnetic molecules. Phys. Rev. B **73**, 235304 (2006)
149. C. Timm, Tunneling through magnetic molecules with arbitrary angle between easy axis and magnetic field. Phys. Rev. B **76**, 014421 (2007)
150. J.E. Grose, E.S. Tam, C. Timm, M. Scheloske, B. Ulgut, J.J. Parks, H.D. Abruna, W. Harneit, D.C. Ralph, Tunnelling spectra of individual magnetic endofullerene molecules. Nat. Mater. **7**, 884–889 (2008)
151. K. Park, M.R. Pederson, C.S. Hellberg, Properties of low-lying excited manifolds in Mn_{12} acetate. Phys. Rev. B **69**, 014416 (2004)
152. K. Park, M.R. Pederson, Effect of extra electrons on the exchange and magnetic anisotropy in the anionic single-molecule magnet Mn_{12}. Phys. Rev. B **70**, 054414 (2004)
153. J. Kortus, C. Massobrio, M. Drillon, First-principles calculations applied to molecular magnetism. J. Comput. Theor. Nanosci. **3**, 11–27 (2006)
154. A.V. Postnikov, J. Kortus, M.R. Pederson, Density functional studies of molecular magnets. Phys. Status Solidi B **243**, 2533–2572 (2006)
155. C. Loose, E. Ruiz, B. Kersting, J. Kortus, Magnetic exchange interaction in triply bridged dinickel(II) complexes. Chem. Phys. Lett. **452**, 38–43 (2008)
156. R. Liu, S.-H. Ke, H.U. Baranger, W. Yang, Organometallic spintronics: dicobaltocene switch. Nano Lett. **5**, 1959–1962 (2005)
157. C. Iacovita, M.V. Rastei, B.W. Heinrich, T. Brumme, J. Kortus, L. Limot, J.P. Bucher, Visualizing the spin of individual cobalt-phthalocyanine molecules. Phys. Rev. Lett. **101**, 116602 (2008)
158. O. Cespedes, M.S. Ferreira, S. Sanvito, M. Kociak, J.M.D. Coey, Contact induced magnetism in carbon nanotubes. J. Phys.: Condens. Matter **16**, L155–L161 (2004)
159. M.S. Ferreira, S. Sanvito, Contact-induced spin polarization in carbon nanotubes. Phys. Rev. B **69**, 035407 (2004)
160. Y. Yamamoto, T. Miura, M. Suzuki, N. Kawamura, H. Miyagawa, T. Nakamura, K. Kobayashi, T. Teranishi, H. Hori, Direct observation of ferromagnetic spin polarization in gold nanoparticles. Phys. Rev. Lett. **93**, 116801 (2004)
161. P. Crespo, R. Litran, T.C. Rojas, M. Multigner, J.M. de la Fuente, J.C. Sanchez-Lopez, M.A. Garcia, A. Hernando, S. Penades, A. Fernandez, Permanent magnetism, magnetic anisotropy, and hysteresis of thiol-capped gold nanoparticles. Phys. Rev. Lett. **93**, 087204 (2004)
162. S.G. Ray, S.S. Daube, G. Leitus, Z. Vager, R. Naaman, Chirality-induced spin-selective properties of self-assembled monolayers of DNA on gold. Phys. Rev. Lett. **96**, 036101 (2006)
163. B.C. Stipe, M.A. Rezaei, W. Ho, A variable-temperature scanning tunneling microscope capable of single-molecule vibrational spectroscopy. Rev. Sci. Instrum. **70**, 137–143 (1999)
164. Y. Kim, T. Pietsch, A. Erbe, W. Belzig, E. Scheer, Benzenedithiol: a broad-range single-channel molecular conductor. Nano Lett. **11**, 3734–3738 (2011)
165. M. Tsutsui, M. Taniguchi, Single molecule electronics and devices. Sensors **12**, 7259–7298 (2012)

166. M. Mayor, H.B. Weber, J. Reichert, M. Elbing, C.V. Haenisch, D. Beckmann, M. Fischer, Electric current through a molecular rod - relevance of the position of anchor groups. Angew. Chem. Int. Ed. **42**, 5834 (2003)
167. A. Bagrets, A. Arnold, F. Evers, Conduction properties of bipyridinium-functionalized molecular wires. J. Am. Chem. Soc. **130**, 9013 (2008)
168. X. Cao, R.J. Hamers, Silicon surfaces as electron acceptors: dative bonding of amines with Si(001) and Si(111) surfaces. J. Am. Chem. Soc. **123**, 10988–10996 (2001)
169. P.G. Piva, G.A. DiLabio, J.L. Pitters, J. Zikovsky, M. Rezeq, S. Dogel, W.A. Hofer, R.A. Wolkow, Field regulation of single-molecule conductivity by a charged surface atom. Nature **435**, 658–661 (2005)
170. F. Anariba, J. Steach, R.L. McCreery, Strong effects of molecular structure on electron transport in carbon/molecule/copper electronic junctions. J. Phys. Chem. B **109**, 11163–11172 (2005)
171. X. Guo, J.P. Small, J.E. Klare, Y. Wang, M.S. Purewal, I.W. Tam, B.H. Hong, R. Caldwell, L. Huang, S. O'Brien, J. Yan, R. Breslow, S.J. Wind, J. Hone, P. Kim, C. Nuckolls, Covalently bridging gaps in single-walled carbon nanotubes with conducting molecules. Science **311**, 356–359 (2006)
172. T. Rakshit, G.-C. Liang, A.W. Ghosh, M.C. Hersam, S. Datta, Molecules on silicon: self-consistent first-principles theory and calibration to experiments. Phys. Rev. B **72**, 125305 (2005)
173. V. Mujica, M.A. Ratner, Semiconductor/molecule transport junctions: an analytical form for the self-energies. Chem. Phys. **326**, 197–203 (2006)
174. J. Chen, M.A. Reed, Electronic transport of molecular systems. Chem. Phys. **281**, 127–145 (2002)
175. Y. Selzer, A. Salomon, D. Cahen, The importance of chemical bonding to the contact for tunneling through alkyl chains. J. Phys. Chem. B **106**, 10432–10439 (2002)
176. S. Lindsay, Molecular wires and devices: advances and issues. Faraday Discuss. **131**, 403–409 (2006)
177. R.M. Metzger, T. Xu, I.R. Peterson, Electrical rectification by a monolayer of hexadecylquinolinium tricyanoquinodimethanide measured between macroscopic gold electrodes. J. Phys. Chem. B **105**, 7280–7290 (2001)
178. A. Salomon, D. Cahen, S.M. Lindsay, J. Tomfohr, V.B. Engelkes, C.D. Frisbie, Comparison of electronic transport. Measurements on organic molecules. Adv. Mater. **15**, 1881–1890 (2003)
179. L.T. Cai, H. Skulason, J.G. Kushmerick, S.K. Pollack, J. Naciri, R. Shashidhar, D.L. Allara, T.E. Mallouk, T.S. Mayer, Nanowire-based molecular monolayer junctions: synthesis, assembly, and electrical characterization. J. Phys. Chem. B **108**, 2827–2832 (2004)
180. N. Gerge-Hackett, M.J. Cabral, T.L. Pernell, L.R. Harriott, J.C. Beanb, C.L.B. Chen, M. Lu, J.M. Tour, Vapor phase deposition of oligo-phenylene-ethynylene molecules for use in molecular electronic devices. J. Vac. Sci. Technol. B **25**, 252–257 (2007)
181. H.B. Akkerman, P.W.M. Blom, D.M. de Leeuw, B. de Boer, Towards molecular electronics with large-area molecular junctions. Nature **441**, 69–72 (2006)
182. H.B. Akkerman, B. de Boer, Electrical conduction through single molecules and self-assembled monolayers. J. Phys.: Condens. Matter **20**, 013001 (2008)
183. B. Xu, P. Zhang, X. Li, N. Tao, Direct conductance measurement of single DNA molecules in aqueous solution. Nano Lett. **4**, 1105–1108 (2004)
184. X.Y. Xiao, D. Brune, J. He, S.M. Lindsay, C.B. Gorman, N.J. Tao, Redox-gated electron transport in electrically wired ferrocene molecules. Chem. Phys. **326**, 138–143 (2006)
185. J.P. Choi, R.W. Murray, Electron self-exchange between $Au_{140}(+/0)$ nanoparticles is faster than that between $Au_{38}(+/0)$ in solid-state, mixed-valent films. J. Am. Chem. Soc. **128**, 10496–10502 (2006)
186. J. Liao, L. Bernard, M. Langer, C. Schonenberger, M. Calame, Reversible formation of molecular junctions in 2D nanoparticle arrays. Adv. Mater. **18**, 2444–2447 (2006)
187. J. Gaudioso, L.J. Lauhon, W. Ho, Vibrationally mediated negative differential resistance in a single molecule. Phys. Rev. Lett. **85**, 1918–1921 (2000)

188. J.R. Hahn, W. Ho, Imaging and vibrational spectroscopy of single pyridine molecules on $Ag(110)$ using a low-temperature scanning tunneling microscope. J. Chem. Phys. **124**, 204708 (2006)
189. J.B. Maddox, U. Harbola, N. Liu, C. Silien, W. Ho, G.C. Bazan, S. Mukamel, Simulation of single molecule inelastic electron tunneling signals in paraphenylene-vinylene oligomers and distyrylbenzene[2.2] paracyclophanes. J. Phys. Chem. A **110**, 6329–6338 (2006)
190. J. Tersoff, D.R. Hamann, Theory of the scanning tunneling microscope. Phys. Rev. B **31**, 805–813 (1985)
191. J. Tersoff, D.R. Hamann, Theory and application for the scanning tunneling microscope. Phys. Rev. Lett. **50**, 1998–2001 (1983)
192. R.A. Kiehl, J.D. Le, P. Candra, R.C. Hoye, T.R. Hoye, Charge storage model for hysteretic negative-differential resistance in metal-molecule-metal junctions. Appl. Phys. Lett. **88**, 172102 (2006)
193. R.P. Berkelaar, H. Sode, T.F. Mocking, A. Kumar, B. Poelsema, H.J.W. Zandvliet, Molecular bridges. J. Phys. Chem. C **115**, 2268 (2011)
194. L. Venkataraman, Y.S. Park, A.C. Whalley, C. Nuckolls, M.S. Hybertsen, M.L. Steigerwald, Electronics and chemistry: varying single-molecule junction conductance using chemical substituents. Nano Lett. **7**, 502–506 (2007)
195. X.D. Cui, A. Primak, X. Zarate, J. Tomfohr, O.F. Sankey, A.L. Moore, T.A. Moore, D. Gust, G. Harris, S.M. Lindsay, Reproducible measurement of single-molecule conductivity. Science **294**, 571–574 (2001)
196. B.Q. Xu, N.J. Tao, Measurement of single molecule conductance by repeated formation of molecular junctions. Science **301**, 1221–1223 (2003)
197. A.S. Blum, J.G. Kushmerick, D.P. Long, C.H. Patterson, J.C. Yang, J.C. Henderson, Y. Yao, J.M. Tour, R. Shashidhar, B.R. Ratna, Molecularly inherent voltage controlled conductance switching. Nat. Mater. **4**, 167–172 (2005)
198. A.S. Blum, J.G. Kushmerick, J.C. Yang, M. Moore, S.K. Pollack, J. Naciri, R. Shashidhar, B.R. Ratna, Charge transport and scaling in molecular wires. J. Phys. Chem. B **108**, 18124–18128 (2004)
199. Y. Selzer, L. Cai, M.A. Cabassi, Y. Yao, J.M. Tour, T.S. Mayer, D.L. Allara, Effect of local environment on molecular conduction: isolated molecule versus self-assembled monolayer. Nano Lett. **5**, 61–65 (2005)
200. S.N. Yaliraki, M.A. Ratner, Molecule-interface coupling effects on electronic transport in molecular wires. J. Chem. Phys. **109**, 5036–5043 (1998)
201. C.D. Lindstrom, X.-Y. Zhu, Photoinduced electron transfer at molecule-metal interfaces. Chem. Rev. **106**, 4281–4300 (2006)
202. L.P. Kadanoff, G. Baum, *Quantum Statistical Mechanics. Green's Function Method in Equilibrium and Nonequilibrium Problems* (Benjamin, Reading, 1962)
203. M. Vagner, Expansions of nonequilibrium Green's functions. Phys. Rev. B **44**, 6104–6117 (1991)
204. E.N. Economou, *Green's Functions in Quantum Physics* (Springer, New York, 2005)
205. N.S. Wingreen, K.W. Jacobsen, J.W. Wilkins, Resonant tunneling with electron-phonon interaction: an exactly solvable model. Phys. Rev. Lett. **61**, 1396–1399 (1988)
206. N.S. Wingreen, K.W. Jacobsen, J.W. Wilkins, Inelastic scattering in resonant tunneling. Phys. Rev. B **40**, 11834–11850 (1989)
207. A. Troisi, M.A. Ratner, Modeling the inelastic electron tunneling spectra of molecular wire junctions. Phys. Rev. B **72**, 033408 (2005)
208. A. Troisi, M.A. Ratner, A. Nitzan, Vibronic effects in off-resonance molecular wire conductance. J. Chem. Phys. **118**, 6072–6082 (2003)
209. L. Yan, Inelastic electron tunneling spectroscopy and vibrational coupling. J. Phys. Chem. A **110**, 13249–13252 (2006)
210. S. Datta, *Electric Transport in Mesoscopic Systems* (Cambridge University Press, Cambridge, 1995)

211. Y. Magarshak, J. Malinsky, A.D. Joran, Diagram techniques for solving Schwinger-Dyson equations: electron transfer pathways in biological molecules. J. Chem. Phys. **95**, 418–431 (1991)
212. J. Bonca, S.A. Trugman, Inelastic quantum transport. Phys. Rev. Lett. **79**, 4874–4877 (1997)
213. H. Ness, Quantum inelastic electron-vibration scattering in molecular wires: Landauer-like versus Green's function approaches and temperature effects. J. Phys.: Condens. Matter **18**, 6307–6328 (2006)
214. Y. Zhou, M. Freitag, J. Hone, C. Staii, A.T. Johnson, N.J. Pinto, A.G. MacDiarmid, Fabrication and electrical characterization of polyaniline-based nanofibers with diameter below 30 nm. Appl. Phys. Lett. **83**, 3800–3802 (2003)
215. A.G. MacDiarmid, Nobel lecture: synthetic metals: a novel role for organic polymers. Rev. Mod. Phys. **73**, 701–712 (2001)
216. J. Joo, Z. Oblakowski, G. Du, J.P. Pouget, E.J. Oh, J.M. Wiesinger, Y. Min, A.G. MacDiarmid, A.J. Epstein, Microwave dielectric response of mesoscopic metallic regions and the intrinsic metallic state of polyaniline. Phys. Rev. B **49**, 2977–2980 (1994)
217. J.P. Pouget, Z. Oblakowski, Y. Nogami, P.A. Albouy, M. Laridjani, E.J. Oh, Y. Min, A.G. MacDiarmid, J. Tsukamoto, T. Ishiguro, A.J. Epstein, Recent structural investigations of metallic polymers. Synth. Met. **65**, 131–140 (1994)
218. M. Pollak, C.J. Adkins, Conduction in granular metals. Philos. Mag. B **65**, 855–860 (1992)
219. V.N. Prigodin, A.J. Epstein, Nature of insulator-metal transition and novel mechanism of charge transport in the metallic state of highly doped electronic polymers. Synth. Met. **125**, 43–53 (2001)
220. C.J. Bolton-Heaton, C.J. Lambert, V.I. Falko, V.N. Prigodin, A.J. Epstein, Distribution of time constants for tunneling through a one-dimensional disordered chain. Phys. Rev. B **60**, 10569–10572 (1999)
221. N.A. Zimbovskaya, A.T. Johnson, N.J. Pinto, Electronic transport mechanism in conducting polymer nanofibers. Phys. Rev. B **72**, 024213 (2005)
222. N.A. Zimbovskaya, Inelastic electron transport in polymer nanofibers. J. Chem. Phys. **123**, 114705 (2008)
223. H. Hang, A.-P. Jauho, *Quantum Kinetics in Transport and Optics in Semiconductors* (Springer, Berlin, 1996)
224. K.-C. Chou, Z.-B. Su, B.-L. Hao, L. Yu, Equilibrium and nonequilibrium formalisms made unified. Phys. Rep. **118**, 1–131 (1984)
225. B.-L. Hao, Closed time path Green's functions and nonlinear response theory. Physica A **109**, 221–236 (1981)
226. E. Wang, U. Heinz, Generalized fluctuation-dissipation theorem for nonlinear response functions. Phys. Rev. D **66**, 025008 (2002)
227. S. Mukamel, *Principles of Nonlinear Optical Spectroscopy* (University Press, New York, 1995)
228. U. Harbola, S. Mukamel, Non-equilibrium superoperator GW-equations. J. Chem. Phys. **124**, 044106–044117 (2006)
229. U. Harbola, S. Mukamel, Superoperator nonequilibrium Green's function theory of many-body systems; applications to charge transfer and transport in open junctions. Phys. Rep. **465**, 191–222 (2008)
230. Y. Meir, N.S. Wingreen, Landauer formula for the current through an interacting electron region. Phys. Rev. Lett. **68**, 2512–2515 (1992)
231. A.P. Jauho, N.S. Wingreen, Y. Meir, Time-dependent transport in interacting and noninteracting resonant-tunneling systems. Phys. Rev. B **50**, 5528–5544 (1994)
232. N.S. Wingreen, Y. Meir, Anderson model out of equilibrium: Noncrossing-approximation approach to transport through a quantum dot. Phys. Rev. B **49**, 11040–11052 (1994)
233. J. Inarrea, G. Platero, A.H. MacDonald, Electronic transport through a double quantum dot in the spin-blockade regime: theoretical models. Phys. Rev. B **76**, 085329 (2007)
234. L.Y. Gorelik, A. Isacsson, M.V. Voinova, B. Kasemo, R.I. Shekhter, M. Jonson, Shuttle mechanism for charge transfer in Coulomb blockade nanostructures. Phys. Rev. Lett. **80**, 4526–4529 (1998)

235. N.M. Chtchelkatchev, W. Belzig, C. Bruder, Charge transport through a single-electron transistor with a mechanically oscillating island. Phys. Rev. B **70**, 193305 (2004)
236. Ya.M. Blanter, O. Usmani, Yu.V. Nazarov, Single-electron tunneling with strong mechanical feedback. Phys. Rev. Lett. **93**, 136802 (2004)
237. C.B. Doiron, W. Belzig, C. Bruder, Electrical transport through a single-electron transistor strongly coupled to an oscillator. Phys. Rev. B **74**, 205336 (2006)
238. K.D. McCarthy, N. Prokofiev, M.T. Tuominen, Incoherent dynamics of vibrating single-molecule transistors. Phys. Rev. B **67**, 245415 (2003)
239. J. Koch, M. Semmelhack, F. von Oppen, A. Nitzan, Current-induced nonequilibrium vibrations in single-molecule devices. Phys. Rev. B **73**, 155306 (2006)
240. J. Koch, F. von Oppen, Franck-Condon blockade and giant Fano factors in transport through single molecules. Phys. Rev. Lett. **94**, 206804 (2005)
241. J. Koch, F. von Oppen, Effects of charge-dependent vibrational frequencies and anharmonicities in transport through molecules. Phys. Rev. B **72**, 113308 (2005)
242. J. Koch, M.E. Raikh, F. von Oppen, Pair tunneling through single molecules. Phys. Rev. Lett. **96**, 056803 (2006)
243. H. Ueba, T. Mii, N. Lorente, B.N.J. Persson, Adsorbate motions induced by inelastic-tunneling current: theoretical scenarios of two-electron processes. J. Chem. Phys. **123**, 084707 (2005)
244. S. Braig, K. Flensberg, Vibrational sidebands and dissipative tunneling in molecular transistors. Phys. Rev. B **68**, 205324 (2003)
245. D.A. Ryndyk, P. D'Amico, K. Richter, Single-spin polaron memory effect in quantum dots and single molecules. Phys. Rev. B **81**, 115333 (2010)
246. M.L. Jones, I.V. Kurnikov, D.N. Beratan, The nature of tunneling pathway and average packing density models for protein-mediated electron transfer. J. Phys. Chem. A **106**, 2002–2006 (2002)
247. J. Tersoff, Theory of semiconductor heterojunctions: the role of quantum dipoles. Phys. Rev. B **30**, 4874–4877 (1984)
248. M.R. Pederson, D.V. Porezag, J. Kortus, D.C. Patton, Strategies for massively parallel local-orbital-based electronic structure methods. Phys. Status Solidi B **217**, 197–218 (2000)
249. N.A. Zimbovskaya, M.R. Pederson, Negative differential resistance in molecular junctions: effect of the electronic structure of the electrodes. Phys. Rev. B **78**, 153105 (2008)
250. R. Landauer, Spatial variation of currents and fields due to localized scatterers in metallic conduction. IBM J. Res. Dev. **32**, 306–316 (1988)
251. A. Wacker, Semiconductor superlattices: a model system for nonlinear transport. Phys. Rep. **357**, 1–111 (2002)
252. A.H. Verbruggen, Fundamental questions in the theory of electromigration. IBM J. Res. Dev. **32**, 93–98 (1988)
253. M. Di Ventra, N.D. Lang, Transport in nanoscale conductors from first principles. Phys. Rev. B **65**, 045402 (2001)
254. F. Kassubek, C.A. Stafford, H. Grabert, Force, charge, and conductance of an ideal metallic nanowire. Phys. Rev. B **59**, 7560–7574 (1999)
255. B.-Y. Choi, S.-J. Kahng, S. Kim, H. Kim, H.W. Kim, Y.J. Song, J. Ihm, Y. Kuk, Conformational molecular switch of the azobenzene molecule: a scanning tunneling microscopy study. Phys. Rev. Lett. **96**, 156106 (2006)
256. A.C. Johnson, C.M. Marcus, M.P. Hanson, A.C. Gossard, Coulomb-modified Fano resonance in a one-lead quantum dot. Phys. Rev. Lett. **93**, 106803 (2004)
257. J.G. Simmons, Generalized formula for the electric tunnel effect between similar electrodes separated by a thin insulating film. J. Appl. Phys. **34**, 1793 (1963)
258. N.A. Zimbovskaya, Electron transport through asymmetric ferroelectric tunnel junctions: current–voltage characteristics. J. Appl. Phys. **106**, 124101 (2009)
259. S.H. Choi, B. Kim, C.D. Frisbie, Electrical resistance of long conjugated molecular wires. Science **320**, 1482 (2008)

260. F. Zahid, M. Paulsson, E. Polizzi, A.W. Ghosh, L. Siddiqui, S. Datta, A self-consistent transport model for molecular conduction based on extended Huckel theory with full three-dimensional electrostatics. J. Chem. Phys. **123**, 064707 (2005)
261. P.F. Bagwell, T.P. Orlando, Landauer's conductance formula and its generalization to finite voltages. Phys. Rev. B **40**, 1456 (1989)
262. A.W. Ghosh, T. Rakshit, S. Datta, Gating of a molecular transistor: electrostatic and conformational. Nano. Lett. **4**, 565 (2004)
263. M.A. Kastner, The single-electron transistor. Rev. Mod. Phys. **64**, 849–858 (1992)
264. L.W. Yu, K.J. Chen, J. Song, J.M. Wang, J. Xu, W. Li, X.F. Huang, Coulomb blockade induced negative differential resistance effect in a self-assembly Si quantum dots array at room temperature. Thin Solid Films **515**, 5466–5470 (2007)
265. F.-R.F. Fan, R.Y. Lai, J. Cornil, Y. Karzazi, J.-L. Bredas, L. Cai, L. Cheng, Y. Yao, D.W. Price, S.M. Dirk, J.M. Tour, A.J. Bard, Electrons are transported through phenylene-ethynylene oligomer monolayers via localized molecular orbitals. J. Am. Chem. Soc. **126**, 2568–2573 (2004)
266. A. Salomon, R. Arad-Yellin, A. Shanzer, A. Karton, D. Cahen, Stable room-temperature molecular negative differential resistance based on molecule-electrode interface chemistry. J. Am. Chem. Soc. **126**, 11648–11657 (2004)
267. M. Grobis, A. Wachowiak, R. Yamachika, M.F. Crommie, Tuning negative differential resistance in a molecular film. Appl. Phys. Lett. **86**, 204102 (2005)
268. J. Repp, G. Meyer, S.M. Stojkovic, A. Gourdon, C. Joachim, Molecules on insulating films: scanning-tunneling microscopy imaging of individual molecular orbitals. Phys. Rev. Lett. **94**, 026803 (2005)
269. E.D. Mentovich, I. Kalifa, A. Tsukernik, A. Caster, N. Rosenberg-Shraga, H. Marom, M. Gozin, S. Richter, Multipeak negative-differential-resistance molecular device. Small **4**, 55–58 (2008)
270. G. Maruccio, P. Marzo, R. Krahne, A. Passaseo, R. Cingolani, R. Rinaldi, Protein conduction and negative differential resistance in large-scale nanojunction arrays. Small **3**, 1184–1188 (2007)
271. Y. Xue, S. Datta, S. Hong, R. Reifenberger, J.I. Henderson, C.P. Kubiak, Negative differential resistance in the scanning-tunneling spectroscopy of organic molecules. Phys. Rev. B **59**, R7852–R7855 (1999)
272. V. Mujica, A. Nitzan, S. Datta, M.A. Ratner, C.P. Kubiak, Molecular wire junctions: tuning the conductance. J. Phys. Chem. B **107**, 91–95 (2003)
273. K.K. Likharev, D.A. Averin, Single-electronics: a correlated transfer of single electrons and cooper pairs in systems of small tunnel junctions, in *Mesoscopic Phenomena in Solids*, ed. by B. Altshuler et al. (Elsevier, Amsterdam, 1991), pp. 173–271
274. C.W.J. Beenakker, H. van Houten, Quantum transport m semiconducting nanostructures, in *Solid State Physics*, vol. 44, ed. by H. Ehrenreich, D. Turnbull (Academic, New York, 1991), pp. 1–228
275. I.O. Kulik, R.I. Shekhter, Kinetic phenomena and charge discreteness effects in granulated media. Zh. Eksp. Teor. Fiz. **68**, 623–629 (1975) [Sov. Phys. JETP **41**, 308–313 (1975)]
276. Y. Hamamoto, T. Kato, Numerical study of the Coulomb blockade in an open quantum dot. J Phys.: Conf. Ser. **150**, 022021 (2009)
277. M. Ezawa, Coulomb blockade in graphene nanodisks. Phys. Rev. B **77**, 155411 (2008)
278. B. Wunsch, T. Stauber, F. Guinea, Electron–electron interactions and charging effects in graphene quantum dots. Phys. Rev. B. **77**, 035316 (2008)
279. H. Park, J. Park, A.K.L. Lim, E.H. Anderson, A.P. Alivisatos, P.L. McEuen, Nanomechanical oscillations in a single-C_{60} transistor. Nature **407**, 57–60 (2000)
280. A.N. Pasupathy, J. Park, C. Chang, A.V. Soldatov, S. Lebedkin, R.C. Bialczak, J.E. Grose, L.A.K. Donev, J.P. Sethna, D.C. Ralph, P.L. McEuen, Vibration-assisted electron tunneling in C140 single-molecule transistors. Nano Lett. **5**, 203–207 (2005)
281. O.D. Miller, B. Muralidharan, N. Kapur, A.W. Ghosh, Rectification by charging-contact induced asymmetry in molecular conductors. Phys. Rev. B **77**, 125427 (2008)

282. B. Muralidharan, A.W. Ghosh, S.K. Pati, S. Datta, Theory of high bias Coulomb blockade through ultra short molecules. IEEE Trans. Nano **6**, 536–544 (2007)
283. B. Muralidharan, S. Datta, Generic model for current collapse in spin blockaded transport. Phys. Rev. B **76**, 035432 (2007)
284. B. Muralidharan, L. Siddiqui, A.W. Ghosh, Role of multiparticle excitations in Coulomb blockaded transport. J. Phys.: Condens. Matter **20**, 374109 (2008)
285. F. Elste, C. Timm, Cotunneling and nonequilibrium magnetization in magnetic molecular monolayers. Phys. Rev. B **75**, 195341 (2007)
286. Y. Nagaoka, Ferromagnetism in a narrow, almost half-filled s band. Phys. Rev. **147**, 392–405 (1966)
287. D. Weinmann, W. Hausler, B. Kramer, Spin blockades in linear and nonlinear transport through quantum dots. Phys. Rev. Lett. **74**, 984–987 (1995)
288. R. Arita, Y. Suwa, K. Kuroki, H. Aoki, Gate-induced band ferromagnetism in an organic polymer. Phys. Rev. Lett. **88**, 127202 (2002)
289. P. Huai, Y. Shimoi, S. Abe, Electronic control of spin alignment in π-conjugated molecular magnets. Phys. Rev. Lett. **90**, 207203 (2003)
290. A. Izuoka, M. Hiraishi, T. Abe, T. Sugawara, K. Sato, T. Takui, Spin alignment in singly oxidized spin-polarized diradical donor: thianthrene bis(nitronyl nitroxide). J. Am. Chem. Soc. **122**, 3234–3235 (2000)
291. J. Zarembowitch, O. Kahn, New J. Chem. **15**, 181–189 (1991)
292. R. Sessoli, D. Gatteschi, A. Caneschi, M.A. Novak, Magnetic bistability in a metal-ion cluster. Nature **365**, 141–143 (1993)
293. O. Kahn, C.J. Martinez, Spin-transition polymers: from molecular materials toward memory devices. Science **279**, 44–48 (1998)
294. S. Hill, R.S. Edwards, N. Aliaga-Alcalde, G. Christou, Quantum coherence in an exchange-coupled dimer of single-molecule magnets. Science **302**, 1015–1018 (2003)
295. A.-L. Barra, P. Debrunner, D. Gatteschi, Ch.E. Schulz, R. Sessoli, Superparamagnetic-like behavior in an octanuclear iron cluster. Europhys. Lett. **35**, 133–136 (1996)
296. W.J. de Haas, J. de Boer, G.J. van den Berg, The electrical resistance of gold, copper and lead at low temperatures. Physica **1**, 1115–1124 (1934)
297. J. Kondo, Resistance minimum in dilute magnetic alloys. Prog. Theor. Phys. **32**, 37–49 (1964)
298. L.I. Glazman, M.E. Raikh, Resonant Kondo transparency of a barrier with quasilocal impurity states. Pisma Zh. Exp. Teor. Fiz. **47**, 378–382 (1988) [JETP Lett. **47**, 452–456 (1988)]
299. T.K. Ng, P.A. Lee, On-site coulomb repulsion and resonant tunneling. Phys. Rev. Lett. **61**, 1768–1771 (1988)
300. D. Goldhaber-Gordon, H. Shtrikman, D. Mahalu, D. Abusch-Magder, U. Meirav, M.A. Kastner, Kondo effect in a single-electron transistor. Nature **391**, 156–159 (1998)
301. J. Schmid, J. Weis, K. Eberl, K.v. Klitzing, Absence of Odd-Even parity behavior for Kondo resonances in quantum dots. Phys. Rev. Lett. **84**, 5824–5827 (2000)
302. F. Simmel, R.H. Blick, J.P. Kotthaus, W. Wegscheider, M. Bichler, Anomalous Kondo effect in a quantum dot at nonzero bias. Phys. Rev. Lett. **83**, 804–807 (1999)
303. A.N. Pasupathy, R.C. Bialczak, J. Martinek, J.E. Grose, L.A.K. Donev, P.L. McEuen, D.C. Ralph, The Kondo effect in the presence of ferromagnetism. Science **306**, 86–89 (2004)
304. J. Martinek, Y. Utsumi, H. Imamura, J. Barnas, S. Maekawa, J. Konig, G. Schon, Kondo effect in quantum dots coupled to ferromagnetic leads. Phys. Rev. Lett. **91**, 127203 (2003)
305. Y. Utsumi, J. Martinek, G. Schon, H. Imamura, S. Maekawa, Nonequilibrium Kondo effect in a quantum dot coupled to ferromagnetic leads. Phys. Rev. B **71**, 245116 (2005)
306. R. Swirkowicz, M. Wilczynski, J. Barnas, Spin-polarized transport through a single-level quantum dot in the Kondo regime. J. Phys.: Condens. Matter **18**, 2291–2304 (2006)
307. M. Krawiec, K.I. Wysokinski, Thermoelectric effects in strongly interacting quantum dot coupled to ferromagnetic leads. Phys. Rev. B **73**, 075307 (2006)
308. R. Swirkowicz, M. Wilczynski, M. Wawrzyniak, J. Barnas, Kondo effect in quantum dots coupled to ferromagnetic leads with noncollinear magnetizations. Phys. Rev. B **73**, 193312 (2006)

309. P. Nordlander, N.S. Wingreen, Y. Meir, D.C. Langreth, Kondo physics in the single-electron transistor with ac driving. Phys. Rev. B **61**, 2146–2150 (2000)
310. P.W. Anderson, A poor man's derivation of scaling laws for the Kondo problem. J. Phys. C: Solid State Phys. **3**, 2436 (1970)
311. A. Rosch, J. Paaske, J. Kroha, P. Wolfle, Nonequilibrium transport through a Kondo dot in a magnetic field: perturbation theory and poor man's scaling. Phys. Rev. Lett. **90**, 076804 (2003)
312. J. Paaske, A. Rosch, P. Wolfle, Nonequilibrium transport through a Kondo dot in a magnetic field: perturbation theory. Phys. Rev. B **69**, 155330 (2004)
313. H. Schmidt, P. Wolfle, Transport through a Kondo quantum dot: functional RG approach. Ann. Phys. (Berlin) **19**, 60 (2010)
314. V. Koerting, J. Paaske, P. Wolfle, Electron transport in the four-lead two-impurity Kondo model: nonequilibrium perturbation theory with almost degenerate levels. Phys. Rev. B **77**, 165122 (2008)
315. D. Secker, S. Wagner, S. Ballmann, R. Hartle, M. Thoss, H.B. Weber, Resonant vibrations, peak broadening, and noise in single molecule contacts: the nature of the first conductance peak. Phys. Rev. Lett. **106**, 136807 (2011)
316. S. Ballmann, W. Hieringer, D. Secker, Q. Zheng, J.A. Gladysz, A. Goerling, H.B. Weber, Molecular wires in single-molecule junctions: charge transport and vibrational excitations. Chem. Phys. Chem. **11**, 2256 (2010)
317. R. Gutierrez, S. Mohapatra, H. Cohen, D. Porath, G. Cuniberti, Inelastic quantum transport in a ladder model: implications for DNA conduction and comparison to experiments on suspended DNA oligomers. Phys. Rev. B **74**, 235105 (2006)
318. S.G. Tikhodeev, H. Ueba, Relation between inelastic electron tunneling and vibrational excitation of single adsorbates on metal surfaces. Phys. Rev. B **70**, 125414 (2004)
319. B.N.J. Persson, A. Baratoff, Inelastic electron tunneling from a metal tip: the contribution from resonant processes. Phys. Rev. Lett. **59**, 339–342 (1987)
320. C.A. Balseiro, P.S. Cornaglia, D.R. Grempel, Electron-phonon correlation effects in molecular transistors. Phys. Rev. B **74**, 235409 (2006)
321. J. Paaske, K. Flensberg, Vibrational sidebands and the Kondo effect in molecular transistors. Phys. Rev. Lett. **94**, 176801 (2005)
322. R. Swirkowicz, M. Wilczynski, M. Wawrzyniak, J. Barnas, Kondo effect in quantum dots coupled to ferromagnetic leads with noncollinear magnetizations: effects due to electron-phonon coupling. J. Phys.: Condens. Matter **20**, 255219 (2008)
323. R.-Q. Wang, Y.-Q. Zhou, B. Wang, D.Y. Xing, Spin-dependent inelastic transport through single-molecule junctions with ferromagnetic electrodes. Phys. Rev. B **75**, 045318 (2007)
324. R.-Q. Wang, Y.-Q. Zhou, D.Y. Xing, Vibration-mediated Kondo effect in a single-molecule quantum dot coupled to ferromagnetic electrodes. J Phys.: Condens. Matter **28**, 045219 (2008)
325. J. Paaske, A. Rosch, P. Wolfle, N. Mason, C.M. Marcus, J. Nygard, Non-equilibrium singlet-triplet Kondo effect in carbon nanotubes. Nat. Phys. **2**, 460–464 (2006)
326. Z.-Z. Chen, H. Lu, R. Lu, B.-F. Zhu, Phonon-assisted Kondo effect in a single-molecule transistor out of equilibrium. J. Phys.: Condens. Matter **18**, 5435 (2006)
327. F. Elste, F. von Oppen, Asymmetric Coulomb blockade and Kondo temperature of single-molecule transistors. New J. Phys. **10**, 065021 (2008)
328. D. Brisker-Klaiman, U. Peskin, Coherent elastic transport contribution to currents through ordered DNA molecular junctions. J. Phys. Chem. C **114**, 19077–19082 (2010)
329. A. Garg, J. Nelson Onuchic, V. Ambegaokar, Effect of friction on electron transfer in biomolecules. J. Chem. Phys. **83**, 4491–4503 (1985)
330. N.A. Zimbovskaya, On the dissipative effects in the electron transport through conducting polymer nanofibers. J. Chem. Phys. **126**, 184901 (2007)
331. Ya.M. Blanter, M. Buttiker, Shot noise in mesoscopic conductors. Phys. Rep. **336**, 1 (2000)
332. C. Flindt, T. Novotny, A.-P. Jauho, Current noise in a vibrating quantum dot array. Phys. Rev. B **70**, 205334 (2004)

333. U. Hanke, Y. Galperin, K.A. Chao, M. Gisselfalt, M. Jonson, R.I. Shekhter, Static and dynamic transport in parity-sensitive systems. Phys. Rev. B **51**, 9084 (1995)
334. M. Galperin, A. Nitzan, M.A. Ratner, Inelastic tunneling effects on noise properties of molecular junctions. Phys. Rev. B **74**, 075326 (2006)
335. D. Mozyrsky, M.B. Hastings, I. Martin, Intermittent polaron dynamics: Born-Oppenheimer approximation out of equilibrium. Phys. Rev. B **73**, 035104 (2006)
336. M. Galperin, A. Nitzan, M.A. Ratner, Inelastic effects in molecular junction transport: scattering and self-consistent calculations for the Seebeck coefficient. Mol. Phys. **106**, 397–404 (2008)
337. A.I. Hochbaum, R. Chen, R. Diaz Delgado, W. Liang, E.C. Garnett, M. Najarian, A. Majumdar, P. Yang, Enhanced thermoelectric performance of rough silicon nanowires. Nature **451**, 163–167 (2008)
338. K. Baheti, J.A. Malen, P. Doak, P. Reddy, S.Y. Jang, T.D. Tilley, A. Majumdar, R.A. Segalman, Probing the chemistry of molecular heterojunctions using thermoelectricity. Nano Lett. **8**, 715–719 (2008)
339. A.I. Boukai, Y. Bunimovich, J. Tahir-Kheli, J.-K. Yu, W.A. Goddard III, J.R. Heath, Silicon nanowires as efficient thermoelectric materials. Nature **451**, 168–171 (2008)
340. K. Schwab, E.A. Henriksen, J.M. Worlock, M.L. Roukes, Measurement of the quantum of thermal conductance. Nature **404**, 974–977 (2000)
341. L.G.C. Rego, G. Kirczenow, Quantized thermal conductance of dielectric quantum wires. Phys. Rev. Lett. **81**, 232–235 (1998)
342. Y. Dubi, M. Di Ventra, Thermoelectric effects in nanoscale junctions. Nano Lett. **9**, 97–101 (2009)
343. T. Markussen, A.-P. Jauho, M. Brandbyge, Electron and phonon transport in silicon nanowires: atomistic approach to thermoelectric properties. Phys. Rev. B **79**, 035415 (2009)
344. A.M. Lunde, K. Flensberg, L.I. Glazman, Interaction-induced resonance in conductance and thermopower of quantum wires. Phys. Rev. Lett. **97**, 256802 (2006)
345. D. Segal, Thermoelectric effect in molecular junctions: a tool for revealing transport mechanisms. Phys. Rev. B **72**, 165426 (2005)
346. B. Wang, Y. Xing, L. Wan, Y. Wei, J. Wang, Oscillatory thermopower of carbon chains: first-principles calculations. Phys. Rev. B **71**, 233406 (2005)
347. Y.-C. Liu, Y.-R. Chen, Y.-C. Chen, Efficiency of energy conversion in thermoelectric nanojunctions. Nano Lett. **9**, 97–101 (2009)
348. C.W.J. Beenakker, A.A.M. Staring, Theory of the thermopower of a quantum dot. Phys. Rev. B **46**, 9667–9676 (1992)
349. Ya.M. Blanter, C. Bruder, R. Fazio, H. Schoeller, Aharonov-Bohm-type oscillations of thermopower in a quantum-dot ring geometry. Phys. Rev. B **55**, 4069–4072 (1997)
350. M. Turek, K.A. Matveev, Cotunneling thermopower of single electron transistors. Phys. Rev. B **65**, 115332 (2002)
351. J. Koch, F. von Oppen, Y. Oreg, E. Sela, Thermopower of single-molecule devices. Phys. Rev. B **70**, 195107 (2004)
352. B. Kubala, J. Konig, Quantum-fluctuation effects on the thermopower of a single-electron transistor. Phys. Rev. B **73**, 195316 (2006)
353. B. Kubala, J. Konig, J. Pekola, Violation of the Wiedemann-Franz law in a single-electron transistor. Phys. Rev. Lett. **100**, 066801 (2008)
354. X. Zianni, Coulomb oscillations in the electron thermal conductance of a dot in the linear regime. Phys. Rev. B **75**, 045344 (2007)
355. D. Boese, R. Fazio, Thermoelectric effects in Kondo correlated quantum dots. Europhys. Lett. **56**, 576–582 (2001)
356. B. Dong, X.L. Lei, Effect of the Kondo correlation on the thermopower in a quantum dot. J. Phys.: Condens. Matter **14**, 11747–11756 (2002)
357. R. Sakano, T. Kita, N. Kawakami, Thermopower of multiorbital Kondo effect via single quantum dot system at finite temperatures. J. Phys. Soc. Jpn. **76**, 074709 (2007)

358. R. Scheibner, H. Buhmann, D. Reuter, M.N. Kiselev, L.W. Molenkamp, Thermopower of a Kondo spin-correlated quantum dot. Phys. Rev. Lett. **95**, 176602 (2005)
359. M. Yoshida, L.N. Oliveira, Thermoelectric effects in quantum dots. Physica B **404**, 3312–3315 (2009)
360. P. Murphy, S. Mukerjee, J. Moore, Optimal thermoelectric figure of merit of a molecular junction. Phys. Rev. B **78**, 161406(R) (2008)
361. C.M. Finch, V.M. Garcia-Suarez, C.J. Lambert, Giant thermopower and figure of merit in single-molecule devices. Phys. Rev. B **79**, 033405 (2009)
362. M. Leijnse, M.R. Wegewijs, K. Flensberg, Nonlinear thermoelectric properties of molecular junctions with vibrational coupling. Phys. Rev. B **82** 045412 (2010)
363. M. Hatami, G.E.W. Bauer, Q. Zhang, P.J. Kelly, Thermoelectric effects in magnetic nanostructures. Phys. Rev. B **79**, 174426 (2009)
364. R. Swirkowicz, M. Wierzbicki, J. Barnas. Thermoelectric effects in transport through quantum dots attached to ferromagnetic leads with noncollinear magnetic moments. Phys. Rev. B **80**, 195409 (2009)
365. T. Seideman, Current-driven dynamics in molecular-scale devices. J. Phys.: Condens. Matter **15**, R521–R549 (2003)
366. N. Lorente, R. Rurali, H. Tang, Single-molecule manipulation and chemistry with the STM. J. Phys.: Condens. Matter **17**, S1049 (2005)
367. Y.-C. Chen, M. Zwolak, M. Di Ventra, Local heating in nanoscale conductors. Nano Lett. **3**, 1691–1694 (2005)
368. C. Li, D. Zhang, X. Liu, S. Han, T. Tang, J. Han, C. Zhou, In_2O_3 nanowires as chemical sensors. Appl. Phys. Lett. **82**, 1613–1615 (2003)
369. J. Repp, G. Meyer, F. Olsson, M. Persson, Controlling the charge state of individual gold adatoms. Science **305**, 493–497 (2004)
370. F.E. Olsson, S. Paavilainen, M. Persson, J. Repp, G. Meyer, Multiple charge states of Ag atoms on ultrathin NaCl films. Phys. Rev. Lett. **98**, 176803 (2007)
371. V.S. Khrapai, S. Ludwig, J.P. Kotthaus, H.P. Tranitz, W. Wegscheider, Double-dot quantum ratchet driven by an independently biased quantum point contact. Phys. Rev. Lett. **97**, 176803 (2006)
372. A. Mitra, I. Aleiner, A.J. Millis, Semiclassical analysis of the nonequilibrium local polaron. Phys. Rev. Lett. **94**, 076404 (2005)
373. P. D'Amico, D.A. Ryndyk, G. Cuniberti, K. Richter, Charge-memory effect in a polaron model: equation-of-motion method for Green functions. New J. Phys. **10**, 085002 (2008)
374. S. Yeganeh, M. Galperin, M.A. Ratner, Switching in molecular transport junctions: polarization response. J. Am. Chem. Soc. **129**, 13313–13320 (2007)
375. R.A. Marcus, On the theory of electron-transfer reactions. VI. Unified treatment for homogeneous and electrode reactions. J. Chem. Phys. **43**, 679–701 (1965)
376. I. Daizadeh, J.N. Gehlen, A.A. Stuchebrukhov, Calculation of electronic tunneling matrix clement in proteins: comparison of exact and approximate one-electron methods for Ru-modified azurin. J. Chem. Phys. **106**, 5658–5666 (1997)
377. W. Kohn, L.J. Sham, Self-consistent equations including exchange and correlation effects. Phys. Rev. **140**, A1133–A1138 (1965)
378. R.G. Parr, W. Yang, *Density-Functional Theory of Atoms and Molecules* (Oxford University Press, New York, 1989)
379. R.O. Jones, O. Gunnarsson, The density functional formalism, its applications and prospects. Rev. Mod. Phys. **61**, 689–746 (1989)
380. W. Kohn, A.D. Becke, R.G. Parr, Density functional theory of electronic structure. J. Phys. Chem. **100**, 12974–12980 (1996)
381. N. Argaman, G. Makov, Density functional theory: an introduction. Am. J. Phys. **68**, 69 (2000)
382. P. Hohenberg, W. Kohn, Inhomogeneous electron gas. Phys. Rev. **136**, B864–B871 (1964)
383. J.P. Perdew, R.G. Parr, M. Levy, J.L. Balduz, Density-functional theory for fractional particle number: derivative discontinuities of the energy. Phys. Rev. Lett. **49**, 1691 (1982)

384. A.A. Quong, M.R. Pederson, J.L. Feldman, First principles determination of the interatomic force-constant tensor of the fullerene molecule. Solid State Commun. **87**, 535–539 (1993)
385. T. Baruah, J. Kortus, M.R. Pederson, R. Wesolowski, J.T. Haraldsen, J.L. Musfeldt, J.M. North, D. Zipse, N.S. Dalal, Understanding the electronic structure, optical, and vibrational properties of the Fe8Br8 single-molecule magnet. Phys. Rev. B **70**, 214410 (2004)
386. T. Baruah, M.R. Pederson, R.R. Zope, Vibrational stability and electronic structure of a B_{80} fullerene. Phys. Rev. B **78**, 045408 (2008)
387. M.R. Pederson, K. Jackson, D.V. Porezag, Z. Hajnal, Th. Frauenheim, Vibrational signatures for low-energy intermediate-sized Si clusters. Phys. Rev. B **54**, 2863–2867 (1996)
388. M.R. Pederson, T. Baruah, Molecular polarizabilities from density-functional theory: from small molecules to light harvesting complexes. Lect. Ser. Comput. Comput. Sci. **3**, 156–167 (2005)
389. T. Baruah, M.R. Pederson, Density functional study on a light-harvesting carotenoid-porphyrin-C60 molecular triad. J. Chem. Phys. **125**, 164706 (2006)
390. R.R. Zope, T. Baruah, M.R. Pederson, B.I. Dunlap, Static dielectric response of icosahedral fullerenes from C_{60} to C_{2160} characterized by an all-electron density functional theory. Phys. Rev. B **77**, 115452 (2008)
391. M.R. Pederson, S.N. Khanna, Electronic structure and magnetism of $Mn_{12}O_{12}$ clusters. Phys. Rev. B **59**, R693–R696 (1999)
392. M.R. Pederson, S.N. Khanna, Magnetic anisotropy barrier for spin tunneling in $Mn_{12}O_{12}$ molecules. Phys. Rev. B **60**, 9566–9572 (1999)
393. R. Car, M. Partinello, Unified approach for molecular dynamics and density-functional theory. Phys. Rev. Lett. **55**, 2471–2474 (1985)
394. R.M. Wentzcovich, J.L. Martins, First principles molecular dynamics of Li: test of a new algorithm. Solid State Commun. **78**, 831–834 (1991)
395. E. Wimmer, H. Krakauer, M. Weinert, A.J. Freeman, Full-potential self-consistent linearized-augmented-plane-wave method for calculating the electronic structure of molecules and surfaces: O_2 molecule. Phys. Rev. B **24**, 864–875 (1981)
396. O.K. Andersen, Linear methods in band theory. Phys. Rev. B **12**, 3060–3083 (1975)
397. M.R. Pederson, N. Bernstein, J. Kortus, Fourth-order magnetic anisotropy and tunnel splittings in Mn_{12} from spin-orbit-vibron interactions. Phys. Rev. Lett. **89**, 097202 (2002)
398. M.R. Pederson, C.C. Lin, All-electron self-consistent variational method for Wannier-type functions: applications to the silicon crystal. Phys. Rev. B **35**, 2273–2283 (1987)
399. M.R. Pederson, K.A. Jackson, Pseudoenergies for simulations on metallic systems. Phys. Rev. B **43**, 7312–7315 (1991)
400. D.V. Porezag, M.R. Pederson, Infrared intensities and Raman-scattering activities within density-functional theory. Phys. Rev. B **54**, 7830–7836 (1996)
401. D.V. Porezag, M.R. Pederson, Optimization of Gaussian basis sets for density-functional calculations. Phys. Rev. A **60**, 2840–2847 (1999)
402. J.P. Perdew, A. Zunger, Self-interaction correction to density-functional approximations for many-electron systems. Phys. Rev. B **23**, 5048–5079 (1981)
403. J.P. Perdew, J.A. Chevary, S.H. Vosko, K.A. Jackson, M.R. Pederson, D.J. Singh, C. Fiolhais, Atoms, molecules, solids, and surfaces: applications of the generalized gradient approximation for exchange and correlation. Phys. Rev. B **46**, 6671–6687 (1992)
404. J.P. Perdew, K. Burke, M. Ernzerhof, Generalized gradient approximation made simple. Phys. Rev. Lett. **77**, 3865–3868 (1996)
405. A.D. Becke, A new mixing of Hartree–Fock and local density-functional theories, J. Chem. Phys. **98**, 1372–1377 (1993)
406. C. Lee, W. Yang, R.G. Parr, Development of the Colle-Salvetti correlation-energy formula into a functional of the electron density. Phys. Rev. B **37**, 785–789 (1988)
407. O. Gunnarsson, B.I. Lundqvist, Exchange and correlation in atoms, molecules, and solids by the spin-density-functional formalism. Phys. Rev. B **13**, 4274–4298 (1976)
408. J.P. Perdew, A. Ruzsinszky, J. Tao, V.N. Staroverov, G.E. Scuseria, G.I. Csonka, Prescription for the design and selection of density functional approximations: more constraint satisfaction with fewer fits. J. Chem. Phys. **123**, 062201 (2005)

409. J.P. Perdew, Accurate density functional for the energy: real-space cutoff of the gradient expansion for the exchange hole. Phys. Rev. Lett. **55**, 1665–1668 (1985)
410. J.P. Perdew, K. Burke, Y. Wang, Generalized gradient approximation for the exchange-correlation hole of a many-electron system. Phys. Rev. B **54**, 16533–16539 (1996)
411. J. Tao, J.P. Perdew, V.N. Staroverov, G.E. Scuseria, Climbing the density functional ladder: nonempirical meta-generalized gradient approximation designed for molecules and solids. Phys. Rev. Lett. **91**, 146401 (2003)
412. V.N. Staroverov, G.E. Scuseria, J. Tao, J.P. Perdew, Tests of a ladder of density functionals for bulk solids and surfaces. Phys. Rev. B **69**, 075102 (2004)
413. J.P. Perdew, J.M. Tao, V.N. Staroverov, G.E. Scuseria, Meta-generalized gradient approximation: explanation of a realistic nonempirical density functional. J. Chem. Phys. **120**, 6898–6911 (2004)
414. W. Kohn, Y. Meir, D.E. Makarov, Van der Waals energies in density functional theory. Phys. Rev. Lett. **80**, 4153–4156 (1998)
415. E.K.U. Gross, W. Kohn, Local density-functional theory of frequency-dependent linear response. Phys. Rev. Lett. **55**, 2850–2852 (1985)
416. J.F. Janak, Proof that $\partial E/\partial ni = \epsilon$ in density-functional theory. Phys. Rev. B **18**, 7165–7168 (1978)
417. L.J. Sham, W. Kohn, One-particle properties of an inhomogeneous interacting electron gas. Phys. Rev. **145**, 561–567 (1966)
418. N. Hadjisavvas, A. Theophilou, Rigorous formulation of Slaters transition-state theory for excited states. Phys. Rev. A **32**, 720–724 (1985)
419. M. Levy, J.P. Perdew, Extrema of the density functional for the energy: excited states from the ground-state theory. Phys. Rev. B **31**, 6264–6272 (1985)
420. T. Baruah, M.R. Pederson, DFT calculations on charge-transfer states of a carotenoid-porphyrin-C_{60} molecular triad. J. Chem. Theor. Comput. **5**, 834–843 (2009)
421. E. Runge, E.K.U. Gross, Density-functional theory for time-dependent systems. Phys. Rev. Lett. **52**, 997–1000 (1984)
422. R. van Leeuwen, Causality and symmetry in time-dependent density-functional theory. Phys. Rev. Lett. **80**, 1280–1283 (1998)
423. M.A.L. Marques, C.A. Ulrich, F. Nogueira, A. Rubio, K. Burke, E.K.U. Gross (eds), *Time-Dependent Density Functional Theory, Lecture Notes in Physics* (Springer, New York, 2006)
424. X. Wang, R.Q. Zhang, S.T. Lee, T.A. Niehaus, Th. Frauenheim, Unusual size dependence of the optical emission gap in small hydrogenated silicon nanoparticles. Appl. Phys. Lett. **90**, 123116 (2007)
425. L. Yang, C.D. Spataru, S.G. Louie, M.Y. Chou, Enhanced electron-hole interaction and optical absorption in a silicon nanowire. Phys. Rev. B **75**, 201304(R) (2007)
426. N.D. Drummond, A.J. Williamson, R.J. Needs, G. Galli, Electron emission from diamondoids: a diffusion quantum Monte Carlo study. Phys. Rev. Lett. **95**, 096801 (2005)
427. G.K. Gueorguiev, J.M. Pacheco, D. Tomanek, Quantum size effects in the polarizability of carbon, fullerenes. Phys. Rev. Lett. **92**, 215501 (2004)
428. Y.H. Hu, E. Ruckenstein, Bond order bond polarizability model for fullerene cages and nanotubes. J. Chem. Phys. **123**, 214708 (2005)
429. E. Westin, A. Rosen, G.T. Velde, E.J. Baerends, Analysis of the polarizability and optical properties of C_{60}. J. Phys. B **29**, 5087–5113 (1996)
430. L.X. Benedict, S.G. Louie, M.L. Cohen, Static polarizabilities of single-wall carbon nanotubes. Phys. Rev. B **52**, 8541–8549 (1995)
431. M.R. Pederson, T. Baruah, P.B. Allen, C. Schmidt, Density-functional-based determination of vibrational polarizabilities in molecules within the double-harmonic approximation: derivation and application. J. Chem. Theor. Comput. **1**, 590–596 (2005)
432. N.G. Szwacki, A. Sadrzadeh, B.I. Yakobson, B_{80} fullerene: an ab initio prediction of geometry, stability, and electronic structure. Phys. Rev. Lett. **98**, 166804 (2007)
433. G. Gopakumar, M.T. Nguyen, A. Ceulemans, The boron buckyball has an unexpected Th symmetry. Chem. Phys. Lett. **450**, 175–177 (2008)

434. H. Tang, S. Ismail-Beigi, Novel precursors for boron nanotubes: the competition of two-center and three-center bonding in boron sheets. Phys. Rev. Lett. **99**, 115501 (2007)
435. J.R. Friedman, M.P. Sarachik, J. Tejada, R. Ziolo, Macroscopic measurement of resonant magnetization tunneling in high-spin molecules. Phys. Rev. Lett. **76**, 3830–3833 (1996)
436. S. Hill, J.A.A.J. Perenboom, N.S. Dalal, T. Hathaway, T. Stalcup, J.S. Brooks, High-sensitivity electron paramagnetic resonance of Mn_{12}-acetate. Phys. Rev. Lett. **80**, 2453–2456 (1998)
437. S. Hill, S. Maccagnano, K. Park, R.M. Achey, J.M. North, N.S. Dalal, Detailed single-crystal EPR line shape measurements for the single-molecule magnets Fe_8Br and Mn_{12}-acetate. Phys. Rev. B **65**, 224410 (2002)
438. E.M. Chudnovsky, D.A. Garanin, Spin tunneling via dislocations in Mn12 acetate crystals. Phys. Rev. Lett. **87**, 187203 (2001)
439. D.A. Garanin, E.M. Chudnovsky, Dislocation-induced spin tunneling in Mn_{12} acetate. Phys. Rev. B **65**, 094423 (2002)
440. A. Cornia, R. Sessoli, L. Sorace, D. Gatteschi, A.L. Barra, C. Daiguebonne, Origin of second-order transverse magnetic anisotropy in Mn_{12}-acetate. Phys. Rev. Lett. **89**, 257201 (2002)
441. S. Hill, R.S. Edwards, S.I. Jones, N.S. Dalal, J.M. North, Definitive spectroscopic determination of the transverse interactions responsible for the magnetic quantum tunneling in Mn_{12}-acetate. Phys. Rev. Lett. **90**, 217204 (2003)
442. E. del Barco, A.D. Kent, N.E. Chakov, L.N. Zakharov, A.L. Rheingold, D.N. Hendrickson, G. Christou, Distribution of internal transverse magnetic fields in a Mn_{12}-based single molecule magnet. Phys. Rev. B **69**, 020411(R) (2004)
443. K. Park, T. Baruah, N. Bernstein, M.R. Pederson, Second-order transverse magnetic anisotropy induced by disorder in the single-molecule magnet Mn_{12}. Phys. Rev. B **69**, 144426 (2004)
444. W. Wernsdorfer, N. Aliaga-Alcalde, D.N. Hendrickson, G. Christou, Exchange-biased quantum tunnelling in a supramolecular dimer of single-molecule magnets. Nature **416**, 406–409 (2002)
445. K. Park, M.R. Pederson, S.L. Richardson, N. Aliaga-Alcalde, G. Christou, Density-functional theory calculation of the intermolecular exchange interaction in the magnetic Mn_4 dimer. Phys. Rev. B **68**, 020405(R) (2003)
446. J. Kortus, C.S. Hellberg, M.R. Pederson, Hamiltonian of the V_{15} spin system from first-principles density-functional calculations. Phys. Rev. Lett. **86**, 3400–3403 (2001)
447. L.H. Yu, Z.K. Keane, J.W. Ciszek, L. Cheng, J.M. Tour, T. Baruah, M.R. Pederson, D. Natelson, Kondo resonances and anomalous gate dependence in the electrical conductivity of single-molecule transistors. Phys. Rev. Lett. **95**, 256803 (2005)
448. D. Wegner, R. Yamachika, X. Zhang, Y. Wang, T. Baruah, M.R. Pederson, B.M. Bartlett, J.R. Long, M.F. Crommie, Tuning molecule-mediated spin coupling in bottom-up-fabricated vanadium-tetracyanoethylene nanostructures. Phys. Rev. Lett. **103**, 087205 (2009)
449. M. Green, *Third Generation Photovolsaics: Advanced Solar Energy Conversion* (Springer, Berlin, 2004)
450. P.V. Kamat, Meeting the clean energy demand: nanostructure architectures for solar energy conversion. J. Phys. Chem. C **111**, 2834–2860 (2007)
451. B. O'Regan, M. Gratzel, A low-cost, high-efficiency solar cell based on dye-sensitized colloidal TiO_2 films. Nature **353**, 737–740 (1991)
452. M. Gratzel, Photoelectrochemical cells. Nature **414**, 338–344 (2001)
453. P.A. Liddell, D. Kuciauskas, J.P. Sumida, B. Nash, D. Nguyen, A.L. Moore, T.A. Moore, D. Gust, Photoinduced charge separation and charge recombination to a triplet state in a carotene-porphyrin-fullerene triad. J. Am. Chem. Soc. **119**, 1400–1405 (1997)
454. D. Carbonera, M. Di Valentin, C. Corvaja, G. Agostini, G. Giacometti, P.A. Liddell, D. Kuciauskas, A.L. Moore, T.A. Moore, D. Gust, EPR investigation of photoinduced radical pair formation and decay to a triplet state in a carotene-porphyrin-fullerene triad. J. Am. Chem. Soc. **120**, 4398–4405 (1998)

455. T. Baruah, M.R. Pederson, Density functional study on a light-harvesting carotenoid-porphyrin-C_{60} molecular triad. J. Chem. Phys. **125**, 164706 (2006)
456. N. Spallanzani, C.A. Rozzi, D. Varsano, T. Baruah, M.R. Pederson, F. Manghi, A. Rubio, Photo-excitation of a light-harvesting supra-molecular triad: a time-dependent DFT study. J. Phys. Chem. B **113**, 5345–5349 (2009)
457. G. Kodis, P.A. Liddell, A.L. Moore, T.A. Moore, D. Gust, Synthesis and photochemistry of a carotene-porphyrin-fullerene model photosynthetic reaction center. Phys. Org. Chem. **17**, 724–734 (2004)
458. S. Vasudevan, N. Kapur, T. He, M. Neurock, J.M. Tour, A.W. Ghosh, Controlling transistor threshold voltages using molecular dipoles. J. Appl. Phys. **105**, 093703 (2009)
459. N.D. Lang, Resistance of atomic wires. Phys. Rev. B **52**, 5335–5342 (1995)
460. K. Hirose, M. Tsukada, First-principles calculation of the electronic structure for a bielectrode junction system under strong field and current. Phys. Rev. B **51**, 5278–5290 (1995)
461. W.G. Aulbur, L. Jonsson, J.W. Wilkins, Quasiparticle calculations in solids original research article. Solid State Phys. **54**, 1–218 (1999)
462. D. Kienle, J.I. Cerda, A.W. Ghosh, Extended Huckel theory for band structure, chemistry, and transport. I. Carbon nanotubes. J. Appl. Phys. **100**, 043714 (2006)
463. D. Kienle, K.H. Bevan, G.-C. Liang, L. Siddiqui, J.I. Cerda, A.W. Ghosh, Extended Huckel theory for band structure, chemistry, and transport. II. Silicon. J. Appl. Phys. **100**, 043715 (2006)
464. Th. Frauenheim, G. Seifert, M. Elsterner, Z. Hajnal, G. Jungnickel, D. Porezag, S. Suhai, R. Scholz, A self-consistent charge density-functional based tight-binding method for predictive materials simulations in physics. Chem. Biol. Phys. Status Solidi B **217**, 41–62 (1999)
465. M. Brandbyge, J.-L. Mozos, P. Ordejon, J. Taylor, K. Stokbro, Density-functional method for nonequilibrium electron transport. Phys. Rev. B **65**, 165401 (2002)
466. A. Arnold, F. Weigend, F. Evers, Quantum chemistry calculations for molecules coupled to reservoirs: formalism, implementation, and application to benzenedithiol. J. Chem. Phys. **126**, 174101 (2007)
467. F. Evers, F. Weigend, M. Koentopp, Conductance of molecular wires and transport calculations based on density-functional theory. Phys. Rev. B **69**, 235411 (2004)
468. G.-C. Liang, A.W. Ghosh, Identifying contact effects in electronic conduction through C_{60} on silicon. Phys. Rev. Lett. **95**, 076403 (2005)
469. N. Lorente, M. Persson, L.J. Lauhon, W. Ho, Symmetry selection rules for vibrationally inelastic tunneling. Phys. Rev. Lett. **86**, 2593–2596 (2001)
470. M.-L. Bocquet, H. Lesnard, N. Lorente, Inelastic spectroscopy identification of STM-induced benzene dehydrogenation. Phys. Rev. Lett. **96**, 096101 (2006)
471. M. Koentopp, K. Burke, F. Evers, Zero-bias molecular electronics: exchange-correlation corrections to Landauer's formula. Phys. Rev. B **73**, 121403 (2006)
472. G. Moore, Cramming more components onto integrated circuits. Electron. Mag. **38**, 114–117 (1965)
473. Y. Li, F. Qian, J. Xiang, C.M. Lieber, Nanowire electronic and optoelectronic devices. Mater. Today **9**, 18–27 (2006)
474. J.P. Colinge, Quantum-wire effects in trigate SOI MOSFETs. Solid-State Electr. **51**, 1153–1160 (2007)
475. C.W. Lee, S.R.N. Yun, C.G. Yu, J.T. Park, J.P. Colinge, Device design guidelines for nano-scale MuGFETs. Solid-State Elect. **51**, 505–510 (2007)
476. D.A. Muller, T. Sorsch, S. Moccio, F.H. Baumann, K. Evans-Lutterodt, G. Timp, The electronic structure at the atomic scale of ultra-thin gate oxides. Nature **399**, 758–760 (1999)
477. J.H. Davis, *The Physics of Low-Dimensional Devices* (Cambridge University Press, Cambridge, 1998)
478. R. Kim, M. Lundstrom, Characteristics features of 1-D ballistic transport in nanowire MOSFETs. IEEE Trans. Nanotechnol. **7**, 787–794 (2008)
479. Y. Xue, M.A. Ratner, Molecular electronics: from physics to computing, in *Nanotechnology: Science and Computation*, ed. by J. Chen, N. Jonoska, G. Rosenberg (Springer, Berlin, 2006)

480. R. Martel, T. Schmidt, H.R. Shea, T. Hertel, Ph. Avouris, Single- and multi-wall carbon nanotube field-effect transistors. Appl. Phys. Lett. **73**, 2447–2449 (1998)
481. S.J. Tans, A.R.M. Verschueren, C. Dekker, Room-temperature transistor based on a single carbon nanotube. Nature **393**, 49–52 (1998)
482. A. Javey, J. Guo, M. Paulsson, Q. Wang, D. Mann, M. Lundstrom, H. Dai, High-field quasiballistic transport in short carbon nanotubes. Phys. Rev. Lett. **92**, 106804 (2004)
483. J.Y. Park, S. Rosenblatt, Y. Yaish, V. Sazonova, H. Ustunel, S. Braig, T.A. Arias, P.W. Brouwer, P.L. McEuen, Electron-phonon scattering in metallic single-walled carbon nanotubes. Nano Lett. **4**, 517–520 (2004)
484. S. Wang, P. Sellin, Pronounced hysteresis and high charge storage stability of single-walled carbon nanotube-baced field-effect transistors. Appl. Phys. Lett. **87**, 133–117 (2005)
485. J. Guo, S. Hasan, A. Javey, G. Bosman, M. Lundstrom, Assessment of high-frequency performance potential of carbon nanotube transistors. IEEE Trans. Nanotechnol. **4**, 715–721 (2005)
486. F. Leonard, J. Tersoff, Role of Fermi-level pinning in nanotube Schottky diodes. Phys. Rev. Lett. **84**, 4693–4696 (2000)
487. Y. Xue, M.A. Ratner, Scaling analysis of Schottky barriers at metal-embedded semiconducting carbon nanotube interfaces. Phys. Rev. B **69**, 161402(R) (2004)
488. S. Auvray, J. Borghetti, M.F. Goffman, A. Filoramo, V. Derycke, J.P. Bourgoin, O. Jost, Carbon nanotube transistor optimization by chemical control of the nanotube-metal interface. Appl. Phys. Lett. **84**, 5106–5108 (2004)
489. J. Appenzeller, J. Knoch, R. Martel, V. Derycke, S.J. Wind, P. Avouris, Carbon nanotube electronics. IEEE Trans. Nanotechnol. **1**, 84–89 (2002)
490. J.U. Lee, P.P. Gipp, C.M. Heller, Carbon nanotube p–n junction diodes. Appl. Phys. Lett. **85**, 145–147 (2004)
491. A. Rakitin, C. Papadopoulos, J.M. Xu, Carbon nanotube self-doping: calculation of the hole carrier concentration. Phys. Rev. B **67**, 033411 (2003)
492. F. Ding, P. Larsson, J.A. Larsson, R. Ahuja, H. Duan, A. Rosen, K. Bolton, The importance of strong carbon-metal adhesion for catalytic nucleation of single-walled carbon nanotubes. Nano Lett. **8**, 463–468 (2008)
493. Y. Wu, Y. Cui, L. Huynh, C.J. Barrelet, D.C. Bell, C.M. Lieber, Controlled growth and structures of molecular-scale silicon nanowires. Nano Lett. **4**, 433–436 (2004)
494. R. Agarwal, Heterointerfaces in semiconductor nanowires. Small **4**, 1872–1893 (2008)
495. Z. Fan, J.G. Lu, Zinc oxide nanostructures: synthesis and properties, J. Nanosci. Nanotechnol. **5**, 1561–1573 (2005)
496. C. Klingshirn, ZnO: from basics towards applications. Phys. Status Solidi B **244**, 3027–3073 (2007)
497. W.-K. Hong, G. Jo, S. Song, J. Maeng, T. Lee, ZnO nanowire field-effect transistors, in *Handbook of Nanophysics; Nanoelectronics and Nanophotonics*, ed. by K.D. Sattler (Taylor and Francis Group, New York, 2010)
498. J. Goldberger, D.J. Sirbuly, M. Law, P. Yang, ZnO nanowire transistors, J. Phys. Chem. B **109**, 9–14 (2005)
499. W.-K. Hong, D.-K. Hwang, I.-K. Park, G. Jo, S. Song, S.-J. Park, T. Lee, B.-J. Kim, E.A. Stach, Realization of highly reproducible ZnO nanowire field-effect transistors with n-channel depletion and enhancement modes. Appl. Phys. Lett. **90**, 243103 (2007)
500. D. Yeom, K. Keem, J. Kang, D.-Y. Jeong, C. Yoon, D. Kim, S. Kim, NOT and NAND logic circuits composed of top-gate ZnO nanowire field-effect transistors with high-k Al_2O_3 gate layers. Nanothechnology **19**, 265202 (2008)
501. K.S. Novoselov, A.K. Geim, S.V. Morozov, D. Jiang, Y. Zhang, S.V. Dubonos, I.V. Grigorieva, A.A. Firsov, Electric field effect in atomically thin carbon films. Science **306**, 666–669 (2004)
502. B. Huard, J.A. Sulpizio, N. Stander, K. Todd, B. Yang, D. Goldhaber-Gordon, Transport measurements across a tunable potential barrier in graphene. Phys. Rev. Lett. **98**, 236803 (2007)

503. H. Cheraghchi, K. Esfarjani, Negative differential resistance in molecular junctions: application to graphene ribbon junctions. Phys. Rev. B **78**, 085123 (2008)
504. B. Xu, X. Xiao, X. Yang, L. Zang, N. Tao, Large gate modulation in the current of a room temperature single molecule transistor. J. Am. Chem. Soc. **127**, 2386–2387 (2005)
505. F. Jackel, M.D. Watson, K. Mullen, J.P. Rabe, Prototypical single-molecule chemical-field-effect transistor with nanometer-sized gates. Phys. Rev. Lett. **92**, 188303 (2004)
506. H.W. Song, Y.S. Kim, Y.H. Jang, H.J. Jeong, M.A. Reed, T.H. Lee, Observation of molecular orbital gating. Nature **462**, 1039–1043 (2009)
507. S. Ballmann, H.B. Weber, An electrostatic gate for mechanically controlled single-molecule junctions. New J. Phys. **14**, 123028 (2012)
508. F. Prins, A. Barreiro, J.W. Ruitenberg, J.S. Seldenthuis, N. Aliaga-Alcalde, L.M.K. Vandersypen, H.S.J. van der Zant, Room-temperature gating of molecular junctions using few-layer graphene nanogap electrodes. Nano Lett. **11**, 4607–4611 (2011)
509. C.A. Martin, D. Ding, H.S.J. van der Zant, J.M. van Ruitenbeek, Lithographic mechanical break junctions for single-molecule measurements in vacuum: possibilities and limitations. New J. Phys. **10**, 065008 (2008)
510. S.-C. Chang, Z. Li, C.N. Lau, B. Larade, R.S. Williams, Investigation of a model molecular-electronic rectifier with an evaporated Ti-metal top contact. Appl. Phys. Lett. **83**, 3198–3200 (2003)
511. H. Haick, J. Ghabboun, D. Cahen, Pd versus Au as evaporated metal contacts to molecules. Appl. Phys. Lett. **8**(6), 042113 (2005)
512. H.B. Akkerman, P.W.M. Blom, D.M. de Leeuw, B. de Boer, Towards molecular electronics with large-area molecular junctions. Nature **441**, 69–72 (2006)
513. D. Velessiotis, A.M. Douvas, S. Athanasiou, B. Nilsson, G. Petersson, U. Sodervall, G. Alestig, P. Argitis, N. Glezos, Molecular junctions made of tungsten-polyoxometalate self-assembled monolayers: towards polyoxometalate-based molecular electronics devices. Microelectr. Eng. **88**, 2775–2777 (2011)
514. R. Dasari, F.J. Ibanez, F.P. Zamborini, Electrochemical fabrication of metal/organic/metal junctions for molecular electronics and sensing applications. Langmuir **27**, 7285–7293 (2011)
515. R.M. Metzger, Unimolecular electrical rectifiers. Chem. Rev. **103**, 3803–3834 (2003)
516. M. Elbing, R. Ochs, M. Koentopp, M. Fischer, C. von Hanisch, F. Weigend, F. Evers, H. Weber, M. Mayor, A single-molecule diode. Proc. Natl. Acad. Sci. USA **102**, 8815 (2005)
517. Diez-Perez, J. Hihath, Y. Lee, L. Yu, L. Adamska, M.A. Kozhushner, I. Oleynik, N. Tao, Rectification and stability of a single molecular diode with controlled orientation. Nat. Chem. **1**, 635–641 (2009)
518. H. Nakamura, Y. Asai, J. Hihath, C. Bruot, N. Tao, Switch of conducting orbital by bias-induced electronic contact asymmetry in a bipyrimidinyl-biphenyl diblock molecule: mechanism to achieve a pn directional molecular diode. Phys. Chem. C **115**, 19931–19938 (2011)
519. M. Geller, A. Marent, D. Bimberg, Nanomemories using self-organized quantum dots, in *Handbook of Nanophysics; Nanoelectronics and Nanophotonics*, ed. by K.D. Sattler (Taylor and Francis Group, New York, 2010)
520. M. Geller, A. Marent, E. Stock, D. Bimberg, V.I. Zubkov, I.S. Shulgunova, A.V. Solomonov, Hole capture into self-organized InGaAs quantum dots. Appl. Phys. Lett. **89**, 232105 (2006)
521. M. Geller, A. Marent, T. Nowozin, D. Bimberg, N. Akcay, N. Oncan, A write time of 6 ns for quantum dot-based memory structures. Appl. Phys. Lett. **92**, 092108 (2008)
522. D.V. Lang, Deep-level transient spectroscopy: a new method to characterize traps in semiconductors. J. Appl. Phys. **45**, 3023–3026 (1974)
523. M. Geller, E. Stock, C. Kapteyn, R.L. Sellin, D. Bimberg, Tunneling emission from self-organized In(Ga)As/GaAs quantum dots observed via time-resolved capacitance measurements. Phys. Rev. B **73**, 205331 (2006)
524. A. Marent, M. Geller, A. Schliwa, D. Feise, K. Potschke, D. Bimberg, N. Akcay, N. Oncan, 10^6 years extrapolated hole storage time in GaSb/AlAs quantum dots, Appl. Phys. Lett. **91**, 242109 (2007)

525. A. Schliwa, M. Winkelnkemper, D. Bimberg, Impact of size, shape, and composition on piezoelectric effects and electronic properties of $In(Ga)As/GaAs$ quantum dots. Phys. Rev. B **76**, 205324 (2007)
526. O. Stier, M. Grundmann, D. Bimberg, Electronic and optical properties of strained quantum dots modeled by 8-band $k \cdot p$ theory. Phys. Rev. B **59**, 5688–5701 (1999)
527. E. Bichoutskaia, A.M. Popov, Y.E. Lozovik, Nanotube-based data storage devices. Mater. Today **11**, 38–43 (2008)
528. O. Wunnicke, Gate capacitance of back-gated nanowire field-effect transistors. Appl. Phys. Lett. **89**, 083102 (2006)
529. M.S. Fuhrer, B.M. Kim, T. Du1rkop, T. Brintlinger, High-mobility nanotube transistor memory. Nano Lett. **2**, 755–759 (2002)
530. M. Radosavljevic, M. Freitag, K.V. Thadani, A.T. Johnson, Nonvolatile molecular memory elements based on ambipolar nanotube field effect transistors. Nano Lett. **2**, 761–764 (2002)
531. J.B. Cui, R. Sordan, M. Burghard, K. Kern, Carbon nanotube memory devices of high charge storage stability. Appl. Phys. Lett. **81**, 3260–3262 (2002)
532. T. Sakurai, T. Yoshimura, S. Akita, N. Fujimura, Y. Nakayama, Single-wall carbon nanotube field effect transistors with non-volatile memory operation. Jpn. J. Appl. Phys. **45** (2006)
533. X. Duan, Y. Huang, C.M. Lieber, Nonvolatile memory and programmable logic from molecule-gated nanowires. Nano Lett. **2**, 487–490 (2002)
534. J. Mannik, B.R. Goldsmith, A. Kane, P.G. Collins, Chemically induced conductance switching in carbon nanotube circuits. Phys. Rev. Lett. **97**, 016601 (2006)
535. B.R. Goldsmith, J.G. Coroneus, V.R. Khalap, A.A. Kane, G.A. Weiss, P.G. Collins, Conductance-controlled point functionalization of single-walled carbon nanotubes. Science **315**, 77–81 (2007)
536. R.S. Chakraborty, S. Narasimhan, S. Bhunia, Hybridization of CMOS with CNT-based nano electromechanical switch for low leakage and robust circuit design using nanoscaled CMOS devices. IEEE Trans. Circ Syst I **54**, 2480–2488 (2007)
537. M.Y.A. Yousif, P. Lundgren, F. Ghavanini, P. Enoksson, S. Bengtsson, CMOS considerations in nanoelectromechanical carbon nanotube-based switches. Nanotechnology **19**, 285204 (2008)
538. T. Rueckes, K. Kim, E. Joselevich, G.Y. Tseng, C.-L. Cheung, C.M. Lieber, Carbon nanotube-based nonvolatile random access memory for molecular computing. Science **289**, (2000) 94–97
539. S.N. Cha, J.E. Jang, Y. Choi, G.A.J. Amaratunga, D.-J. Kang, D.G. Hasko, J.E. Jung, J.M. Kim, Fabrication of a nanoelectromechanical switch using a suspended carbon nanotube. Appl. Phys. Lett. **86**, 083105 (2005)
540. M. Dequesnes, S.V. Rotkin, N.R. Aluru, Calculation of pull-in voltages for nanoelectromechanical switches. Nanotechnology **13**, 120–131 (2002)
541. V.V. Deshpande, H.-Y. Chiu, H.W.Ch. Postma, C. Miko, L. Forro, M. Bockrath, Carbon nanotube linear bearing nanoswitches. Nano Lett. **6**, 1092–1095 (2006)
542. L. Maslov, Concept of nonvolatile memory based on multiwall carbon nanotubes. Nanotechnology **17**, 2475–2482 (2006)
543. J.W. Kang, Q. Jiang, Electrostatically telescoping nanotube nonvolatile memory device. Nanotechnology **18**, 095705 (2007)
544. Y.-K. Kwon, D. Tomanek, S. Iijima, "Bucky Shuttle" memory device: synthetic approach and molecular dynamics simulations. Phys. Rev. Lett. **82**, 1470–1473 (1999)
545. Z.J. Donhauser, B.A. Mantooth, K.F. Kelly, L.A. Bumm, J.D. Monnell, J.J. Stapleton, D.L. Allara, J.M. Tour, P.S. Weiss, Conductance switching in single molecules through conformational changes. Science **292**, 2303–2307 (2001)
546. G.K. Ramachandran, T.J. Hopson, A.M. Rawlett, L.A. Nagahara, A. Primak, S.M. Lindsay, A bond-fluctuation mechanism for stochastic switching in wired molecules. Science **300**, 1413–1416 (2003)
547. D. Dulic, F. Pump, S. Campidelli, P. Lavie, G. Cuniberti, A. Filoramo, Controlled stability of molecular junctions. Angew. Chem. **121**, 8423–8426 (2009)

548. E.G. Emberly, G. Kirczenow, The smallest molecular switch. Phys. Rev. Lett. **91**, 188301 (2003)
549. G. Li, A. Mishchenko, Z. Li, I. Pobelov, Th. Wandlowski, X.Q. Li, F. Wurthner, A. Bagrets, F. Evers, Electrochemical gate-controlled electron transport of redox-active single perylene bisimide molecular junctions. J. Phys.: Condens. Matter **20**, 374122 (2008)
550. Y. Wada, T. Uda, M. Lutwyche, S. Kondo, S. Heike, A proposal of nanoscale devices based on atom/molecule switching. J. Appl. Phys. **74**, 7321–7328 (1993)
551. C. Joachim, J.K. Gimzewski, H. Tang, Physical principles of the single-C_{60} transistor effect. Phys. Rev. B **58**, 16407–16417 (1998)
552. Ch. Loppacher, M. Guggisberg, O. Pfeiffer, E. Meyer, M. Bammerlin, R. Luthi, R. Schlittler, J.K. Gimzewski, H. Tang, C. Joachim, Direct determination of the energy required to operate a single molecule switch. Phys. Rev. Lett. **90**, 066107 (2003)
553. V. Meded, A. Bagrets, A. Arnold, F. Evers, Molecular switch controlled by pulsed bias voltages. Small **5**, 2218–2223 (2009)
554. M. Taniguchi, M. Tsutsui, K. Yokota, T. Kawai, Mechanically-controllable single molecule switch based on configuration specific electrical conductivity of metal–molecule-metal junctions. Chem. Sci. **1**, 247–253 (2010)
555. C. Li, D. Zhang, X. Liu, S. Han, T. Tang, C. Zhou, W. Fan, J. Koehne, J. Han, M. Meyyappan, A.M. Rawlett, D.W. Price, J.M. Tour, Fabrication approach for molecular memory arrays. Appl. Phys. Lett. **82**, 645–647 (2003)
556. R. Jorn, T. Seideman, Implications and applications of current-induced dynamics in molecular junctions. Acc. Chem. Res. **43**, 1186–1194 (2010)
557. E. Lortscher, J.W. Ciszek, J. Tour, H. Riel, Reversible and controllable switching of a single-molecule junction. Small **2**, 973–977 (2006)
558. Y. Chen, D.A.A. Ohlberg, X. Li, D.R. Stewart, R.S. Williams, J.O. Jeppesen, K.A. Nielsen, J.F. Stoddart, D.L. Olynick, E. Anderson, Nanoscale molecular-switch devices fabricated by imprint lithography. Appl. Phys. Lett. **82**, 1610–1612 (2003)
559. J.E. Green, J.W. Choi, A. Boukai, Y. Bunimovich, E. Johnston-Halperin, E. DeIonno, Y. Luo, B.A. Sheriff, K. Xu, Y.S. Shin, H.-R. Tseng, J.F. Stoddart, J.R. Heath, A 160-kilobit molecular electronic memory patterned at 10^{11} bits per square centimetre. Nature **445**, 414–417 (2007)
560. D.R. Stewart, D.A.A. Ohlberg, P.A. Beck, Y. Chen, R.S. Williams, Molecule-independent electrical switching in Pt/organic monolayer/Ti devices. Nano Lett. **4**, 133–136 (2004)
561. D. Vuillaume, Molecular nanoelectronics. Proc. IEEE **98**, 2111–2123 (2010)
562. A.J. Kronemeijer, H.B. Akkerman, T. Kudernac, B.J. van Wees, B.L. Feringa, P.W.M. Blom, B. de Boer, Reversible conductance switching in molecular devices. Adv. Mater. **20**, 1467–1473 (2008)
563. A.S. Kumar, T. Ye, T. Takami, B.-C. Yu, A.K. Flatt, J.M. Tour, P.S. Weiss, Reversible photo-switching of single azobenzene molecules in controlled nanoscale environments. Nano Lett. **8**, 1644–1648 (2008)
564. J.M. Mativetsky, G. Pace, M. Elbing, M.A. Rampi, M. Mayor, P. Samori, Azobenzenes as light-controlled molecular electronic switches in nanoscale metal-molecule-metal junctions. J. Am. Chem. Soc. **130**, 9192–9193 (2008)
565. X. Zhang, Y. Wen, Y. Li, G. Li, S. Du, H. Guo, L. Yang, L. Jiang, H. Gao, Y. Song, Molecularly controlled modulation of conductance on azobenzene monolayer-modified silicon surfaces. J. Phys. Chem. C **112**, 8288–8293 (2008)
566. D. Dulic, S.J. van der Molen, T. Kudernac, H.T. Jonkman, J.J.D. de Jong, T.N. Bowden, J. van Esch, B.L. Feringa, B.J. van Wees, One-way optoelectronic switching of photochromic molecules on gold. Phys. Rev. Lett. **91**, 207402 (2003)
567. D. Nozaki, G. Cuniberti, Silicon-based molecular switch junctions. Nano Res. **2**, 648–659 (2009)
568. M. Zhuang, M. Ernzerhof, Mechanism of a molecular electronic photo switch. Phys. Rev. B **72**, 073104 (2005)
569. J. Li, G. Speyer, O.F. Sankey, Conduction switching of photochromic molecules. Phys. Rev. Lett. **93**, 248302 (2004)

570. C. Bertarelli, M.C. Gallazzi, F. Stellacci, G. Zerbi, S. Stagira, M. Nisoli, S. De Silvestri, Ultrafast photoinduced ring-closure dynamics of a diarylethene polymer. Chem. Phys. Lett. **359**, 278–282 (2002)
571. D. Dulic, S.J. van der Molen, T. Kudernac, H.T. Jonkman, J.J.D. de Jong, T.N. Bowden, J. van Esch, B.L. Feringa, B.J. van Wees, One-way optoelectronic switching of photochromic molecules on gold. Phys. Rev. Lett. **91**, 207402 (2003)
572. N. Katsonis, T. Kudernac, M. Walko, S. Jan van der Molen, B.J. van Wees, B.L. Feringa, Reversible conductance switching of single diarylethenes on a gold surface. Adv. Mater. **18**, 1397–1400 (2006)
573. A.C. Whalley, M.L. Steigerwald, X. Guo, C. Nuckolls, Reversible switching in molecular electronic devices. J. Am. Chem. Soc. **129**, 12590–12591 (2007)
574. K. Smaali, S. Lenfant, S. Karpe, M. Oafrain, P. Blanchard, D. Deresmes, S.Godey, A. Rochefort, J. Roncali, D. Vuillaume, High on-off conductance switching ratio in optically-driven self-assembled conjugated molecular systems. ACS Nano **4**, 2411–2421 (2010)
575. Q. Li, G. Mathur, M. Homsi, S. Surthi, V. Misra, V. Malinovskii, K.-H. Schweikart, L. Yu, J.S. Lindsey, Z. Liu, R.B. Dabke, A. Yasseri, D.F. Bocian, W.G. Kuhr, Capacitance and conductance characterization of ferrocene-containing self-assembled monolayers on silicon surfaces for memory applications. Appl. Phys. Lett. **81**, 1494–1496 (2002)
576. K.M. Roth, A.A. Yasseri, Z. Liu, R.B. Dabke, V. Malinovskii, K.-H. Schweikart, L. Yu, H. Tiznado, F. Zaera, J.S. Lindsey, W.G. Kuhr, D.F. Bocian, Measurements of electron-transfer rates of charge-storage molecular monolayers on Si(100). Toward hybrid molecular/semiconductor information storage devices. J. Am. Chem. Soc. **125**, 505–517 (2003)
577. Z. Liu, A.A. Yasseri, J.S. Lindsey, D.F. Bocian, Molecular memories that survive silicon device processing and real-world operation. Science **302**, 1543–1545 (2003)
578. M.N. Leuenberger, D. Loss, Spin tunneling and phonon-assisted relaxation in Mn_{12}-acetate. Phys. Rev. B **61**, 1286–1302 (2000)
579. E.S. Snow, F.K. Perkins, E.J. Houser, S.C. Badescu, T.L. Reinecke, Chemical detection with a single-walled carbon nanotube capacitor. Science **307**, 1942–1945 (2005)
580. J. Kong, N.R. Franklin, C. Zhou, M.G. Chapline, S. Peng, K. Cho, H. Dai, Nanotube molecular wires as chemical sensors. Science **287**, 622–625 (2000)
581. M. Freitag, A.T. Johnson, S.V. Kalinin, D.A. Bonnell, Role of single defects in electronic transport through carbon nanotube field-effect transistors. Phys. Rev. Lett. **89**, 216801 (2002)
582. C. Staii, A.T. Johnson, DNA-decorated carbon nanotubes for chemical sensing. Nano Lett. **5**, 1774–1778 (2005)
583. R.R. Breaker, Natural and engineered nucleic acids as tools to explore biology. Nature **432**, 838–845 (2004)
584. J.T. Robinson, F.K. Perkins, E.S. Snow, Z. Wei, P.E. Sheehan, Reduced graphene oxide molecular sensors. Nano Lett. **8**, 3137–3140 (2008)
585. S. Gilje, S. Han, M. Wang, K.L. Wang, R.B. Kaner, A chemical route to graphene for device applications. Nano Lett. **7**, 3394–3398 (2007)
586. Y.-M. Lin, Ph. Avouris, Strong suppression of electrical noise in bilayer graphene nano devices. Nano Lett. **8**, 2119–2125 (2008)
587. A. Modi, N. Koratkar, E. Lass, B. Wei, P.M. Ajayan, Miniaturized gas ionization sensors using carbon nanotubes. Nature **424**, 171–174 (2003)
588. Y. Zhang, J. Liu, C. Zhu, Novel gas ionization sensors using carbon nanotubes. Sensor Lett. **8**, 219–227 (2010)
589. Y. Cui, Q. Wei, H. Park, C.M. Lieber, Nanowire nanosensors for highly-sensitive, selective and integrated detection of biological and chemical species. Science **293**, 1289–1292 (2001)
590. C.Z. Li, H.X. He, A. Bogozi, J.S. Bunch, N.J. Tao, Molecular detection based on conductance quantization of nanowires. Appl. Phys. Lett. **76**, 1333–1335 (2000)
591. A. Tao, F. Kim, C. Hess, J. Goldberger, R. He, Y. Sun, Y. Xia, P. Yang, Langmuir–Blodgett silver nanowire monolayers for molecular sensing using surface-enhanced Raman spectroscopy. Nano Lett. **3**, 1229–1233 (2003)

592. Z. Li, Y. Chen, X. Li, T.I. Kamins, K. Nauka, R.S. Williams, Sequence-specific label-free DNA sensors based on silicon nanowires from northwestern. Nano Lett. **4**, 245–247 (2004)
593. Z. Fan, J.G. Lu, Gate-refreshable nanowire chemical sensors. Appl. Phys. Lett. **86**, 123510 (2005)
594. A. Kolmakov, D.O. Klenov, Y. Lilach, S. Stemmer, M. Moskovits, Enhanced gas sensing by individual SnO_2 nanowires and nanobelts functionalized with Pd catalyst particles. Nano Lett. **5**, 667–673 (2005)
595. T.I. Kamins, S. Sharma, A.A. Yasseri, Z. Li, J. Straznicky, Metal-catalysed, bridging nanowires as vapour sensors and concept for their use in a sensor system. Nanotechnology **17**, S291–S297 (2006)
596. A.T.C. Johnson, C. Staii, M. Chen, S. Khamis, R. Johnson, M.L. Klein, A. Gelperin, DNA-decorated carbon nanotubes for chemical sensing. Phys. Status Solidi B **243**, 3252–3256 (2006)
597. Q. Kuang, C. Lao, Z.L. Wang, Z.X. Xie, L.S. Zheng, High-sensitivity humidity sensor based on single SnO2 nanowire. J. Am. Chem. Soc. **129**, 6070–6071 (2007)
598. B. Li, L. Shang, M.S. Marcus, T.L. Clare, E. Perkins, R.J. Hamers, Chemoselective nanowire fuses: chemically induced cleavage and electrical detection of carbon nanofiber bridges. Small **4**, 795–801 (2008)
599. X.-J. Huang, Y.-K. Choi, Chemical sensors based on nanostructured materials. Sens. Actuators B **122**, 659–671 (2007)
600. Q. Wan, Q.H. Li, Y.J. Chen, T.H. Wang, X.L. He, J.P. Li, C.L. Lin, Fabrication and ethanol sensing characteristics of ZnO nanowire gas sensors. Appl. Phys. Lett. **84**, 3654–3656 (2004)
601. N.S. Ramgir, I.S. Mulla, K.P. Vijayamohanan, A room temperature nitric oxide sensor actualized from Ru-doped SnO_2 nanowires. Sens. Actuators B **107**, 708–715 (2005)
602. C. Li, D. Zhang, B. Lei, S. Han, X. Liu, C. Zhou, Surface treatment and doping dependence of In_2O_3 nanowires as ammonia sensors. J. Phys. Chem. B **107**, 12451–12455 (2003)
603. Y. Zhang, A. Kolmakov, S. Chretien, H. Metiu, M. Moskovits, Control of catalytic reactions at the surface of a metal oxide nanowire by manipulating electron density inside it. Nano Lett. **4**, 403–407 (2004)
604. F. Favier, E. Walter, M.P. Zach, T. Benter, R.M. Penner, Hydrogen sensors and hydrogen-activated switches were fabricated from arrays of mesoscopic palladium wires. Science **293**, 2227–2231 (2001)
605. A.S. Blum, C.M. Soto, K.E. Sapsford, C.D. Wilson, M.H. Moore, B.R. Ratna, Molecular electronics based nanosensors on a viral scaffold. Biosens. Bioelectron. **26**, 2852–2857 (2011)
606. A.P. de Silva, H.Q.N. Gunaratne, T. Gunnlaugsson, A.J.M. Huxley, C.P. McCoy, J.T. Rademacher, T.E. Rice, Signaling recognition events with fluorescent sensors and switches. Chem. Rev. **97**, 1515–1566 (1997)
607. C.W. Rogers, M.O. Wolf, Luminescent molecular sensors based on analyte coordination to transition metal complexes. Coord. Chem. Rev. **233–234**, 341–350 (2002)
608. J.N. Demas, B.A. DeGraff, Applications of luminescent transition platinum group metal complexes to sensor technology and molecular probes. Coord. Chem. Rev. **211**, 317–351 (2001)
609. B. Valeur, I. Leray, Design principles of fluorescent molecular sensors for cation recognition. Coord. Chem. Rev. **205**, 3–40 (2000)
610. A.P. de Silva, D.B. Fox, A.J.M. Huxley, T.S. Moody, Combining luminescence, coordination and electron transfer for signalling purposes. Coord. Chem. Rev. **205**, 41–57 (2000)
611. R.E. Gawley, S. Pinet, C.M. Cardona, P.K. Datta, T. Ren, W.C. Guida, J. Nydick, R.M. Leblanc, Chemosensors for the marine toxin saxitoxin. J. Am. Chem. Soc. **124**, 13448–13453 (2002)
612. R.E. Gawley, H. Mao, M.M. Haque, J.B. Thorne, J.S. Pharr, Visible fluorescence chemosensor for saxitoxin. J. Org. Chem. **72**, 2187–2191 (2007)
613. Y. Dubi, M. Di Ventra, Heat flow and thermoelectricity in atomic and molecular junctions. Rev. Mod. Phys. **83**, 131–155 (2011)

614. L.D. Hicks, M.S. Dresselhaus, Effect of quantum wells on the thermoelectric figure of merit. Phys. Rev. B **47**, 2727–1273 (1993)
615. C.J. Vineis, A. Shakouri, A. Majumdar, M.G. Kanatzidis, Nanostructured thermoelectrics: big efficiency gains from small features. Adv. Mater. **22**, 3970 (2010)
616. R. Venkatasubramanian, E. Siivola, T. Colpitts, B. O'Quinn, Thin-film thermoelectric devices with high room-temperature figures of merit. Nature **413**, 597–602 (2001)
617. R. Rurali, Colloquium: structural, electronic, and transport properties of silicon nanowires. Rev. Mod. Phys. **82**, 427–449 (2010)
618. M. Paulsson, S. Datta, Thermoelectric effect in molecular electronics. Phys. Rev. B **67**, 241403(R) (2003)
619. J.A. Malen, P. Doak, K. Baheti, T.D. Tilley, R.A. Segalman, A. Majumdar, Identifying the length dependence of orbital alignment and contact coupling in molecular heterojunctions. Nano Lett. **9**, 1164–1169 (2009)
620. J.R. Widawsky, P. Darancet, J.B. Neaton, L. Venkataraman, Simultaneous determination of conductance and thermopower of single molecule junctions. Nano Lett. **12**, 354–358 (2012)
621. S.K. Yee, J.A. Malen, A. Majumdar, R.A. Segalman, Thermoelectricity in fullerene-metal heterojunctions. Nano Lett. **11**, 4089–4094 (2011)
622. M. Zwolak, M. Di Ventra, Colloquium: physical approaches to DNA sequencing and detection. Rev. Mod. Phys. **80**, 141–165 (2008)
623. E. Macia, Thermoelectric power and electrical conductance of DNA based molecular junctions. Nanotechnology **16**, S254–S260 (2005)
624. E. Macia, DNA-based thermoelectric devices: a theoretical prospective. Phys. Rev. B **75**, 035130 (2007)
625. Z. Wang, J.A. Carter, A. Lagutchev, Y.K. Koh, N.-H. Seong, D.G. Cahill, D.D. Dlott, Ultrafast flash thermal conductance of molecular chains. Science **317**, 787–789 (2007)
626. M. Tsutsui, M. Taniguchi, K. Yokota, T. Kawai, Roles of lattice cooling on local heating in metal–molecule-metal junctions. Appl. Phys. Lett. **96**, 103110 (2010)
627. F. Liu, K.L. Wang, Correlated random telegraph signal and low-frequency noise in carbon nanotube transistors. Nano Lett. **8**, 147–151 (2008)
628. J. Chan, B. Burke, K. Evans, K.A. Williams, S. Vasudevan, M. Liu, J. Campbell, A.W. Ghosh, Reversal of current blockade in nanotube-based field effect transistors through multiple trap correlations. Phys. Rev. B **80**, 033402 (2009)
629. N.P. Guisinger, M.E. Greene, R. Basu, A.S. Baluch, M.C. Hersam, Room temperature negative differential resistance through individual organic molecules on silicon surfaces. Nano Lett. **4**, 55–59 (2004)
630. S. Kubatkin, A. Danilov, M. Hjort, J. Cornil, J.L. Bredas, N. Stuhr-Hansen, P. Hedega, T. Bjornholm, Single-electron transistor of a single organic molecule with access to several redox states. Nature **425**, 698–701 (2003)
631. M. Tsutsui, M. Taniguchi, Vibrational spectroscopy of single-molecule junctions by direct current measurements. J. Appl. Phys. **113**, 084301 (2013)

Index

A
Activation energy, 243
Advanced Green's functions, 39–41, 43–45, 66, 150, 171, 226
Approximation
  Born-Oppenheimer, 1, 184, 191, 321
  Condon, 23
  Hartree, 13, 120, 187, 188, 196, 200, 207, 222
  Hartree-Fock, 120, 196
  non-crossing, 66, 70, 128, 241
Atomic-size contact, 29, 30, 34, 35
Atomic structure, 82, 191

B
B80, 210–212
Bipolaronic attraction, 149
Bistable system, 250, 251
Bose-Einstein distribution function, 70, 71, 77, 154
Break junctions, 29, 32, 34, 236, 240
Buttiker model, 56, 57, 63, 139, 274

C
Capacitive forces, 245, 246
Capture potential barrier, 242
Carbon nanotubes, 18, 20, 23, 108, 133, 148, 157, 233, 234, 242–247, 256–258
Charge reservoirs, 15
Charging energy, 11–14
Charge transfer, 1, 2, 10, 11, 29, 79–84, 86, 87, 178, 217, 220, 221, 234, 269–271
Chemical potentials of electrodes, 81, 177
Chemoselective sensor, 255–262
Conducting polymers, 18, 62, 63
Conduction electrons, 118, 119, 123, 168
Conduction mechanisms, 18, 158
Conduction-based sensors, 257
Conformational changes in molecules, 85, 251, 252
Contact time, 17, 18, 153, 154
CNT-based nanoelectromechanical memory cells, 245, 246
Coulomb blockade, 11–16, 67, 70, 105–115, 118–121, 148–153, 170, 216, 227, 270, 273
Current-induced charge, 85, 86
Current rectification, 10, 11, 84–90, 240
CWNT crossbar array, 245

D
Defects density, 256
Density matrix, 43–45, 80, 114, 224–227
Differential conductance, 36, 116, 124, 125, 135, 147, 151–153, 166, 216
Differential noise, 165–167
Digital signal generation, 259
Dipole polarizability, 208, 210, 211
Direct vibrational relaxation, 135, 137, 138
Dissipative effects, 17, 18, 153, 154, 161
Dissipative electron transport, 154, 158
Donor-bridge-acceptor units, 218
Doping, 2, 234, 258, 267

E
Easy axis anisotropies of magnetic molecules, 118
Effective potential, 184, 188, 204–206
Electric charge, 2, 10, 53, 80–82, 84–90, 116, 117, 178, 242, 254, 255, 271

N.A. Zimbovskaya, *Transport Properties of Molecular Junctions*, Springer Tracts in Modern Physics 254, DOI 10.1007/978-1-4614-8011-2,

Electrostatic potential distribution, 81, 84, 89, 102, 104
Electromigration, 29, 34, 85, 237, 243, 247
Electron-electron interactions, 5–16, 67, 75–77, 79, 118, 148, 171, 184, 187, 190, 194, 197, 204, 274
Electron-phonon interactions, 1, 7, 18–23, 37, 43, 54, 56, 64, 66, 67, 70, 72, 75, 77, 78, 109, 121, 126, 144, 148, 153, 164, 167, 169, 179, 227, 228, 230, 263, 268, 274, 275
Electron-vibron interactions, 20, 22, 74, 75, 133, 139, 143–145, 150, 164–167, 273, 274
Electron reservoirs, 28, 41, 52, 53, 103, 104
Electrical breakdown, 257
Electrode-electrolyte interfaces, 236
Electron transfer reactions, 60, 81, 153, 178–182, 220, 262
Electron tunneling, 18, 58, 62, 97, 99, 104, 112, 118, 123, 126, 136, 139–148, 158, 163, 178, 238, 245, 251
Emission energies, 261
Energy functional, 186–191, 203
Electron-phonon coupling strengths, 55
Electron density of states (DOS), 11, 72, 83, 84, 121–124, 127, 143, 151, 254, 263
Electron transmission, 3, 5, 7, 17, 18, 52–56, 59–63, 80, 83, 84, 91–93, 95, 99–101, 104, 106, 133, 135, 139, 141, 143, 144, 155, 156, 161–163, 170, 178, 181, 227, 229, 230, 263
Electron-vibron coupling, 151, 152, 165, 167, 177–179, 214
Equilibrium Fermi energy, 2, 83, 84, 103, 165
Exchange-correlation energy, 118, 192–196, 202
Excited electronic states, 200
Extended molecule, 80

F
Fano factors, 165, 166
Fano resonances, 90–93, 146, 172, 173, 264
Ferromagnetic leads, 25, 27, 118, 122, 123, 151, 174, 216
Field effect transistors, vi
Fermi distribution functions, 52, 74, 225
Finite-field method, x, 209, 211
Fourier transforms, 150, 169
Franck-Condon blockade, 77, 167, 274
Franck-Condon factor, 178, 180
Franck-Condon matrix elements, 152
Fullerene, 120, 207–212, 216, 220, 221, 227, 228, 234, 237, 247, 248, 251, 252, 256, 267

G
Gate electrode, 10, 31, 32, 34, 36, 99, 105, 117, 125, 126, 237, 245
Gate voltage, 12, 24, 32, 34, 99–102, 105–107, 117, 170–172, 216, 236–238, 244, 245, 251, 258, 270, 273
Gaussian orbitals, 191, 210
Generalized gradient approximation (GGA), 194, 196, 222
Graphene nanoribbons, 236
Graphene oxide, 256, 257
Graphene sheets, 233, 238, 256
Ground-state energy, 189, 190, 196–199

H
Hartree term, 13, 207
Heat conductance, 168
Heat transfer, 19, 68–175, 267, 268, 273
Highest occupied molecular orbital (HOMO), 2, 21, 34, 62, 81, 85, 88, 114, 161, 192–194, 201, 218–223, 238, 240, 241, 254, 261, 263–267
Hopping transmission, 18
Hopping transport, vii, 18, 157
Hydrocarbon chains, 6, 83, 84, 104
Hyperfine interactions, 24
Hysteresis in molecular junctions, 175

I
Inelastic transport channel, 146
Ionization energy, 51, 139, 144, 220
Ionization sensor, 257, 258

J
Joule heating, 169

K
Keldysh equations, 65, 150, 226
Kohn-Sham equations, 181, 183–192, 204, 206, 209
Kohn-Sham orbitals, 189–191, 200, 202, 210, 213, 217, 219
Kondo effect, viii, x, 118–127

Kondo peak, 15, 70, 74, 119–126, 148, 150, 151
Kondo temperature, 119, 125, 126, 130, 171, 216

L
Lagrange multiplier, 190
Landauer expression for electron current, 52–55, 63, 179, 228, 230
Landauer-Buttiker formalism, 52
Lesser Green's functions, 44, 54, 80, 109, 226
Linear response theory, 209, 266
Light-harvesting, 212, 217–221, 275
Local-density approximation (LDA), 192, 193, 222, 223, 227
Local spin-density approximation (LSDA), 193–196, 222
Localization energy, 243
Long range electron transfer reactions, 178–182
Lowest unoccupied orbital (LUMO), 2, 21, 34, 88, 113, 114, 161, 218, 219, 240, 241, 254, 263–266
Luminophores, 261
Luminescence, 261, 262
Luminescent gas sensors, 261

M
Magnetic quantum number, 28, 115
Magnetic molecules, 28, 115–118, 212, 214, 217, 255
Magnetic relaxation time, 118
Microscopic quantum tunneling (MQT), 213
Magnetic anisotropy barrier (MAB), 213, 214
Many-electron wave function, 184
Master equations, 14, 109, 110, 112, 114
Memory elements, 117
Metal-molecule-metal (MMM) junctions, 8, 14, 17, 22–24, 28–32, 34, 37–47, 83, 85, 87–90, 94, 99, 109, 118, 119, 126, 148, 163, 170, 173, 221–230, 236, 239, 240, 248, 249, 254, 264–268, 273–275
Metal-oxide-silicon field effect transistors (MOSFET) 231, 232, 234, 272
Metastable electron levels, 133
Molecular conductance, ix, 18, 22, 23, 26, 35, 37, 102, 133, 150, 158, 179, 248, 251, 261, 275
Mn12, 28, 115–117, 212–215, 217, 255
Mn4 dimer, 215
Molecular magnets, 212–217, 273
Molecular networks, 32, 83
Molecular recognition, 255, 260, 262
Molecular spin-multiplets, 116
Molecular targeting, 256
Molecular orbital, ix, 1–7, 13, 20, 34, 36, 54, 73, 83, 88, 110, 133, 218, 238, 240, 261, 263, 266, 271
Molecular spin, 24–26, 28, 116, 117
Molecular switching device, 248
Molecular vibrations, 20, 21, 36, 67, 75, 133, 135, 148–153, 167
Molecular-based electronic devices, 233
Molecular spin state, 117
Monomers, 215, 248
Monomer-monomer exchange interaction, 215

N
Nagaoka mechanism, 117
Nanoelectromechanical systems, 245
Natural length of the device, 232
Negative differential resistance (NDR), 11, 102–105, 111, 113–117, 175–177, 271, 272
Nonequilibrium states, 137
Nonequilibrium Green's functions formalism (NEGF), 15, 55, 56, 64–72, 78, 79, 90, 105–109, 111, 119–121, 135, 139, 145, 146, 150, 164, 168, 228, 230, 241, 274

O
Occupation numbers, 106–108, 200, 201, 220
Occupation probabilities, 14, 73, 110, 111, 114, 137, 274
Optimized molecular geometry, 191
Organometallic compounds, 216

P
Phonon-induced broadening, 135, 147, 156
Phonon-induced peaks in the transmission, 135, 162
Photoabsorption, 204, 218, 219
Polaron, vii, 21, 134, 144, 148, 149, 169, 175–177, 180, 181
PBE-GGA approximation, 196
Phonon density of states, 71, 72, 142
Photoabsorption spectra, 204, 218
Photoconversion units, 217
Photoinduced charge separation, 218
Photoinduced charge transfer reactions, 220
Photoinduced electron transfer, 262

Photoswitching devices, 252
Poisson equation, 10, 80
Polarization fluctuations, 178, 181
Polaronic shift, 21, 134, 144, 149, 180
Polymer chains, 62, 154
Polaronic conduction, vii
Preferable pathways for electron transport, 7

Q
Quantum dot, vii, viii, 12, 15, 90, 93, 102, 105, 107, 109, 110, 112, 114, 115, 119, 120, 126–132, 149–151, 170, 174, 217, 233, 242–247, 263, 273
Quantum switching, 176
Quantum tunneling, 36, 213, 214, 232

R
Random phase approximation (RPA), 208, 209
Raman spectra, 190
Resistivity dipole, 85
Retarded Green's functions, 40–42, 44, 45, 48, 66, 67, 80, 91, 92, 106, 121, 141, 158, 224, 225
Retention time, 242, 251, 255

S
Scanning tunneling microscope (STM), 26, 29, 33, 34, 36, 83, 90, 145, 146, 228, 236, 237, 248, 249, 252, 267
Schrodinger equation, 42, 80, 184, 188, 199, 204, 205
Seebeck effect, 174
Self-assembled monolayers, 31, 38, 239
Self-energy, 21, 42–48, 51, 54, 64, 66, 69–72, 103, 106, 109, 139, 142, 143, 154, 157, 169, 179, 180, 201, 223, 225–227
Self-consistent field model, 13–15
Short-channel effects, 231, 232
Short-range adhesive forces, 246
Shottky barriers, 234
Single-electron excitations, 202
Single-electron Hamiltonian, 203
Silicon nanowires, 235
Slater's determinant, 202, 203, 205, 217
Spin-orbit interactions, 23–27, 118, 122, 213
Spin-switching MMM junction, 117
Single-molecule magnets, 212
Solar cells, 217, 218
Spin-orbit-vibron coupling, 214
Spin blockade, 74, 116
Spin dephasing, 24
Spin Seebeck effect, 174
Spin selection rules, 116, 118
Spin valve, 24–26, 28
Stochastic switching, 23, 175, 248
Surface adsorption, 258
Surface-roughness effects, 236
Switching time, 176
Spontaneous polarization, 244
Storage time, 242–244, 255
Strong vibrational excitation, 133

T
Thermal conductance, 172–175, 267, 273
Thermal phonons, 20, 21, 64, 142, 154, 158–163, 242
Thermoelectric current, 264
Thermopower, 36, 168–175, 263–267, 273
Thomas-Fermi model, 193
Transition rates, 14, 73, 74, 76, 77, 107, 110–113, 116, 119, 137, 138, 152
Trapping centers, 244
Tunnel transmission, 18
Tunneling transport, 85

V
V15, 215, 216
Van der Waals forces, 101, 196, 246
Vibration-induced effects, 20, 133
Vibrational frequencies, 135, 136, 211
Vibrational modes, 20, 21, 37, 55, 109, 134, 153, 177, 190, 210
Vibrational phonons, 21, 22, 64, 133, 144, 145, 148, 149, 165
Vibrational relaxation, 77, 135, 137, 138
Vibrational stability, 210, 211

W
Wiedemann-Franz law, 172, 263
Write time, 242

Z
ZnO nanowires, 235, 236

MIX
Papier aus verantwortungsvollen Quellen
Paper from responsible sources
FSC® C105338

FSC
www.fsc.org

If you have any concerns about our products,
you can contact us on
ProductSafety@springernature.com

In case Publisher is established outside the EU,
the EU authorized representative is:
Springer Nature Customer Service Center GmbH
Europaplatz 3, 69115 Heidelberg, Germany

Printed by Libri Plureos GmbH
in Hamburg, Germany